AF328674

FUJI

Fuji

A MOUNTAIN IN THE MAKING

ANDREW W. BERNSTEIN

PRINCETON UNIVERSITY PRESS

PRINCETON & OXFORD

Published by Princeton University Press
41 William Street, Princeton, New Jersey 08540
99 Banbury Road, Oxford OX2 6JX

press.princeton.edu

GPSR Authorized Representative: Easy Access System Europe - Mustamäe tee 50, 10621 Tallinn, Estonia, gpsr.requests@easproject.com

All Rights Reserved

Library of Congress Cataloging-in-Publication Data

Names: Bernstein, Andrew, 1968– author
Title: Fuji : a mountain in the making / Andrew W. Bernstein.
Description: Princeton : Princeton University Press, 2025. | Includes
    bibliographical references and index.
Identifiers: LCCN 2024054178 (print) | LCCN 2024054179 (ebook) | ISBN
    9780691256290 hardback | ISBN 9780691256306 ebook
Subjects: LCSH: Fuji, Mount (Japan)—History | Fuji, Mount (Japan)—In
    literature | Fuji, Mount (Japan)—In art | Mountain
    worship—Japan—Fuji, Mount | National characteristics, Japanese |
    BISAC: HISTORY / Asia / Japan | SCIENCE / History
Classification: LCC DS894.495.F85 B47 2025 (print) | LCC DS894.495.F85
    (ebook) | DDC 952/.166—dc23/eng/20250521
LC record available at https://lccn.loc.gov/2024054178
LC ebook record available at https://lccn.loc.gov/2024054179

ISBN 9780691256290
ISBN (e-book) 9780691256306

British Library Cataloging-in-Publication Data is available

Editorial: Priya Nelson, Emma Wagh
Production Editorial: Elizabeth Byrd
Jacket: Karl Spurzem
Production: Danielle Amatucci
Publicity: Alyssa Sanford (US), Carmen Jimenez (UK)
Copyeditor: Stephen Beitel

Jacket image: Katsushika Hokusai, *South Wind, Clear Sky* (*Gaifū kaisei*), also known as *Red Fuji*, from the series *Thirty-six Views of Mount Fuji* (*Fugaku sanjūrokkei*), ca. 1830–32. Courtesy of The Metropolitan Museum of Art, New York / Rogers Fund, 1914.

Studies of the Weatherhead East Asian Institute, Columbia University
The Studies of the Weatherhead East Asian Institute of Columbia University
were inaugurated in 1962 to bring to a wider public the results of significant
new research on modern and contemporary East Asia.

Printed in the United States of America

10 9 8 7 6 5 4 3 2 1

To Philip

CONTENTS

Plates

ILLUSTRATION CREDITS

I AM GRATEFUL to the institutions and individuals who provided permissions for, and/or copies of, the following items either directly or by making them publicly available.

Figures

2.2. Prince Shōtoku flies over Fuji. *Illustrated Biography of Prince Shōtoku (Shōtoku taishi eden)*. Fifteenth century. Tanzan Shrine, Sakurai. One of four painted silk scrolls. 147 × 81 cm. Alamy stock photo.

4.2. Fujizuka at Senjūkawada Sengen Shrine, Tokyo. Photo by Philip Ellway.

4.3. Fujizuka at Teppōzu Inari Shrine, Tokyo. Photo by Philip Ellway.

4.5. Illustration of Kamiyoshida. In Nagashima Taigyō, *True View Illustrations of Mount Fuji (Fujisan shinkei no zu)*. 1847. Reprinted in Nagashima Taigyō, *Fujisan shinkei no zu: Edo jidai sankei emaki*, ed. Okada Hiroshi (Tokyo: Meicho Shuppan, 1985), pp. 8–9.

4.8. View of Ōsawa Failure from a distance. Photo by Watanabe Michihito.

4.9. Illustration of Fuji's summit. In Nagashima Taigyō, *True View Illustrations of Mount Fuji (Fujisan shinkei no zu)*. 1847. Reprinted in Nagashima Taigyō, *Fujisan shinkei no zu: Edo jidai sankei emaki*, ed. Okada Hiroshi (Tokyo: Meicho Shuppan, 1985), pp. 72–73.

5.1. Sesshū Tōyō. *Mount Fuji, Miho, and Seikenji Temple (Fujisan, Miho, Seikenji zu)*. Sixteenth century. Hanging scroll, ink on paper. 43 × 102 cm. Eisei Bunko Museum, Tokyo. Alamy stock photo.

5.2. Kanō Tan'yū. *Mount Fuji, Mount Yuwang, and Jinshan Temple* (*Fujisan, Ikuōzan, Kinzanji*). Seventeenth century. Three painted silk scrolls. Each scroll 104.8 × 54.8 cm. Private collection.

5.3. Ariwara no Narihira passes Fuji. In *Yamato Library: Teaching One Hundred Poems by One Hundred Poets* (*Oshie hyakunin isshu Yamato bunko*). 1829. Image found in Emori Ichirō, *Edo jidai josei seikatsu ezu daijiten*, vol. 5, *Shiki, dōshokubutsu, meisho* (1993; repr., Tokyo: Ōzorasha, 2005), pp. 136–37.

5.4. Suzuki Harunobu. *Lucky Dream for the New Year: Mt. Fuji, Falcon, and Eggplants.* ca. 1768–1769. Color woodblock print. Image: 10 1/4 × 8 5/16 in.; Sheet: 11 1/4 × 8 7/16 in. Los Angeles County Museum of Art, Gift of Barbara S. Bowman. M.2015.298.72. Photo © Museum Associates/LACMA.

5.5. Shiba Kōkan. *Seven League Beach, Kamakura, Sagami Province* (*Sōshū Kamakura Shichiri-ga-hama zu*). 1796. Oil painting on two-fold screen. 95.7 × 178.4 cm. Kobe City Museum. Image from Wikimedia Commons.

5.6. Katsushika Hokusai. *The Appearance of Mount Hōei* (*Hōeizan shutsugen*), from the series *One Hundred Views of Mount Fuji*. 1835. Woodblock print in book. 24.6 × 15.88 cm. The Miriam and Ira D. Wallach Division of Art, Prints and Photographs: Print Collection, The New York Public Library Digital Collections.

5.7. Katsushika Hokusai. *Fuji and Yatsugatake in Shinshū* (*Shinshū Yatsugatake no Fuji*), from the series *One Hundred Views of Mount Fuji*. 1835. Woodblock print in book. 24.6 × 15.88 cm. The Miriam and Ira D. Wallach Division of Art, Prints and Photographs: Print Collection, The New York Public Library Digital Collections.

5.8. Ishikawa Ryūsen. Portion of *Map of the Seas, Mountains, and Lands of Japan* (*Nihon kaisan chōrikuzu*). 1691. Woodblock print. 82 × 168 cm. Japanese Historical Maps, University of California, Berkeley Library Digital Collections.

5.9. Katsushika Hokusai. *Famous Places of the Tōkaidō at a Glance* (*Tōkaidō meisho ichiran*). Woodblock print. 1818. 43.9 × 58.3 cm. Japanese Maps of the Tokugawa era. Rare Books and Special Collections, University of British Columbia Library. G7962.T6 E635 1818 K2.

5.10. Katsushika Hokusai. *Fuji and Foreign Embassy (Raichō no Fuji)*, from the series *One Hundred Views of Mount Fuji*. 1835. Woodblock print in book. 24.6 × 15.88 cm. The Miriam and Ira D. Wallach Division of Art, Prints and Photographs: Print Collection, The New York Public Library Digital Collections.

5.11. Matsumoto Yasuoki (Gengendō). *Detailed Copperplate Map of Japan (Dōsen Nihon yochi saizu)*. 1835. Woodblock print. 29 × 27 cm. Japanese Historical Maps, University of California, Berkeley Library Digital Collections.

6.1. Alcock goes to Fuji. In Kanagaki Robun, *Humorous Pilgrimage to Fuji (Kokkei Fuji mairi)*. 1860. Image found in Ronald Toby, *Engaging the Other: "Japan" and Its Alter Egos, 1550–1850* (Leiden: Brill, 2019), p. 308.

6.2. Storm blows Alcock and his men down Fuji. Anonymous broadsheet. 1860. Image found in Kinoshita Naoyuki et al., *Nyūsu no tanjō: Kawaraban to shinbun nishikie no jōhō sekai* (Tokyo: Tōkyō Daigaku Shuppankai, 1999), p. 89.

6.3. Kanō Tōsen. *Cranes Flying over Mount Fuji (Fuji hikaku zu)*. 1859. Painting on silk scroll. 130.2 × 56.3 cm. Mount Fuji World Heritage Centre, Shizuoka.

6.4. Altar with Buddhist statues rescued from Fuji's summit at Fuji Takasago Sake Brewery. Photo by Philip Ellway.

7.1. Hand-painted kimono. Meiji period. Andrea Aranow Textile Design Collection. Courtesy of Textile Hive.

7.2. Nonaka Chiyoko and Itaru, and the Fuji Weather Observatory. In Ochiai Naobumi, *Takane no yuki* (Tokyo: Meiji Shoin, 1896), front matter.

7.6. "Japan's greatest mountain." In *National Language Reader for Elementary Schools (Kokugo dokuhon, jinjō shōgakkō yō)*, vol. 2. 1900. Image found in Kaigo Tokiomi, ed., *Nihon kyōkasho taikei kindaihen*, vol. 6, *Kokugo* 3 (Tokyo: Kodansha, 1964), p. 211.

7.7. Bunkichi creates a mini-Fuji in his family garden. In *Elementary Studies Reader (Jinjō shōgaku dokuhon)*, vol. 4. 1903. Image found in Kaigo Tokiomi, ed., *Nihon kyōkasho taikei kindaihen*, vol. 6, *Kokugo* 3 (Tokyo: Kodansha, 1964), p. 434.

7.8. "Tarō drew a picture of a battleship. Hanako drew a picture of Mount Fuji." In *Elementary National Language Reader* (*Shōgaku kokugo dokuhon*), vol. 1. 1933. Image found in Kaigo Tokiomi, ed., *Nihon kyōkasho taikei kindaihen*, vol. 7, *Kokugo* 4 (Tokyo: Kōdansha, 1963), p. 563.

7.9. Yokoyama Taikan. *Mount Fuji* (*Fujisan*). 1940. Color on paper, hanging scroll. 73.5 × 94.0 cm. Tokyo Fuji Art Museum.

7.10. Postcard commemorating the twenty-six-hundredth anniversary of the founding of the empire by the mythical Emperor Jimmu. 1940. Private collection of Matsushima Jin.

7.11. B-29 bomber over Mount Fuji. 1945. United States Army Air Forces. Wikimedia Commons.

7.12. Fujisan Hongū Sengen Taisha in Fujinomiya. Photo by Philip Ellway.

8.2. Entrance to cave in Aokigahara forest where locals stored silkworm eggs. Photo by Philip Ellway.

8.3. Fuji and tea fields. Alamy stock photo.

8.9. Spring burning of Nashigahara grassland. Photo by Watanabe Michihito.

Plates

1. Kanō Motonobu. *Fuji Pilgrimage Mandala* (*Fuji sankei mandara*). Sixteenth century. Ink and colors on silk. 186.6 × 118.2 cm. Fujisan Hongū Sengen Taisha, Fujinomiya.

2. *Fuji Pilgrimage Mandala* (*Fuji sankei mandara*). Sixteenth century. Ink and colors on silk. 91.5 × 67.3 cm. Fujisan Hongū Sengen Taisha, Fujinomiya.

3. Utagawa Sadahide. *Mount Fuji Womb Cave Pilgrimage* (*Fujisan tainai meguri no zu*). 1858. Three woodblock prints. Each print 35 × 24 cm. Private collection of D. Max Moerman.

4. Utagawa Sadahide. *Portrayal of Mount Fuji* (*Fujisan no zu*). 1848. Woodblock print. 91.4 × 96.5 cm. Japanese Maps of the Tokugawa Era. Rare Books and Special Collections, University of British Columbia Library. G7962.K3 1848 S2.

5. Shiomi Masanari. *Ariwara no Narihara at Mt. Fuji* (inrō/ojime/netsuke ensemble). Early nineteenth century. [Inrō] Gold and iroye togidashi

on black lacquer ground, with mother-of-pearl inlays; [ojime] shakudō and sentoku; [netsuke] ivory with staining, a) inrō: 3 3/16 × 2 × 15/16 in.; b) ojime: 9/16 × 9/16 × 5/16 in.; c) netsuke: 1 5/16 × 1 5/16 × 1 5/16 in. Los Angeles County Museum of Art, gift of Miss Bella Mabury. M.39.2.325a-c. Photo © Museum Associates/ LACMA.

6. Box for Inkstone and Writing Implements (*Suzuribako*) with Geese against Mount Fuji in Moonlight and (inner lid) with Plovers by the Seashore. Early to mid-nineteenth century. Black lacquer ground with gold and silver maki-e. 9 9/16 in. (24.3 cm) × 8 5/8 in. (21.9 cm) × 1 13/16 in. (4.6 cm). Bequest of Stephen Whitney Phoenix, 1881. Metropolitan Museum of Art, New York.

7. Katsushika Hokusai. *South Wind, Clear Sky* (*Gaifū kaisei*), also known as *Red Fuji*, from the series *Thirty-Six Views of Mount Fuji* (*Fugaku sanjūrokkei*). ca. 1830–1832. Woodblock print; ink and color on paper. 9 5/8 × 14 in. (24.4 × 35.6 cm). Rogers Fund, 1914. Metropolitan Museum of Art, New York.

8. Katsushika Hokusai. *Under the Wave off Kanagawa* (*Kanagawa oki nami ura*), also known as *The Great Wave*, from the series *Thirty-Six Views of Mount Fuji* (*Fugaku sanjūrokkei*). ca. 1830–1832. Woodblock print; ink and color on paper. 10 1/8 × 14 15/16 in. (25.7 × 37.9 cm). H. O. Havemeyer Collection, Bequest of Mrs. H. O. Havemeyer, 1929. Metropolitan Museum of Art, New York.

9. Katsushika Hokusai. *Mitsui Shop at Surugachō in Edo* (*Edo Surugachō Mitsui mise ryaku zu*), from the series *Thirty-Six Views of Mount Fuji* (*Fugaku sanjūrokkei*). ca. 1830–1832. Woodblock print; ink and color on paper. 10 × 15 in. (25.4 × 38.1 cm). Rogers Fund, 1922. Metropolitan Museum of Art, New York.

10. Utagawa Hiroshige. *New Fuji, Meguro* (*Meguro, Shin Fuji*), from the series *One Hundred Famous Views of Edo* (*Meisho Edo hyakkei*). 1857. Color woodblock print. 35.6 × 24.4 cm (14 × 9 9/16 in.). Frederick W. Gookin Collection, Art Institute of Chicago.

11. Utagawa Hiroshige. *Surugachō*, from the series *One Hundred Famous Views of Edo* (*Meisho Edo hyakkei*). 1856. Color woodblock print. 36.3 × 24.1 cm. Library of Congress Prints and Photographs Division, Washington, DC.

12. Akiyama Einen. *Map of the Thirteen Provinces from which to View Fuji* (*Fujimi jūsanshū yochi no zenzu*). 1843. Woodblock print. 147.2 × 170.7 cm. Japanese Maps of the Tokugawa Era. Rare Books and Special Collections, University of British Columbia Library. G7962. H6 E635 1843 A5.

13. *Map of the World and its Peoples* (*Bankoku jinbutsu zue*). Late Tokugawa period. Color woodblock print. 38 × 52 cm. Gifu Prefectural Library.

15. Mount Fuji World Heritage Centre, Shizuoka (in Fujinomiya). Photo by Philip Ellway.

NAMES, DATES, & TRANSLITERATION

IT IS CONVENTIONAL for Japanese names to be written with the family name first. There are exceptions, however. Sometimes when a Japanese author publishes in English, the given name appears before the family name. In the notes and bibliography, I adhere to the order that appears in each publication. As is Japanese custom, I refer to major historical personages such as daimyo and shoguns by their given names after providing full names and use the professional names adopted by Japanese artists, writers, and religious figures.

Japan followed a lunisolar calendar until 1873, so dates before that point are indicated by the day of the lunar month—for example, "seventh day of the fourth month." This rule applies except in cases where a geologist or historian has provided dates based on the Gregorian solar calendar, such as those used to chart the progress of the Hōei eruption of 1707.

Macrons designate long vowels. They are omitted for familiar places and terms like Tokyo and Shinto.

Only the first word in the title of a Japanese publication is capitalized unless it is a proper name (Fuji, for instance). All of the words in the name of a Japanese publisher or organization are capitalized except for particles.

A MOUNTAIN of debt has accumulated over the many years it has taken to produce this book. With the knowledge that a debt this large can never be repaid in full, I thank those who have kindly provided support along the way.

To write a history of this chronological and topical scope would not have been possible without the guidance of numerous experts in fields ranging from geology to art history. My profound thanks go to David Feldman, Carol Gluck, Andrew Gordon, Julia Adeney Thomas, Anne Walthall, and Kären Wigen for giving feedback on drafts of the entire manuscript; and to Gina Barnes, David Fedman (not to be confused with David Feldman), David Lurie, Margaret McKean, Miyazaki Fumiko, Kenneth Ruoff, Elizabeth Safran, Saitō Haruo, Janine Anderson Sawada, Timon Screech, Henry Smith, Sarah Thal, and Pamela Winfield for commenting on drafts of one or more chapters. Miyazaki, Smith, Thal, Thomas, Walthall, and Wigen also provided various forms of assistance over the course of my research. I would like to express special gratitude to Carol Gluck for her wise counsel and unflagging support every step of the way.

My colleagues at Lewis and Clark College have been a tremendous help from start to finish. Thank you to Dean of the College Bruce Suttmeier and History Department members David Campion, Nancy Gallman, Susan Glosser, Maureen Healy, Reiko Hillyer, Jane Hunter, Benjamin Westervelt, and Elliott Young for both steadfast friendship and valuable advice. I am also grateful to my colleagues in the Asian Studies Program for their support—in particular to Dawn Odell for her keen art historian's eye and to Satomi Hayashi Newsom for help with translations. Thanks to Amy Baskin, Justin Counts, Kelly Delfatti, Scott Feickert, Daena Goldsmith, Jessica Kleiss, James Proctor, Debra Richman, Margaret Salstrom, Susan Smith, Alison Walcott, Rishona Zimring, and the staff of Lewis and Clark's Aubrey R. Watzek Library for their assistance. I am especially grateful to Elizabeth Safran, who partnered with me to develop and run the Mount Fuji Study Abroad Program for Lewis and Clark students in the summers of 2014 and 2017. It was a pleasure to work with a colleague as magnanimous, astute, and patient as Liz. Thanks, too, to Blythe Knott, Larry Meyers, and the rest of the Overseas Studies and Off-Campus Programs team; and last but not least, to the students who participated in the program. Their boundless curiosity and enthusiasm made it a joy to lead them up, down, and around Fuji.

Many other individuals and organizations have provided help. Aoki Naoko, Watanabe Michihito, and Watanabe Sadamoto deserve special mention for generously sharing their time and knowledge with both me and the students on the Mount Fuji Study Abroad Program. Watanabe Michihito also contributed two of the photographs used in this book. Financial support for research, writing, and publication was provided by the Fulbright-Hays and Fulbright Scholar programs, the National Endowment for the Humanities, Lewis and Clark College, and the Weatherhead East Asian Institute. Obinata Sumio, Okamoto Koichi, and Tsurumi Tarō arranged an institutional home at Waseda University for the times I conducted research in Japan. Other people who provided suggestions, introductions, or other types of assistance include Jessamyn Abel, Barbara Ambros, Anma Sō, Aramaki Shigeo, Kara Batdorff, Philip Brown, Chiba Tatsurō, Logan Drain, Byron Earhart, Wendy Frantz, Marco Gottardo, Harashida Minoru, Hōjō Hiroshi, Honma Chie, Mark Hudson, Ida Kiyoshige, Inoue Takuya, Richard Jaffe, Thomas Jones, Koyama Masato, Matsushima Jin, Bryan Miller, Ian Miller, Max Moerman, Micah Muscolino, Kate Wildman Nakai, Nishikawa Osamu, Nishiki Kōichi, Ōtsuka Sōzō, Sakakibara Sayoko, Jordan Sand, Sano Mitsuru, Caleb Sayan, Shimada Naoyuki, Shishino Fumio, David Sprague, Yui Suzuki, Takamura Fujiyoshi, Tanabe Shirō, Tanaka Kei, John Treat, Tsunematsu Kae, Keith Vincent, Brett Walker, and Watai Hideyo.

I am also grateful for assistance provided by Fukui Hirokazu, Kōda Yoshitaka, Moriwaki Tetsuhiko, and Watanabe Shin at Fujisan Hongū Sengen Taisha; by Komatsu Yukiko and Watanabe Toyohiro at Groundwork Mishima; and by researchers, officials, and/or staff at the Fujisan Club, Japan–United States Friendship Commission, Waseda University Library, Institute for Agro-Environmental Sciences, Mount Fuji Research Institute, Fujisan Museum, Mount Fuji and Princess Kaguya Museum, Fujisan World Heritage Center (Yamanashi Prefecture), Mount Fuji World Heritage Centre (Shizuoka Prefecture), Onshirin Kumiai, Fuji Takasago Sake Brewery, Combined Arms Training Center Camp Fuji, Mount Fuji Sabō Office, Tokyo Fire Department, Japanese Ministry of the Environment, and Japanese Ministry of Agriculture, Forestry, and Fisheries. I would also like to thank the many others who contributed their time and expertise to the Mount Fuji Study Abroad Program.

Thank you to Priya Nelson, senior editor at Princeton University Press, for believing in this book from its inception, and to the infinitely patient Emma Wagh, Elizabeth Byrd, and Michelle Scott for shepherding it to completion. I am grateful to Shane Kelley for his expertly created maps and to Ariana King at the Weatherhead East Asian Institute for her help in the publication process. Parts of chapters 7 and 8 are adapted from two earlier publications: "Whose Fuji? Religion, Region, and State in the Fight for a National Symbol," *Monumenta Nipponica* 63, no. 1 (Summer 2008): 51–99; and "Weathering Fuji: Marriage, Meteorology, and the Meiji Bodyscape," in *Japan at Nature's Edge: The Environmental Context of a Global Power*, edited by Ian Jared Miller, Julia Adeney Thomas, and Brett L. Walker (Honolulu: University of Hawai'i Press, 2013).

Finally, a tremendous thank-you to the family and friends whose love and support have sustained me over the years. Thanks most of all to Philip for putting up with me—and Fuji—in good times and bad. This book is dedicated to you.

FUJI

Introduction

MOUNT FUJI commands attention. Part of the "Ring of Fire" that circles the Pacific, the volcano rises an impressive 3,776 meters (12,389 ft) above sea level. Visible to millions in Tokyo, its iconic silhouette is recognizable around the world. At the same time, industries and military bases located on and around Fuji connect it to economic and political networks with global reach. The mountain also has a rich and sometimes turbulent history as a major pilgrimage site, aesthetic inspiration, and national symbol. Other mountains may be as famous, but none has the combination of material and cultural attributes that makes Fuji such a dominating presence.

This book is a "biography" of the mountain—with a particular focus on its relationship to humans—that stretches from its geological origins to today. To use the term "biography" gives Fuji pride of place in its own story, not only as a mountain but as an active volcano.[1] Volcanoes are often depicted as having human qualities. Stories associate them with deities like the Hawaiian goddess Pele or the Roman god Vulcanus, the origin of the English word "volcano." Volcanoes take on human features in scientific literature too. The introduction to the *Encyclopedia of Volcanoes* states that "each volcano seems to have its distinct 'personality,'" and "just as the medical doctor analyzes the various fluids of the human body, the geochemist samples and analyzes the magmatic liquids that issue from Earth's volcanoes."[2] It is also common to refer to life cycles in which volcanoes are born, grow old, and die. In the words of a character in Endō Shūsaku's novel *Volcano* (*Kazan*), "A volcano resembles human life. In youth it gives rein to passions, and burns with fire. It spurts out lava. But when it grows old, it assumes the burden of those past evil deeds, and it turns quiet as a grave."[3]

Fuji's life has been relatively brief. What volcanologists call "Old Fuji" (Ko Fuji) dates to about one hundred thousand years ago, while "New Fuji"

(Shin Fuji), the volcano that now covers Old Fuji and two older volcanic cones (called "edifices" by geologists), started to take shape only around seventeen thousand years ago. It became especially active about six thousand years later, at the end of the last ice age.[4] Thus the Fuji we see today did not precede the presence of humans but grew up among them. At times Fuji oozed slowly moving lava; at others it erupted with explosive force. There were long periods when it continuously emitted steam and other periods—the last few hundred years, for example—when it seemed to go completely quiet. These episodes of volcanic activity set the conditions for human interactions with Fuji, in the short and long term, in physical and imagined ways, in close proximity and from afar.

Volcanoes produce substantial differences in soil, elevation, and access to water in a relatively small area, so they clearly reveal the ability of humans to occupy and ultimately transform a wide range of habitats. At certain sites around Fuji, springs and rivers provide reliable supplies of fresh water to rice paddies, shrines, and paper factories; on a young lava flow, where water quickly escapes into the permeable rock and makes it unsuitable for agriculture or heavy industry, people once took advantage of the cool and constant temperatures in lava caves to store silkworm eggs.[5] In another example of human resourcefulness, people living at the northern foot of Fuji in the Tokugawa period (1600–1868) diverted debris from occasional slush flows (*yukishiro*) and deposited it on a nearby lava flow to create arable land.[6]

Because eruptions can abruptly alter the nearby landscape—and in dramatic cases, even the global climate—volcanoes also reveal the vulnerability of the technologies and institutions humans use to manipulate different environments. The massive eruption of Fuji in 1707 not only buried nearby villages but, because of prevailing winds, deposited up to sixteen centimeters of ash fifty kilometers to the east and up to eight centimeters of ash over a hundred kilometers away.[7] The ash clogged rivers and canals, wreaking havoc on elaborate irrigation systems that had dramatically increased the output of rice over the previous few centuries and, in turn, increased the number of people living in harm's way. The result was a famine that took more lives than the eruption itself had.

Volcanic eruptions also highlight the uneven distribution of this vulnerability. Some disasters strike the haves and have-nots in equal measure—the eruption of Vesuvius in 79 CE buried both patricians and plebeians of Herculaneum and Pompeii—but more often than not, suffering and relief correlate with political influence and material wealth. The villages most devastated by debris in the eruption of Fuji in 1707 were located at higher elevations, where the value of agricultural output was far lower than in lowland villages. Yet the lowland communities were the ones that received the lion's share of government relief since they produced far more agricultural tribute than those in the poorer uplands.

By the time it erupted in 1707, Fuji had acquired a privileged place in the collective Japanese imagination. Its popularity would only increase over time, first in Japan and then abroad, to the point that Fuji is now one of the most famous sites in the world. Representations of Fuji grew to enjoy lives of their own, and, like the icons of saints, became objects of desire in and of themselves. Today these objects include everything from protective talismans and woodblock prints to Fuji-shaped sponges and erasers.

However one views Fuji, whether directly or through representations, certain features are bound to dominate while others remain obscure. The writer Lafcadio Hearn (1850–1904), who enthusiastically introduced Japanese culture to Western audiences, reflected on this phenomenon in his late nineteenth-century account of climbing the mountain. When describing Fuji at a distance he praises the beauty of its symmetrical shape, which brings to mind "some exquisite sloping of shoulders toward the neck."[8] But in recounting his ascent he portrays the mountain as "a frightful extinct heap of visible ashes and cinders and slaggy lava," where "most of the green has disappeared. Likewise all of the illusion. . . . Above—miles above—the snow patches glare and gleam against that blackness,—hideously. I think of a gleam of white teeth I once saw in a skull,—a woman's skull,—otherwise burnt to a sooty crisp." He concludes, "So one of the fairest, if not the fairest of earthly visions, resolves itself into a spectacle of horror and death. . . . But have not all human ideals of beauty, like the beauty of Fuji seen from afar, been created by forces of death and pain?—are not all, in their kind, but composites of death, beheld in retrospective through the magical haze of inherited memory?"[9]

Any history privileges some views over others, but by treating Fuji as an actor in, and product of, both the physical world and the human imagination, I hope to dispel some of the "magical haze" that surrounds the volcano. This means breaching the wall erected between humans and nature and drawing widely on different disciplines, from geology to art history. With that in mind, I conducted my research not only in libraries and archives but also on (and under) the ground, including two summers of Lewis and Clark College's Mount Fuji Study Abroad Program that I led with a geologist, Elizabeth Safran, in 2014 and 2017. We took soil cores, counted tree species, participated in religious rituals, visited a toilet paper factory, and, of course, climbed to the summit.[10]

This book proceeds chronologically, but the story of Fuji here does not unfold in a straight line from the deeper to the recent past. Nor does it confuse trajectory with inevitability. There are moments when history might have taken a different course. Given the focus on Fuji's changing relations with humans, this book touches only briefly on developments predating the Upper Paleolithic (the Late Stone Age), when groups of *Homo sapiens* established themselves in the vicinity of Old Fuji.[11] Most of the book deals with the past few centuries, when human relations with Fuji multiplied and intensified at a rapid pace and Fuji became a mountain of global importance.

Our history starts with Fuji's geological origins and the earliest humans who lived near the volcano. It then delves into the beliefs and aesthetic sensibilities of Japan's ancient imperial court, for which the volcano was both a poetic inspiration and a deity to be feared. Fuji proved to be an especially violent god, erupting at least seven times between the late eighth and late eleventh centuries CE. The most dramatic eruption took place in 864–866, producing massive amounts of lava that engulfed thirty-five square kilometers of Fuji's northwestern skirt. Yet the court and its representatives feared Fuji mainly as a god of epidemics, situating the volcano in a network of disease that linked microbes, humans, highways, and rituals performed to appease the angry deity. At the same time, Fuji appeared in literature and art as a realm of Daoist immortals, which towered above the mere flesh-and-blood humans below.

After quieting in the twelfth century, Fuji became a deity more beneficent than destructive. First a home to mountain ascetics, it gradually evolved into a destination for pilgrims seeking divine favor. In the process, Fuji's imagined role in the spread of disease was radically transformed, thanks in large part to the ascetic Kakugyō Tōbutsu (1541?-1646). He and his followers worshipped Fuji as the "Original Father and Mother" (Moto no Chichihaha) of the cosmos and viewed the parental god as a source of healing rather than illness. Other divinities clustered around the mountain, including buddhas and bodhisattvas, the dragon goddess Benzaiten, and the "shining princess" Kaguya-hime. Pilgrims who came to Fuji could appeal for help from as few or as many of them as they liked.

Although worshippers now turned to Fuji the sacred mountain for assistance, Fuji the fierce volcano was not done wreaking havoc. The eruption of 1707 buried homes and fields and clogged waterways with ash, causing floods for decades to come. Over the previous century, Japan had experienced enormous growth in both population and agricultural output, the latter driven by new farming techniques and extensive irrigation systems. The eruption and its aftermath starkly revealed the vulnerability of a complex infrastructure designed to produce crops for subsistence, taxation, and sale to distant markets. Efforts by the Tokugawa regime to provide relief laid bare the challenges and contradictions of a political order in which the shogun shared power—and responsibilities—with the daimyo (lords) of several hundred domains operating within an increasingly national context.

Fuji worship and pilgrimage expanded steadily in the decades after the eruption, especially in the city of Edo, the seat of the shogunal government. Joining worship groups known as *Fujikō*, devotees were inspired by figures like the highly influential leader Jikigyō Miroku (1671–1733). Following in the footsteps of Kakugyō, he identified Fuji with the Original Father and Mother and articulated a cosmic—and dynamic—"bodyscape," as I think of it, in which the body of Fuji sustained the bodies of humans and other living things through its production of water and rice. Inspired in large part by Jikigyō, who dramatically fasted to death on Fuji in an ultimate act of devotion, Fujikō proliferated in the eighteenth and nineteenth centuries. These groups came together to worship on a regular basis, and some members also made the trek to Fuji during the summer months, supporting a pilgrimage economy that generated not only profits but also competition and often discord among locals who participated in it.

While Jikigyō and his followers transformed Fuji into an entity of popular worship, others incorporated the mountain into daily life through fiction, poetry, art, and consumer objects ranging from dishes to combs. As a result, sacred Fuji became a familiar friend. The people of Edo, who could see the mountain from the city on a clear day, laid special claim to it, but as time passed, Fuji entered the everyday lives of people throughout Japan, becoming a national possession—and obsession.

The link between Fuji and a sense of nationhood grew more pronounced in the face of the threat posed by the Western powers in the mid-nineteenth century. Some Japanese sought to nationalize Fuji by turning it into an object of purely Shinto—that is, native—worship. Their efforts intensified after the Meiji Restoration of 1868, when a new imperial regime replaced the Tokugawa shogunate and initially looked to unify Japan by Shintoizing it. The promotion of Shinto met with mixed success, however, and ideologues in and out of government soon found other ways to foster national sentiment within the "national body" (*kokutai*) headed by the emperor. Fuji played a symbolic role in this ideological project, appearing in schoolbooks designed to teach schoolchildren to be productive and loyal subjects. At the same time, Japanese were encouraged to climb Fuji and other mountains not for religious purposes but to foster personal health and, by extension, national vigor.

Not that Fuji was viewed by everyone in the same way. Its power as a national icon lay in its capacity to mean different things to different people, much like the nation itself. After World War II, these differences played out in a political and legal battle over who owned Fuji's summit: a Shinto shrine at the base of the mountain or the state. This conflict had its roots in local competition that dated back centuries, but it took on national significance because participants saw control of the summit of Fuji, symbol of Japan, as an indication of the sort of country Japan was and would become.

From the late 1800s, Fuji also became a prominent site of scientific inquiry. The husband-and-wife team Nonaka Itaru and Chiyoko achieved fame, for example, by attempting (and failing) to spend the winter of 1895–1896 on the summit to take meteorological readings. They aimed to make Fuji's summit into a place for empirical measurement, but the thin air and severe cold took their toll, and their mission was cut short when a rescue party took them down the mountain in December. The advance of modern science may have deepened the conceptual divide between humans and nature, but the Nonakas' grueling experience—and the publicity surrounding it—attest to the ways in which the creation of this divide has involved living, breathing people entangled with both the objects of their study and the expectations of their fellow humans.

As Fuji developed into a national symbol and scientific object, it became increasingly enmeshed in commodified networks, first nationally, then globally. The production of tea, textiles, and paper depended, and continues to depend, on the steady supply of water at the base of the mountain. Industrial growth created environmental problems along with economic prosperity, however. The most notorious was the pollution of nearby Suruga Bay, which had grown so dire by the 1960s that it was featured in a Godzilla film. Despite efforts to clean up the bay, the bodies of sharks and other marine life are still contaminated with industrial chemicals.

Environmental impacts have been anything but straightforward, as the militarization of common lands at the base of Fuji from the 1890s demonstrates. Initially used by the army of imperial Japan, since World War II these common lands have hosted both the Japanese and American armed forces. Military leaders found the region attractive not only because it was close to Tokyo but also because of its vast grasslands, which were maintained *as* grasslands by locals who regularly mowed and burned them. Previously used to feed horses, grow edible plants, and thatch roofs, today they are burned to preserve a terrain suitable for artillery practice and other military exercises, which in turn generates rental fees from the government for common land associations. Were it not for those fees, the associations would likely stop burning the grasslands, since there is little demand today for horses or thatch. Elsewhere in Japan former grasslands once maintained by burning or mowing have turned into forest—or golf courses. It might seem unproblematic to allow nature to take its course and let the grasslands around Fuji disappear into forest as well. But should this occur, endangered butterfly species and other organisms that depend on grassland plants would disappear, too. The curious fact is that these organisms have the militarization of Fuji to thank for the maintenance of their habitat and thus their survival. In an epoch so dominated by *Homo sapiens* that it is frequently called the "Anthropocene," ecological entanglements of this sort are multiplying at a rapid pace.

Fuji may be famous throughout the world, but its contradictions, like its summit, are often obscured by mist. Designated a UNESCO World Heritage Site in 2013, it is depicted as stable, peaceful, and even timeless. Yet it has erupted many times in the past and will almost surely erupt again. It is an awe-inspiring example of natural beauty, but its water supports polluting industries and its lower slopes are home to artillery ranges. It is one of Japan's most prominent symbols of national unity, yet its history is full of economic, political, and ideological conflict. What follows is therefore a biography of multiple lives, multiple Fujis. It is also a parable—one that shows how fantasies about nature can be destructive as well as creative, for humans and nonhumans alike.

1

Active Volcano, Lofty Mountain

Rising between the lands of Kai
and Suruga,
    where the waves draw near,
is Fuji's lofty peak.
It thwarts the very clouds
        from their path.
Even the birds
    cannot reach its summit
        on their wings.
There, the snow drowns the flame
and the flame melts the snow.
I cannot speak of it,
I cannot name it,
this occultly dwelling god!
It envelops the waters
that we call a sea,
    the Sea of Se.[1]
The river called Fuji
that men must cross
is but a stream of this mountain.
God who dwells there,
defender of the sun-source,
    the land of Yamato,
treasure of a mountain!
Though I gaze
on Fuji's lofty peak
        in Suruga,
I never tire.[2]

Usually attributed to low-ranking aristocrat Takahashi Mushimaro, and appearing in the *Collection of Ten Thousand Leaves* (*Man'yōshū*), a canonical anthology of Japanese poetry composed in the seventh and eighth centuries, this poem celebrates Fuji in its singular, volcanic glory.[3] Bringing together the polar opposites of fire and snow, Fuji's divine power is as uncanny as it is impressive. To emphasize the boundlessness of this power, the characters originally used to write "Fuji" in Mushimaro's poem are 不盡, meaning "never exhausted." Other poems in the *Collection of Ten Thousand Leaves* convey Fuji's majestic status with the characters 不二, "peerless," and 布時, "timeless."[4]

While Mushimaro exalts Fuji as an awe-inspiring divinity, he also situates it within a bureaucratic geography. The poem starts with a reference to the provinces (*kuni*) of Kai and Suruga and ends with another mention of Suruga, locating Fuji in a terrain organized in service to a court-centered state (see map 1.1). It identifies Fuji as "defender of the sun-source, the land of Yamato," which not only places the divine volcano in the territory claimed by that state but also enlists its divine power to protect it.

At the time the poem was written in the early eighth century, the land of Yamato stretched from the southwestern island of Kyushu to northeastern Honshu and was ruled by a dynasty from its court in Nara, located several hundred kilometers to the west of Fuji and thus well out of its destructive reach. This physical distance is important to keep in mind when considering the place of Fuji in the Yamato imagination. The vast majority of aristocrats living in the capital of Nara never laid eyes on the volcano, and although Mushimaro's poem exalts Fuji as a protective deity, Fuji was not a cultic center for the Yamato court as was Mount Miwa (on the outskirts of Nara), Ise (on the Kii Peninsula southeast of Nara), or Izumo (in western Honshu). In fact, Fuji does not even make an appearance in the *Record of Ancient Matters* (*Kojiki*) or the *Chronicles of Japan* (*Nihon shoki* or *Nihongi*), the two ancient myth-histories completed in the early eighth century to bolster the legitimacy of the Yamato dynasty. And although thirteen poems in the *Collection of Ten Thousand Leaves* mention Fuji (this includes "envoy" stanzas that accompany longer poems), they make up only a small number of the anthology's mountain-centric poems, with twice as many focused on Mount Tsukuba alone.[5] This is not to diminish Fuji's place in ancient Japan, only to put it in perspective. It is all too easy to project the present status of Fuji as a national icon back in time, but those responsible for the written record of ancient Japan regarded Fuji as just one divine force among many that could be brought into an alliance with the Yamato court.

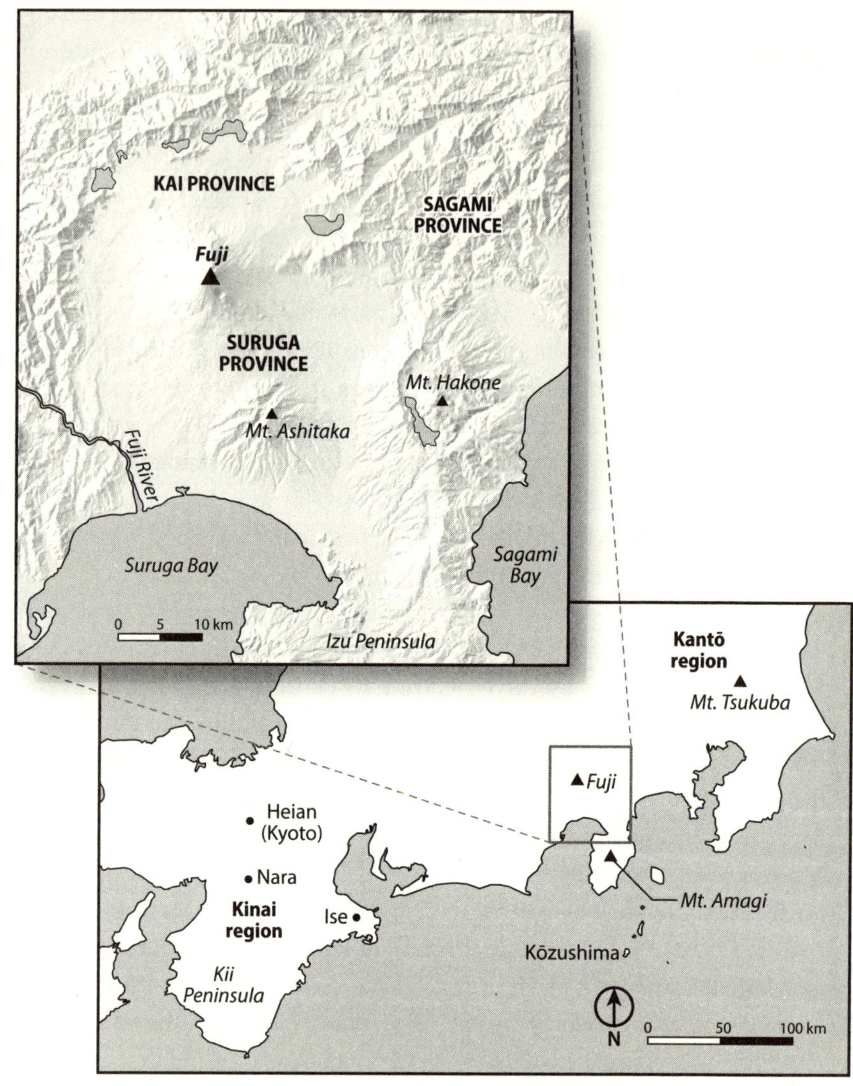

MAP 1.1. Central Japan with close-up of Fuji region.

The following pages trace the processes, both human and nonhuman, that eventually drew Fuji into that alliance. Mushimaro's poem foregrounds Fuji's nature as a volcano by describing its snow-melting flames and tremendous height and emphasizes its agency by referring to it as a god. This chapter, too, treats Fuji not as a static place where the history of ancient Japan unfolded but as an active—sometimes very active—force in the making of that history, one reaching back millennia before such a thing as the Yamato court existed. Travel into that distant past shows how Fuji evolved not apart from humans but together with them, generating a terrain that provided both physical sustenance and food for the imagination.

Fuji and Humans in Motion

Fuji is located on the largest island in an archipelago now known to English speakers as Japan and to Japanese speakers as *Nihon*. The phrase "*now* known" serves as a reminder that the national framing of this geographical space is a relatively recent invention. The term "Japan" appears in the pages that follow, but only as shorthand to designate a series of islands (the main four being Hokkaido, Honshu, Shikoku, and Kyushu) whose geological boundaries overlap conveniently, not inevitably, with later national ones.

Japan sits where the Pacific, Philippine, Eurasian, and Okhotsk (or North American) plates converge, making it one of the most tectonically active segments of the "Ring of Fire" that encircles the Pacific. Responsible for the volcanoes forming this ring is a process called subduction, in which certain plates bend and sink, or "subduct," under others. The subducting plate is always capped by oceanic crust, which, unlike continental crust, is dense enough to be forced into the asthenosphere, the semisolid, slowly flowing upper mantle just beneath. As the subducting plate sinks under either a continental plate or a neighboring oceanic one, it creates a deep underwater trench where devastating earthquakes are generated. Subduction also introduces volatile substances such as water and carbon dioxide into the asthenosphere, lowering its melting point and thus allowing pockets of liquid magma to form. Magma then finds its way to the surface (where it is called lava), and with that, a volcano is born.[6]

Fuji stands at the triple junction of the Eurasian, Okhotsk/North American, and Philippine plates, with the Pacific Plate sinking under the Philippine Plate just to the east (see map 1.2). Although at a slight remove, the subducting Pacific Plate drives the generation of magma constituting Fuji.[7] It is also responsible for the volcanic arc—or chain of volcanoes—that produced the Izu Peninsula and the Bonin and Mariana islands stretching south of Japan into the Pacific. The arc began colliding with what is now the Japanese archipelago's main island of Honshu about fifteen million years ago (when much of the archipelago was under water), thus forming the geological "basement" that underlies Fuji and its environs.[8]

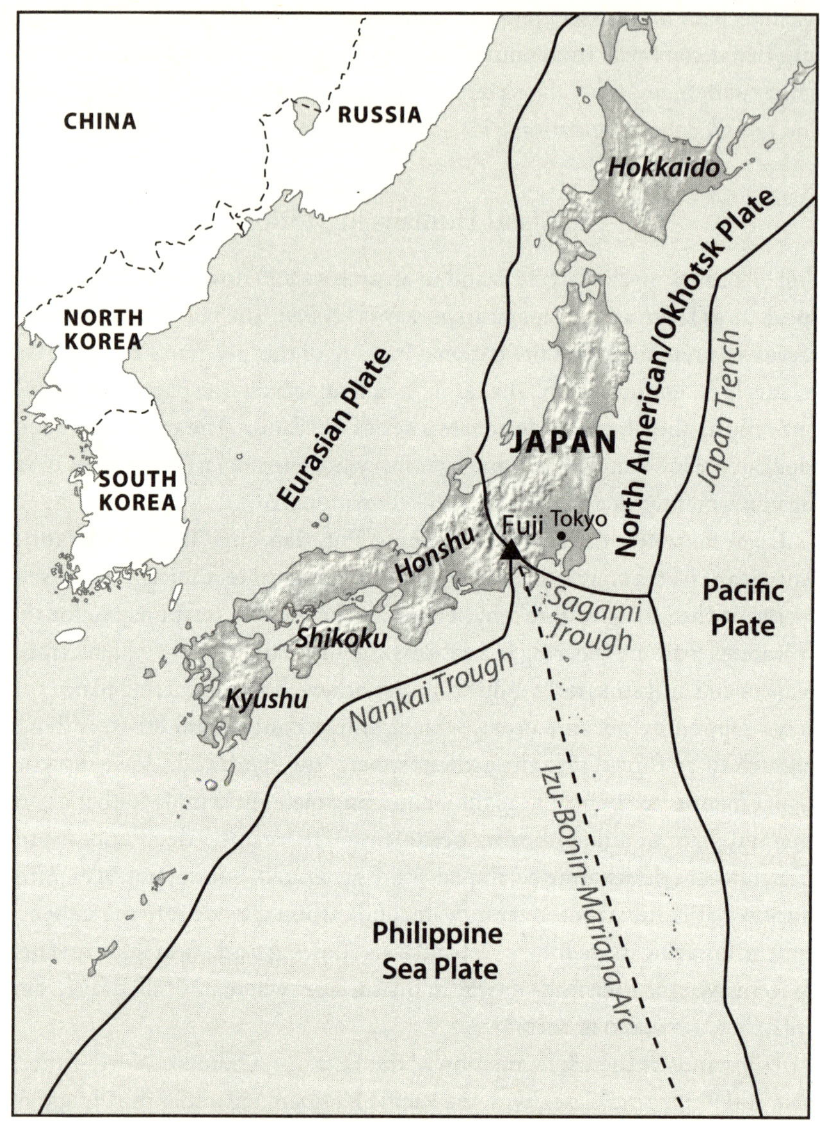

MAP 1.2. Tectonic plates in Northeast Asia.

Typical of volcanoes in subduction zones, Fuji is a "stratovolcano," built by alternating eruptions of flowing lava on the one hand and airborne tephra—material ejected by volcanic eruptions ranging in size from ash particles to boulders—on the other. Having produced unusually large amounts of magma in the geologically short time span of one hundred thousand years—its

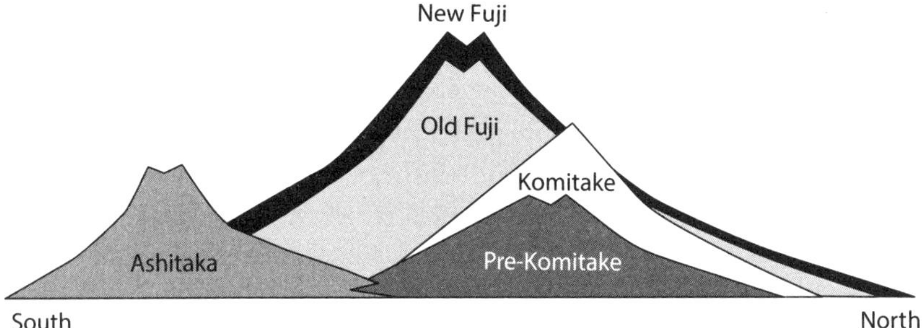

FIGURE 1.1. Cross-section showing Fuji, pre-Fuji, and Ashitaka volcanoes. Adapted from diagram in Aramaki, "Fujisan no tokushoku," p. 157.

average eruption rate of four to six cubic kilometers per thousand years is far greater than the rate of 0.01 to 0.1 cubic kilometers per thousand years for other volcanoes along the Izu-Bonin-Mariana Arc—Fuji now reaches 3,776 meters (12,389 ft) high, making it by far the tallest volcano, and indeed the tallest mountain, in Japan.[9]

The volcanic edifice—or cone—that we see today is what volcanologists call "New Fuji" (Shin Fuji), which appears to have started taking shape about seventeen thousand years ago and became especially active from approximately eleven thousand years ago (explaining why many consider the latter date to be the starting point for New Fuji). Buried beneath New Fuji is "Old Fuji" (Ko Fuji), which began growing about one hundred thousand years ago.[10] Some argue that the difference between the two is negligible, but New and Old Fuji are widely considered to be distinct volcanoes because their mineral ratios differ.[11] Even older than Old Fuji is Komitake, which protrudes from the northeastern slope of New Fuji. Deep-core drilling in 2001–2002 revealed yet another, "pre-Komitake" volcano below that (see fig. 1.1). New Fuji therefore owes its height not only to its own eruptions but also to those of its several predecessors.[12]

New Fuji developed in five stages. In Stage One (17,000–8,000 years ago), large amounts of lava flowed continuously from both the summit and lateral vents (eruptive openings on the volcano's flanks). It was during this stage that New Fuji distinguished itself from Old Fuji. Stage Two (8,000–5,600 years ago) was defined by smaller, intermittent eruptions that produced airborne tephra; while Stage Three (5,600–3,500 years ago) was characterized by an increased number of intense, explosive eruptions and midscale lava flows. In Stage Four (3,500–2,200 years ago), the crater at the summit frequently and explosively erupted, while during Stage Five (2,200 years ago to the present), eruptions of lava and airborne tephra shifted to lateral vents.[13]

Even when not erupting, Fuji was constantly on the move. Rain and melting snow continually reconfigured the material heaped on and around the volcano by triggering debris flows and "slush flows" known locally as *yukishiro*. Composed of snow, volcanic debris, and water, *yukishiro* usually strike in the early winter or spring.[14] Fuji's chronic restlessness, combined with periodic eruptions, explains why the height of the timberline varies by hundreds of meters from one part of the mountain to the next. It is also why the subalpine zone (roughly 1,800 to 2,850 meters in elevation) is still dominated by pioneer species hardy enough to colonize harsh and rapidly shifting environments, and why numerous small plants as well as the creeping pine (*haimatsu*) common in the Japanese Alps are absent.[15] Because Fuji is so young and relatively isolated, the Japanese monkeys (*macaca fuscata*) that thrive in mountainous regions of Japan are not found on its slopes (ironic perhaps, in that according to legend Fuji emerged in the year of the monkey).[16]

So while admirers have often portrayed Fuji as timeless, the volcano has always been a product and driver of change. For tens of thousands of years, humans and nonhumans have had to contend with its eruptions along with its frequent slush and debris flows, making Fuji an active participant in history, not merely a passive backdrop for it.

The earliest evidence of humans living near Fuji—in fact, anywhere in Japan—has been unearthed at the southeastern foot of Mount Ashitaka and the western foot of nearby Mount Hakone, and dates to what archaeologists identify as Japan's Early Upper Paleolithic (38,000–30,000 years ago). Ashitaka is a volcano immediately south of Fuji that went extinct about one hundred thousand years ago. Hakone is a volcano located just to the east of Ashitaka that continues to release sulfurous gas to this day (see map 1.1).

The Ashitaka-Hakone region was attractive to Paleolithic humans thanks to the availability of food and water and relatively easy access to obsidian, a substance ideal for knapping, or shaping, sharp tools. Warmed by the nearby waters of the Pacific, the climate supported temperate forests in which hunter-gatherers could find a greater quantity and diversity of edible plants and animals than in most of Pleistocene Japan (2.58 million to 11,700 years ago).[17] We know they lived in this region because they dug trapping pits that archaeologists have excavated over the past several decades. The size and shape of these pits suggest they were used to capture midsized deer and boar rather than the elephants and giant Yabe deer that also roamed the area. Using tephrochronology, a technique that estimates the age of objects found in, under, or between deposits of tephra by linking the geochemical composition of those deposits to particular eruptions, archaeologists have dated them to between thirty-four and thirty-one thousand years ago, making them among the oldest trapping pits in the world.[18]

In addition to the pits, archaeologists have identified a large number of Paleolithic tools made of obsidian, or "volcanic glass," which can be traced to particular sources. Because each source of obsidian has a unique geochemical "portrait" or "signature," obsidian artifacts provide clear evidence of transport from one place to another—a boon for archaeologists seeking to understand procurement and exchange networks.[19]

The obsidian tools discovered near Fuji are important evidence of the movements and behaviors of Upper Paleolithic humans in the Japanese archipelago. There are two reasons for this: first, dating as far back as thirty-eight thousand years ago, they are among the earliest tools discovered in Japan; second, the encampments in which they were found sat at the crossroads of obsidian procurement and transportation routes that extended east to the Kantō Plain, north to the mountainous central region of Honshu, and south to the island of Kōzushima, today located about fifty kilometers offshore from the Izu Peninsula (see map 1.1).[20] These routes connected sources of obsidian to sites that were in some cases located up to two hundred kilometers away.[21] Toolmakers wanted to use obsidian as free of inclusions (rock fragments trapped in cooling obsidian) as possible, so the higher the quality of the material, the greater the distance traveled to obtain it. The obsidian found at Hakone and at Mount Amagi (on the Izu Peninsula) is inferior to that from Kōzushima and several places in the highlands to the north, so the former was used almost exclusively within a radius of one hundred kilometers while the latter traveled twice that distance.[22]

Fuji did not provide the obsidian used by humans who lived nearby. Obsidian is formed by rhyolitic lava, which has a silica content ranging anywhere from 68 to 77 percent. Fuji is an outlier among stratovolcanoes because it mainly produces basalt, a fine-grained rock with a silica content of only about 50 percent.[23] The greater the silica content, the more viscous the magma, and the more viscous the magma, the more liable it is to trap gases that then explode when it reaches the crust's surface.[24] This means rhyolitic magma can be highly explosive, producing large amounts of airborne tephra, while less viscous basaltic magma tends to produce lava flows that can build massive but gentle-sloping "shield" volcanoes like those found in Hawaii. Stratovolcanoes consist of rocks formed from different types of magma, but most commonly those rocks are "intermediate" in composition, meaning that their silica content—anywhere from 52 to 63 percent—falls between that for rhyolite and basalt. This explains why stratovolcanoes are more explosive than entirely basaltic shield volcanoes, and why Fuji, a frequently explosive stratovolcano in the Ring of Fire, is unusual in that it mostly consists of basalt.[25]

Fuji may not have produced obsidian, but its size, shape, and location provided access to it. The mountain's gently sloping skirts offered relatively easy passage from Suruga Bay to routes leading through valleys and over mountain passes to the obsidian located in the highlands to the north. In addition, Fuji's height encourages the development of distinctive cloud formations at its summit—such as the "cap cloud" that in spring and summer usually indicates rain is on the way[26]—thus signaling to travelers what weather they could expect. Forecasting the weather would have been important to those traveling by sea to and from Kōzushima;[27] and although we have no proof that they looked to Fuji for guidance, we can presume they did. Even in the age of satellite forecasts, people still look to Fuji to see what weather is in store. In an interview given in 2005, veteran fisherman Inaki Noboru recalled how his grandfather taught him to monitor Fuji's clouds when fishing in Suruga Bay: "Often he'd stop work suddenly and turn the boat toward port, saying the wind was going to blow hard and make high waves. I'd look up and see that the top of the cloud above Mount Fuji had an up-and-down wavy shape. In time, without consciously trying to do so, I knew how to forecast the weather from the shape of the clouds on the mountain." He added that, although he listens to weather forecasts, "the weatherman is often wrong. Sometimes you can get a more accurate forecast by looking at Fuji's clouds."[28]

Humans are storytellers, so just as surely as Paleolithic people looked to Fuji to gauge the weather, they must have swapped tales about it. Maybe they created myths that paired Fuji's eruptions with storms at sea. Perhaps they featured the volcano in sagas with themes of love and revenge, death and survival. We will never know if this was the case, but anyone who saw Fuji rising in the distance would have had something to say about it, making Fuji's viewshed at the same time a "storyshed" in which mobile groups of humans exchanged tales as well as things. These people are often described as "prehistoric," but the pits they dug, the tools they made, and the stories they told made them not the forerunners of history but participants in it. A lack of written records does not mean a lack of history. It just means we cannot see it as fully as we would like.

Neolithic Fuji

The humans who first migrated to Japan 40,000–38,000 years ago—probably by boat from the Korean Peninsula—lived in a climate relatively warm by Pleistocene standards.[29] It then cooled significantly from about 29,000 years ago as the planet headed toward the Last Glacial Maximum, the period in which the most recent continental ice sheets reached their greatest extent. Although temperature trends and the advance and retreat of ice sheets differed according to region, the maximum lasted from approximately 26,500 to 19,000 years ago.[30] At its height, temperatures on the southern side of Fuji were on average about seven to eight degrees Celsius lower than they are today, making the climate similar to what is currently found in Hokkaido.[31]

With more water locked up in the form of ice, sea levels dropped to the point that Hokkaido was joined to the continent, and a strait less than two kilometers wide separated it from the rest of Japan.[32] Archaeological discoveries, as well as the genetic analysis of skeletal remains and living populations, show that humans in northeast Eurasia looking for milder climes took advantage of this land bridge to reach the Japanese archipelago.[33] To what extent this new wave of migrants intermingled with existing populations is unclear. So is the degree to which they contributed to the extinction of Japan's Pleistocene megafauna, which, depending on the specific region and species, vanished during or soon after the end of the Last Glacial Maximum.[34]

We do know that starting between about seventeen to sixteen thousand years ago, the descendants of these migrants developed pottery and other Neolithic (New Stone Age) technologies that profoundly altered the ways humans manipulated and experienced the world around them. This was particularly true from about eleven thousand seven hundred years ago, when the chilly Pleistocene gave way to the much warmer Holocene, the climatically hospitable epoch in which everything we associate with "civilization" developed.

The Neolithic technologies and ways of life that took shape in Japan toward the end of the Pleistocene and flourished in the early Holocene are collectively known as Jōmon culture, "Jōmon" referring to patterns created on pottery by pressing cords onto wet clay. Hunter-gatherers of the Jōmon period led more sedentary lives than those of their Paleolithic forebears. This was thanks in large part to the food-filled deciduous forests that spread throughout Holocene Japan as well as human manipulation of local ecologies. Jōmon people collected a rich array of plants, roots, and nuts produced by broadleaf woodlands; and while they did not develop grain-based agriculture, they actively

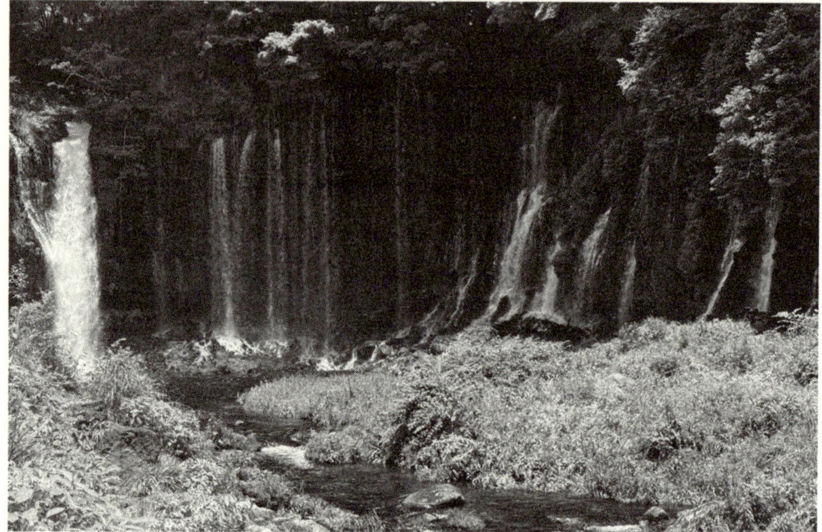

FIGURE 1.2. Shiraito (White Thread) Falls in Fujinomiya. Photo by author.

encouraged the growth of beans, burdock, chestnut trees, and other species.[35] Like Stone Age counterparts across the world, they also lit fires to produce grassland and open woodland environments optimal for hunting game and gathering edible plants, including lands at the base of Fuji.[36]

The Jōmon people who lived around Fuji built settlements where they could take advantage of its abundant water. Rain and snowmelt quickly sink into Fuji's highly permeable lava and tephra, meaning its upper reaches lack water close to the surface. But conducted via layers of lava fragments ("clinkers") at the boundaries of lava flows, it becomes available at lower elevations as springs, spring-fed rivers, ponds, lakes, and waterfalls (see fig. 1.2). It can take decades for water to percolate through these layers before emerging at the foot of the mountain, the amount of time differing according to where on Fuji it originates.[37] Numerous springs are located at the ends of lava flows around Fuji, with large concentrations to the south in what are today the cities of Fujinomiya, Fuji, and Mishima (see map 1.3).[38] These springs supply a number of small- to moderate-sized rivers, although not the Fuji River, which, despite its name, originates in the mountains to the north. Immediately to the north of the volcano are the renowned "Fuji five lakes" of Motosu, Shōji, Sai, Kawaguchi, and Yamanaka. Recent studies suggest that their water comes primarily not from Fuji but from nearby mountains.[39]

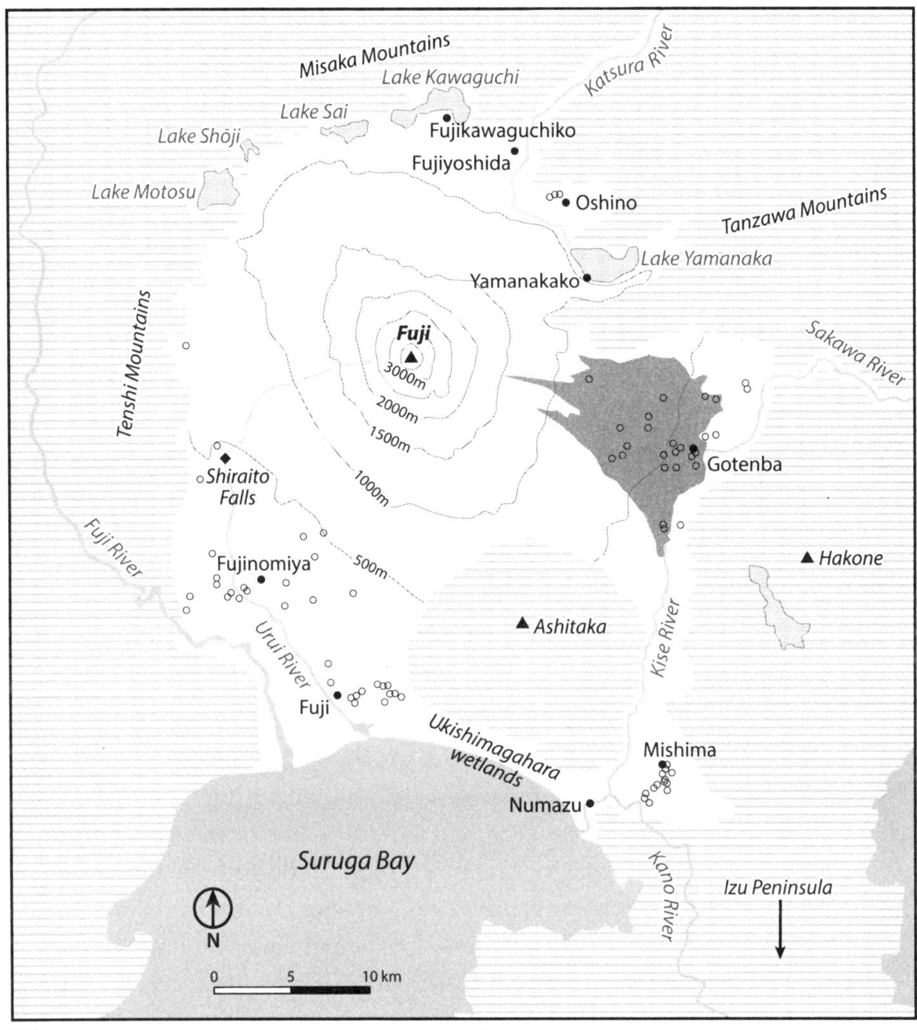

MAP 1.3. Fuji groundwater system. Adapted from spring distribution map in Tsuchi, "Fujisan chikasui, yūsui," p. 377.

Jōmon settlements have been found along river terraces to Fuji's south and near springs at higher elevations to its north.[40] Three of the better-studied sites are at Ikenomoto, Tenmanzawa, and Sengo. Ikenomoto is located in Fujiyoshida City, Sengo in Fujinomiya City, and Tenmanzawa in Fuji City. At all three locations archaeologists have uncovered ceramics, stone tools, and the footprints of pit dwellings characteristic of Jōmon villages throughout central and eastern Japan, with the ones in Ikenomoto dating as far back as ten thousand years. These dwellings consisted of pits covered with roofs of wood and/or thatch to ward off the elements, and the fact that Jōmon people spent the time and energy to construct them is one sign among many that they stayed put for greater lengths of time than their Paleolithic forerunners.[41]

This does not mean they resided in only one spot. Although these hunter-gatherers might have spent much or almost all of their time at a particular settlement, they established outposts to take advantage of resources distributed across different locations and seasons as well.[42] While almost two-thirds of the stone tools found at Tenmanzawa appear to have been used to dig up roots, the majority at Sengo are arrowheads, indicating that hunting predominated there.[43] Jōmon people also developed extensive exchange networks to acquire what they could not find close at hand. Procurement of high-quality obsidian suitable for arrowheads and finely worked tools intensified during and after the Paleolithic-Jōmon transition, leading to large-scale mining and systematized trade networks. Continuing a pattern from the Paleolithic period, obsidian from Kōzushima and several sites in the highlands of central Honshu was especially prized, with exchange routes running along the Fuji River and Fuji's eastern skirt.[44]

In addition to supplying their everyday material needs, Jōmon people created structures and objects that appear to have been explicitly devoted to ceremonial purposes. At Sengo, Tenmanzawa, and other sites around Fuji, archaeologists have uncovered configurations of stone, often circular, that demarcate what appear to have been ritual spaces.[45] Sometimes associated with burials and sometimes not, these configurations can be found throughout central and eastern Japan, with most dating from the mid- to late Jōmon period.[46] The specific practices associated with each are a mystery, but the orientation of stones at Tenmanzawa and other sites within view of Fuji suggests the mountain may have been a ritual focal point, both literally and figuratively.[47] At Sengo, circular arrangements of stone are accompanied by graves, so they might have served as places where people aimed to assist the dead and/or beseech them to help the living. Because hunting took place at Sengo, people may also have used them to conduct rituals designed to increase the odds of catching prey. On the opposite side of Fuji at Ikenomoto, archaeologists have found traces of two Jōmon pit dwellings distinguished by their stone floors and the raised terraces surrounding them. Since they are located next to a burial ground, they too may have been involved in death practices of some kind.[48]

Ritual objects provide further material to speculate about Jōmon belief systems. Among the most common items to be discovered at Jōmon sites across Japan, including Tenmanzawa and other places near Fuji, are clay figurines (*dogū*) with mostly female attributes that appear in many cases to have been intentionally smashed. Archaeologists often interpret them to be fertility symbols, while some point to a Japanese folk practice of destroying dolls at times of epidemic or disaster to argue that the figurines were similarly destroyed in rituals to cure illness.[49] Stone rods (*sekibō*) have been found at Jōmon sites as well. Whereas the figurines mainly seem to be female, the rods, because of their phallic shape, are thought to signify male energy; and some scholars argue that they, like the figurines, were used as fertility symbols. Because of their association with males, other Jōmon experts propose they were used in hunting rites, a theory that might explain why one was found at Sengo.[50]

Just as the artifacts found near Fuji are representative of Jōmon culture, the decline of settlements around the mountain toward the end of the Jōmon period reflects a broader demographic trend. Jōmon culture originated in an "incipient" phase approximately 15,630–11,200 years ago and ended between 2,400 and 2,300 years ago.[51] From the early to mid-Jōmon period, the population of central to northeastern Honshu grew considerably, increasing from 15,100 people to 238,600, according to the widely cited estimates of Jōmon scholar Koyama Shuzo. By the late Jōmon period, that number had plunged to as low as 58,300.[52] What caused this drop in population is uncertain, but a cooling climate seems to have been a significant reason.[53]

Fuji's volcanic activity contributed to this decline on a local level. During Stage Four (3,500–2,200 years ago) of New Fuji's development, its activity intensified, making life much more hazardous for those within its destructive reach. Around 3,500 years ago, Fuji buried the Ikenomoto site under a layer of tephra.[54] With the help of earthquakes as well as eruptions of Hakone and Amagi, it also wreaked destruction to the south and east, where it caused a decline in settlements severe even by late Jōmon standards.[55] To make things worse, around 2,900 years ago, Fuji's southeastern flank collapsed, creating a massive debris avalanche. Much of it over thirty meters in depth, the Gotenba (or Gotemba) avalanche deposit forms a hillock-filled terrain covering fifty-three square kilometers between five hundred and one thousand meters above sea level (see map 1.3). Following the collapse, secondary mud and debris flows spread to the south and east, creating layers many meters thick—in one location reaching a depth of forty-eight meters. Traveling down the Sakawa

River to the east, fluvial deposits reached as far as Sagami Bay, and following the Kise River to the south, nearly made it to Suruga Bay too.[56]

While settlements decreased in size and number from central Honshu to Hokkaido in the late Jōmon period, in thinly populated western Japan they remained steady, and at the close of the Jōmon period, they shot upwards. Why western Japan had held less appeal for so long is unknown, but the most important factor was probably the dominance of evergreen broadleaf forests, which yield less edible vegetation than deciduous ones.[57] Between about 3,000 and 2,600 years ago, waves of migration from the continent brought a new way to secure food: agriculture. Introduced to northern Kyushu from the Korean Peninsula, this form of ecological manipulation generated calories on an unprecedented scale, fueling population growth in western Japan and thereby placing Fuji to the east of the archipelago's newly emerging demographic and cultural heart.

This shift represented a dramatic reversal from Jōmon times, when that heart was located in eastern Japan, and communities around Fuji, judging from their pottery styles, homes, ritual objects, and obsidian trade routes, were oriented to the Kantō and the highlands to the north, not the sparsely populated regions to the west.[58] In the centuries that followed, people inhabiting the communities that ringed Fuji remained culturally and economically tied to the east, although western Japan generated the forces of change—agriculture foremost among them—that would increasingly transform their lives.

Paddies, Tombs, and the Yamato

The arrival of agriculturalists from the Korean Peninsula initiated the Yayoi period (named for an archaeological site in the Yayoi neighborhood of present-day Tokyo), which, depending on the archaeological benchmarks used, began anywhere from 1,000 to 600 BCE and ended around 300 CE.[59] In some regions the newcomers from Korea may have entirely displaced existing Jōmon groups, but archaeological discoveries and genetic analysis show that a certain degree of intermingling must have occurred, creating a cultural hybrid that first took shape in northern Kyushu and western Honshu and spread from there.[60]

The adoption of large-scale agriculture transformed Japan. Plants had already been domesticated during the Jōmon period, but none could provide enough calories and nutrition to lure people away from a diet based mainly on collecting food to one that depended on producing it.[61] In contrast, the impressive yields generated by Yayoi rice paddies, and to a lesser extent the more modest harvests of dry-field farming, induced communities in Japan to adopt a predominantly agricultural lifestyle, which in turn led to substantial population growth. Those who adopted agriculture did not instantly or completely abandon hunting and gathering, but as time passed, Yayoi food production radically altered how people organized and imagined their lives.

With the increase in population came greater competition for resources and the need to defend them—as the Yayoi period's ditch-enclosed settlements and fortifications attest—leading to the emergence of warrior-elites who lived, and were buried, apart from commoners.[62] Placed in their graves were items indicating both their warrior status (whether or not earned on the field of battle) and their ability to take advantage of long-distance trade. These included daggers, mirrors, bells, and other grave goods made from bronze, which were either imported in finished form from Korea or China or made from materials originating there.[63] As the Yayoi period wore on, elites were also buried with weapons made from iron, a material far stronger (and deadlier) than bronze.[64]

The Yayoi culture that originated in northern Kyushu and western Honshu gradually spread east to the Kantō and beyond, but this does not mean it uniformly unfurled from west to east or that it was always adopted whole. Ecological conditions in particular shaped to what degree and in what form its key component, agriculture, supplemented or replaced hunting and gathering; and this was especially true in a volcanic landscape like Fuji's, where differences in elevation, water supply, and soil quality were extreme.

Volcanoes can be both a blessing and a curse as far as agriculture is concerned. On the upside, they can provide bountiful supplies of water. On the downside, volcanic material is often so permeable that it fails to retain sufficient amounts of water close to the surface. And while volcanic soils known as andosols or andisols are rich in certain nutrients, they also contain high amounts of aluminum that bond with phosphorus, making the phosphorus unavailable to plants that depend on it to grow.[65] This is the case for the "Kantō loam," which was produced largely from the ash spewed by Mount Fuji and Mount Asama. Found throughout the Kantō region, it requires copious amounts of fertilizer to support the production of rice and other crops.[66]

The terrain and climate in the immediate vicinity of Fuji could prove especially difficult for farmers. Intensive agriculture was possible at lower elevations in places with the right kind of soil and access to water, but locations higher up were problematic because of steep, unstable slopes and cold temperatures. This was particularly true on the northern side of the mountain.

Always looming, of course, was the danger of a major eruption that could lead to catastrophic losses and render once-fertile land unusable for generations.

Despite this danger, by the end of the early Yayoi period, rice paddy cultivation had taken hold in and around the Ukishimagahara wetlands located between Suruga Bay and the southern skirts of Fuji and Ashitaka (see map 1.3).[67] About ten thousand years ago, the bay extended up to the lower skirts of each volcano. But material carried by the Fuji, Kano, and Kise rivers, as well as streams originating on Ashitaka, built sand reefs that eventually blocked the bay, creating a lagoon that, thanks to continued sediment deposition, was surrounded by marsh by the time rice farmers came on the scene.[68] Judging from the location of their settlements, Yayoi farmers also utilized the bottomlands of the Kise and Kano rivers.[69] Early Yayoi sites discovered on the northern side of Fuji—several by Lake Kawaguchi (one of them on an island in the lake) as well as the Oshino springs region—were also located near areas suitable for wet-field rice farming. It would have been possible only on a limited scale in these places, however, requiring inhabitants to rely heavily on hunting and gathering.[70]

Even later in the Yayoi period, paddy cultivation in areas immediately adjoining Fuji failed to assume the prominence that it did in more amenable regions of Japan, including the Kōfu Basin just to the north.[71] Torrential rains make the bottomlands alongside rivers highly vulnerable to flooding, and this is truer still for alluvial wetlands located by the coast, where storms bring devastating tidal surges in addition to heavy rain. Following a pattern seen around Japan, farmers to the south of Fuji used the more manageable headwaters and tributaries of major rivers to create irrigation ponds that could make growing rice more reliable, but despite the production of more predictable yields, it appears that most communities living around Fuji did not subsist on rice. Excavated remains of fishing gear show that people near the Suruga coast, for example, depended largely on seafood.[72] Wherever they were located, settlements around Fuji remained relatively modest in size throughout the Yayoi period when compared with their counterparts in northern Kyushu and western Honshu, where rice paddy farming was far more developed.

In the middle of the third century CE, the Yayoi period transitioned into the Kofun period, named after large tumuli constructed for the burial of elites. Lasting to the middle of the sixth century (or, depending on one's periodization, the mid-seventh century), this period witnessed an unprecedented consolidation of power that eventually resulted in the formation of the dynasty known as the Yamato.

Often surrounded by *haniwa*—clay figurines of people, animals, tools, and so forth—and containing weapons and prestige goods made both in the archipelago and on the continent, tombs were initially constructed in shapes that differed somewhat according to locale. From the start, the grandest were built in the Kinai region of western Honshu (see map 1.1)—today home to the cities of Kyoto, Nara, Osaka, and Kobe—and were formed in a distinctive "round keyhole" style that combined a circular mound with a flared tail. Over time this style spread to other parts of Japan, providing scholars a way to map the growth of a "supraregional league of chieftains" (Joan Piggott's term) that initially cohered, it seems, more through ritual diffusion than military subordination.[73]

Resources on and immediately around Fuji were limited, suggesting why no major tombs were constructed at the base of its northern, western, or eastern flanks.[74] Power holders in the rice-growing lowlands near Suruga Bay, however, did have the means to build impressive tumuli.[75] The oldest, dating to the Yayoi-Kofun period transition in the third century, is the Marugaito tomb, located in today's Fujinomiya City. A relatively modest twenty-six meters long and consisting of a square mound and tail surrounded by a trench, it sits alongside a cluster of smaller graves and is situated so that its axis points toward Fuji, apparently associating the grandeur of the mountain with the stature of the deceased.[76] The second-oldest tumulus, located to the east in Numazu City and sixty-two meters in length, also has a square mound, but the third-oldest, Numazu's fifty-three-meter-long Shinmeizuka tomb (like the other two, dating to the third century), is shaped like a round keyhole, closely resembling its counterparts in the Kinai region several hundred kilometers to the west.[77]

In the fourth century, three more tumuli for local magnates were built south of Fuji—the largest, the ninety-meter-long Sengen tomb in Fuji City, with a square platform; and the two others, a seventy-meter-long tumulus also located in Fuji City and a sixty-meter-long one in Mishima City, in the Kinai's round keyhole style. None were constructed from the end of the fourth to the end of the fifth century, however, resulting in an anomalous one-hundred-year gap at the height of Japan's Kofun period.[78] Perhaps political changes were responsible, but it appears that Fuji's volcanic activity might have been to blame.[79]

Construction of large tumuli in the region picked up again in the early sixth century but, following a pattern seen throughout much of the archipelago, soon came to an end.[80] While keyhole tumuli built from the third to sixth centuries demonstrate the influence of rulers in the Kinai region, their decline and replacement by clusters of small tombs is explained by outright domination by those rulers, who squelched mortuary competition while continuing to build large tombs of their own.[81]

By the end of the seventh century, the construction of large tumuli ceased throughout all of Japan—even in the Kinai, where the Yamato lineage of rulers had created a dynastic state with more systematic and far-reaching ways of asserting dominance. Under Yamato leadership, labor that had once gone to building tombs was steered toward other purposes, such as constructing roads, irrigation systems, and monumental buildings. This same dynasty (which has continued in one form or another to the present) also sponsored the development of Buddhist temples, which took over the memorialization of the dead.[82]

The foundation of the Yamato dynasty took shape from the fifth to seventh centuries as paramount rulers not only extended their influence in the archipelago but actively participated in the wars on the Korean Peninsula. These wars generated a stream of refugees who brought to Japan technologies and ideas that the dynasty and its allies used to build a court-centered, bureaucratic state that could stave off threats originating both in the archipelago and on the continent.[83]

Chinese dynasties had long treated rulers in the Japanese archipelago as tributary kings of the land of Wa, but by the early eighth century, the Yamato monarchs had redefined the terms of engagement by naming their realm *Nihon*. The characters for *Nihon* (日本) were also pronounced more lyrically as *hi-no-moto*, translated as "sun-source" in Mushimaro's poem that appears at the beginning of this chapter. The new name worked on two levels. First, equating Japan with the source of the sun implicitly acknowledged that it was located to the east of China, so the term made geographic sense in a Sinocentric world. Second, it reinforced the heaven-sanctioned position of the Yamato lineage, which had by the end of the seventh century appropriated the sun goddess Amaterasu as its ancestral deity.[84] Around the same time, Yamato rulers also adopted the title of *tennō*, meaning "heavenly sovereign" and commonly translated into English as "emperor."[85]

The Yamato court instituted a host of other measures in the seventh and eighth centuries both to enhance its status and manage the realm, such as establishing formal court ranks, conducting censuses, and promulgating legal codes. They also constructed a series of *tennō*-worthy capitals in the area around Nara that were ceremonial as well as administrative centers, providing a range of sites to conduct rituals designed to justify, protect, and enhance Yamato rule. Beyond the capital region, a network of major roads linked together provinces supervised by governors appointed by the court.[86] Straddling the boundary between Kai Province to the north and Suruga Province to the south, Fuji loomed over the Tōkaidō (Eastern Sea Road) that connected the capital region to eastern Japan.

To harness the worship of state-endorsed *kami*—powerful beings who took the form of ancestral deities, remarkable features in the landscape such as Fuji, and a host of other awe-inspiring entities—the Yamato established a Council of Divinities (Jingikan) in the capital and sponsored cultic centers in the provinces, the most prominent being the one devoted to Amaterasu at Ise. The dynasty also created a ranking system for kami and the shrines devoted to them that closely tracked political fortunes. Finally, it commissioned the official myth-histories in which the progenitor kami of the Japanese Islands as well as the Yamato lineage and its courtier allies figure prominently. These different institutions and belief systems involving kami tend to be lumped together under the term "Shinto" (Way of the gods), which is often considered the "native religion" of Japan. At this point in history, however, the multiple forms of kami worship conducted by both elites and commoners lacked the theological and ritual coherence denoted by the word "religion." The Yamato myth-histories commonly cited as foundational texts for Shinto, the *Record of Ancient Matters* and *Chronicles of Japan*, are replete with Chinese rhetoric and beliefs, making it difficult if not impossible to sort out the indigenous from the foreign.[87] Thus the Yamato and their affiliates developed beliefs and rituals that were distinct to Japan but fundamentally shaped by influences from the continent. This was equally true for the beliefs and rituals surrounding Fuji.

The Lofty Mountain of Suruga

The *Record of Ancient Matters* and *Chronicles of Japan* are filled with tales of kami and humans performing extraordinary deeds at different places in Japan, including prominent mountains, so a reader might expect towering Fuji to make at least a cameo appearance or perhaps even enjoy a starring role. Instead it is entirely absent, evidence that what is now considered the symbol of Japan was far from central to the ideological efforts of the fledgling ancient state.

The closest we get to Fuji in the *Record of Ancient Matters* is through the tale of Yamato-takeru, the mythological prince dispatched by his equally mythological father to bring unruly kami and peoples beyond the Kinai region to heel. At the start of his expedition to conquer Azuma, a term referring to the lands of eastern and northern Honshu, Yamato-takeru is ambushed by a local ruler who sets fire to a grassy plain in Sagami, a province just to the east of Fuji (see map 1.1). Yamato-takeru then starts his own fire to counter the one set by his enemy.[88] This episode may allude to the real-life practice of burning grasslands to encourage fresh growth. Having survived his trial by fire, Yamato-takeru sets about conquering the inhabitants of Azuma known as the Emishi, a people considered "barbaric" largely by virtue of their opposition to Yamato rule.[89] After subduing these barbarians, Yamato-takeru heads back toward the capital by crossing the Ashigara Pass immediately to the east of Fuji and then entering the province of Kai. This means he must have traveled along Fuji's lower flanks, but while the *Record of Ancient Matters* highlights the Ashigara Pass, the route used by Yamato forces (represented in the figure of Yamato-takeru) to quell the barbarous Emishi of Azuma, it ignores the majestic volcano towering over it.

Fuji is also absent in the telling of the Yamato-takeru story in the *Chronicles of Japan*, despite the fact that, in this version, the attempt to kill the prince by fire takes place in Suruga, the province encompassing Fuji's southern flank.[90] Another place where it could appear is in an account of a man living near the Fuji River who was executed in 644 CE for inciting the frenzied worship of an insect believed to be a kami. Here the name "Fuji" is used, but in connection with the river, not the mountain looming over it.[91]

It is easier to note that Fuji is missing from the *Record of Ancient Matters* and *Chronicles of Japan* than it is to explain why. Perhaps it is a case of "out of sight, out of mind." The authors of the myth-histories might have had no use for a volcano in the faraway east as they worked to legitimize the Kinai-based royal house and its allies, suggesting that the failure to mention it was a passive rather than active choice. Whatever the reason, Fuji is nowhere mentioned in the two foundational texts composed to justify and celebrate Yamato rule.

The earliest surviving trace of Fuji in writing appears in the *Hitachi Province Gazetteer* (*Hitachi fudoki*), one of a number of local gazetteers (*fudoki*) produced in the early eighth century to introduce the court to different parts of the realm. Hitachi Province was located in the northeastern Kantō region, and one of the folktales appearing in its *fudoki* unfavorably contrasts the province's beloved Mount Tsukuba (see map 1.1) with cold and forbidding Fuji. According to the story, the great ancestor kami, in the course of paying visits to different deities, asks Fuji if it could spend the night. The Fuji deity turns the ancestor kami away, saying, "It is the first tasting of the early grain, and the house is taboo. This one day I cannot admit you." Here we see a reference to the custom of secluding the women of a household ahead of the annual harvest festival. There was also a court-sponsored variant of this festival (the *niinamesai*) in which the emperor went into seclusion. Despite the Fuji deity's excuse, the ancestral kami replies, "How is it that you do not give lodging to your parent? As long as you live, on the mountain where you dwell snow and frost shall fall winter and summer, cold shall pile upon cold, the people shall not climb up, and there shall be none to make offerings of food and drink." After cursing Fuji, the ancestor kami travels to the more hospitable Tsukuba, where the local deity provides refreshment. As a reward, the ancestor kami blesses the mountain, declaring:

> Oh, well beloved,
> my progeny,
> oh, lofty,
> shrine of gods!
> Even with heaven and earth,
> together with sun and moon,
> the people shall gather to celebrate
> with food and drink abundant,
> age after age never failing,
> day after day more flourishing:
> for a thousand autumns and ten thousand years
> their sports and pleasures never spent.

The story ends by noting that this is why crowds of people dance and feast at Tsukuba but no one climbs snowy Fuji.[92] It is perhaps unsurprising that a folktale originating in the vicinity of Tsukuba would praise that mountain at Fuji's expense. Regrettably the *fudoki* for Suruga and Kai are lost, though one can imagine that Fuji's location between the two provinces might have resulted in a more generous depiction.

Unlike the *Hitachi Province Gazetteer*, which condemns Fuji as unfilial, Mushimaro's poem at the start of this chapter praises the mountain as a defender and treasure of the Yamato realm. What the two works share in common is an emphasis on the unapproachable nature of Fuji, which soars so high that, as the poem puts it, "even the birds cannot reach its summit on their wings." This portrayal of Fuji as distant and otherworldly stands in stark contrast to the poems that Mushimaro composed about Mount Tsukuba.[93] Reflecting a general trend in the depiction of Tsukuba—and of other peaks appearing in the *Collection of Ten Thousand Leaves*—humans in these works make intimate contact with the mountain by climbing it. Composed to memorialize Mushimaro's climb with revenue officer Ōtomo Tabito, one of them begins:

As you, honoured Lord, travelled all the way,
    Wishing to see the twin-peaked Tsukuba
In the land of Hitachi,
I have brought you to the summit to enjoy the view,
Climbing, out of breath, by the roots of trees,
And wiping away beads of sweat in the scorching
    sun.[94]

Rather than look upward and remain awestruck from below, the author and his companion climb Tsukuba "to enjoy the view." Mushimaro also emphasizes a sensual engagement that goes beyond the detached act of seeing to include the primal, bodily experience of grasping roots and sweating in the sun.

The interplay between Tsukuba and human bodies also features prominently in Mushimaro's "Climbing Mount Tsukuba on the day of *kagai*." The term *kagai* (also known as *utagaki*) refers to ritualized outdoor gatherings in which men and women engaged in poetry contests and paired off to have sex with the blessing of the local kami. According to the *Hitachi Province Gazetteer*, Tsukuba was a famous spot for kagai, which took place in both the spring and fall.[95] Mushimaro's poem on the Tsukuba kagai conveys an orgiastic spirit free of ordinary constraints:

On Mount Tsukuba where eagles dwell,
    By the founts of Mohakitsu,
Maidens and men, in troops assembling,
Hold a *kagai*, vying in poetry;
I will seek company with others' wives,
Let others woo my own;
The gods that dominate this mountain
Have allowed such freedom since of old;
This day regard us not
With reproachful eyes,
Nor say a word of blame.[96]

This depiction of Tsukuba as a place whose kami invited freewheeling, joyful, and intimate passions could not be more different from that of Mushimaro's Fuji, towering defender of the realm.

In his famous paean to Fuji in the *Collection of Ten Thousand Leaves*, celebrated poet Yamabe Akahito also emphasizes grandeur at the expense of intimacy:

Since the time
when heaven and earth split apart,
Fuji's lofty peak
has stood in the land of Suruga,
    high and noble,
    like a very god.

As I gaze up to it
through the fields of heaven,
I see it hides the light
of the sky-traversing sun,
and the very gleam of the moon
is invisible in its shadow.
It thwarts the white clouds
    from their path,
and snow falls on its summit
outside the bounds of time.
Let us speak of it
and recount it to the ages—
Fuji's lofty peak![97]

Here Fuji exceeds the bounds of both time and space. It touches heaven and dates to the time "when heaven and earth split apart," effectively situating it at the start of history rather than in the course of it. Although the characters used to write "Fuji" in the first verse appear to be only phonetic in value, those used at the end of the poem are the same that appear in Mushimaro's—不盡 (never exhausted)—thus emphasizing Fuji's timeless nature. The fact that snow appears on Fuji's summit for almost the entire year puts Fuji "outside the bounds of time" in a cyclical sense as well. This season-defying quality distinguishes Fuji from most of the places celebrated in the *Collection of Ten Thousand Leaves*. Although several poems in the anthology feature mountains other than Fuji where snow can be found in the spring and summer, the majority of poems employ seasonally appropriate images—cherry blossoms in springtime, for example—to situate those places within the cyclical order of the cosmos.

The poems by Akahito and Mushimaro create a gap between humans and Fuji that can be bridged only by a worshipful gaze. With the exception of the short envoys that accompany those poems, the remaining verses in the *Collection of Ten Thousand Leaves* that mention Fuji are anonymous love poems that focus on yearning rather than awe. One reads:

> If we lose this chance,
> in the darkness of the wooded slopes
> of Fuji rising to the heavens,
> there is little prospect
> our meeting
> will come to pass.[98]

This poem refers to Fuji's tremendous height, but rather than gaze at the summit, the poet waits longingly in the dark woods for a lover who will most likely not appear. Two other poems in the collection use Fuji's volcanic activity as a way to express thwarted desire, establishing a trope that would remain popular into the future:

> Unable to meet my beloved,
> must I burn forever
> like the lofty peak of Fuji in Suruga?[99]

> Regrettable it would be
> if gossip ruined
> my beloved's name and my own,
> so like the lofty peak of Fuji,
> we continually burn within.[100]

By comparing the steam rising from its summit to the passion of a frustrated lover, these love poems transform Fuji into a metaphorical kindred spirit. And yet, unlike Tsukuba, where humans climb, sing, and have sex on its slopes, Fuji remains physically distant and is associated with painful longing rather than bodily pleasure. Future poets would continue to connect the steam rising from its summit with human passion long after the steam had ceased—as good a demonstration as any that the fantasies people construct about nonhuman things like mountains take on lives that may be inspired and shaped by those things but are hardly bound to them.

FROM THE EARLIEST human habitation of the Japanese Islands to the development of the Japanese state tens of thousands of years later, Fuji actively and sometimes violently participated in the construction of people's imaginations and material lives. They may not have left written traces, but the Paleolithic hunter-gatherers who trapped animals and transported obsidian most likely had stories to tell about the towering and sometimes fearsome volcano. The same holds true for both the Jōmon people who constructed pit dwellings and ritual sites at Fuji's base and the Yayoi farmers and tomb builders who followed them. All of these groups inhabited worlds in which Fuji, plants, animals, and other nonhumans were entangled with human beliefs and deeds.

With the rise of the ancient Yamato court came the creation of written records that reveal how Fuji was perceived among literate Japanese, most of whom had never laid eyes on the majestic peak but were nevertheless familiar with it. These prominently included Takahashi Mushimaro and Yamabe Akahito, state functionaries well trained in the art of deference and praise, who glorified Fuji as a kami so august that it exceeded the bounds of time and space. Mushimaro took the extra step of declaring it to be "defender of the sun-source, the land of Yamato."

Treated in isolation, Mushimaro's portrayal of Fuji as guardian of the realm can give the impression that the mountain was central to the ideology of the court, especially when viewed with an eye trained to see it as a symbol of Japan. But the main myth-histories of the Yamato monarchs make no mention of Fuji; and although the mountain was singled out by Akahito and Mushimaro as a mysterious and powerful kami, it was only one of many sacred places celebrated in the *Collection of Ten Thousand Leaves*. The most important of those locales, such as Mount Miwa or the mountains of Yoshino just south of Nara, were located not in the east, but in the western regions where the Yamato dynasty took root. In fact, Mushimaro's use of *hi-no-moto* ("sun-source") in his poem on Fuji can potentially be interpreted not only as a reference to all of *Nihon* but also to the fact that Fuji, in relation to the court, was located to the east. According to the Chinese cosmology adopted by the Yamato court, the northeast was a dangerous direction associated with demonic forces—a notion reinforced by the fact that northeastern Honshu was the land of the rebellious Emishi—so the Fuji appearing in Mushimaro's poem might be imagined as a frontier kami defending the realm from the margins rather than from the center.

Fuji would later prove to be a highly temperamental and demanding god who could destroy as well as protect. It is now time to turn to the ways in which Japanese developed a wide array of beliefs, practices, and institutions in response to the two opposing sides of Fuji: fearsome on the one hand, benevolent on the other. As they did, they would draw heavily on Daoism, Buddhism, and other continental influences, reinforcing time and again that Fuji was as much an Asian as a Japanese mountain.

2

From Angry God to Parent
of the World

At Fuji's northwestern base sprawls the thickly forested Aokigahara lava field, which covers thirty square kilometers and is up to 135 meters thick.[1] Growing on its rocky surface is a "sea of trees" (*jukai*) dominated by cedar and cypress, their gnarled, monstrous-looking roots grabbing hold wherever they can (see fig. 2.1). Unfortunately the Aokigahara is now a popular site for people looking to end their lives, giving it a sinister reputation as a "suicide forest."[2]

This uncanny landscape owes its existence to the Jōgan eruption, Fuji's most effusive in the last 4,500 years, which started in the summer of 864 CE and seems to have come to an end by early 866 (a small portion of the Aokigahara lava field, the Kōriana lava group 1, had already formed between 838 and 864).[3] Although the eruption lasted for a year and a half, most of the lava was produced during the first two months, with flows taking different but overlapping routes as they engulfed everything they encountered, including people's homes. The lava also filled in part of one lake, Motosu, and split another, Se (or Se-no-umi), into the two lakes now known as Sai and Shōji (see map 2.1).[4] In an account of the eruption submitted to the imperial court, the governor of Kai Province reported that Fuji had "set the hills on fire, destroying grass and trees, pouring out molten rock, and filling in lakes Motosu and Se in Yatsushiro District. The water was so hot that it steamed and all the fish and turtles died. Commoners' homes were buried along with the lakes, and of the remaining homes, countless were abandoned."[5]

This Fuji is entirely different from the one portrayed in the *Collection of Ten Thousand Leaves*. Several poems in the anthology associate its rising steam with romantic longing, but its volcanism is viewed as constant rather than violent. Takahashi Mushimaro's poem acknowledges Fuji's volcanic activity as

34

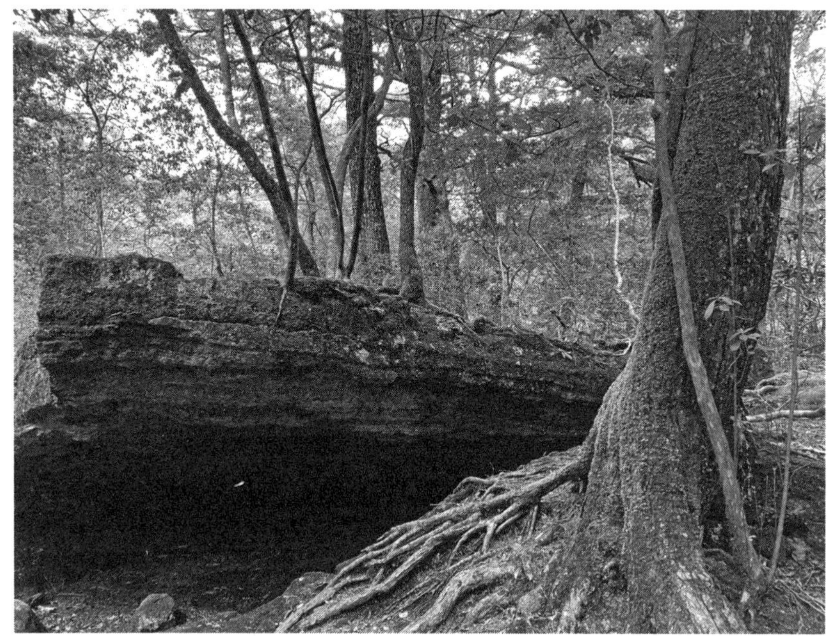

FIGURE 2.1. Inside Aokigahara "sea of trees." Photo by author.

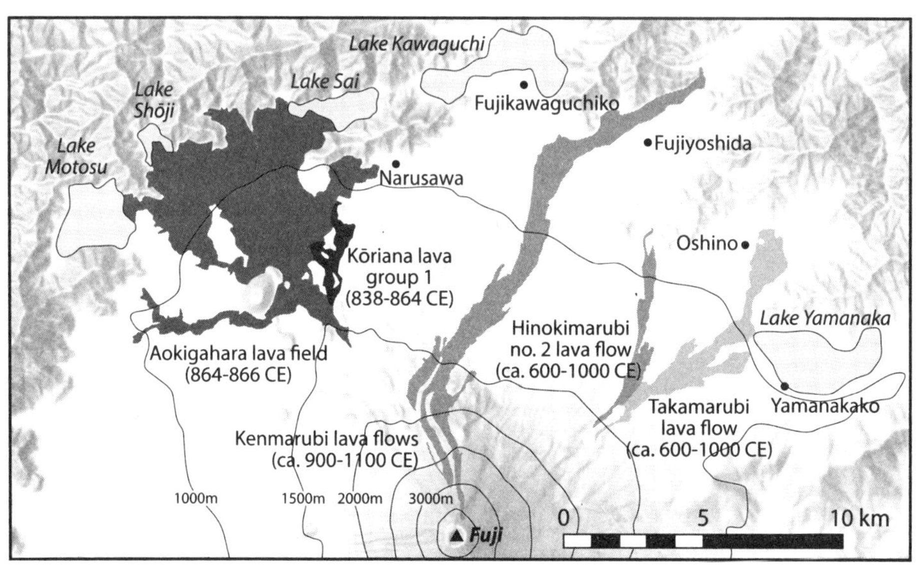

MAP 2.1. Major lava flows on northern side of Fuji (ca. 600–1100 CE). For a comprehensive map of lava flows and other eruption products on and around Fuji, see Takada et al., *Geological Map of Fuji Volcano*.

well, but by observing that "the snow drowns the flame and the flame melts the snow," it portrays that activity as an ongoing collaboration of opposing forces that is predictable and in balance. With his emphasis on constancy rather than volatility, Mushimaro, like fellow *Ten Thousand Leaves* poet Yamabe Akahito, depicts a mountain that does not violently intrude into the world but rises mysteriously and grandly above it. As the Kai governor's account shows, between 864 and 866, Fuji showed itself to be anything but.

Seeking to understand the motives of the angry kami and to appease it, the Yamato court mobilized a ritual apparatus that was centuries in the making. First it ordered the performance of an augury known as *kiboku* (plastromancy), a practice originating in China that involved heating the underside of a tortoise shell (the plastron) and interpreting the resulting cracks.[6] In this case the cracks revealed that the priests of the shrine in Suruga Province dedicated to the Fuji deity, who was known at this point as Sengen (or Asama) Myōjin, had been delinquent in their duties. Based on the divination, the court ordered the governors of both Suruga and Kai provinces (the Jōgan eruption mainly affected Kai) to have rituals performed to mollify the violent kami.[7]

Fuji continued to command attention both ritually and in writing over the centuries to come. It erupted at least four more times during the Heian period (794–1185), reinforcing its status as a deity to be respected and feared, while the steam rising from its summit bolstered the association with smoldering love first seen in the *Collection of Ten Thousand Leaves*. While these divergent views of Fuji—as destructive god on the one hand and inspiration for love poetry on the other—might seem contradictory, they are not so far apart when one considers that both burning passion and violent eruptions involve suffering and danger. In addition, they demonstrate that there was plenty of latitude for people to project their own fears and desires onto the volcano, which was also viewed as a realm of immortals.

After the Heian period, Fuji quieted down for the next few hundred years, and during that time, de facto power shifted from the imperial court to a series of military regimes. The first of these shogunates (*bakufu*) was established by Shogun Minamoto no Yoritomo (1147–1199) in the town of Kamakura. Located in the southern Kantō region, Kamakura offered everyday views of Fuji, while the mountain's foothills provided terrain for hunts involving thousands of samurai, effectively linking the mountain to the authority of the bakufu. There was also a sharp increase in travel between Kamakura and western Japan, exposing more people than ever to firsthand views of Fuji. These included prominent writers who elevated the mountain to unprecedented heights in Japanese literature.

Meanwhile, ritual institutions first built for local mountain ascetics and priests came to host travelers who journeyed to Fuji as pilgrims. Buddhism, along with forms of mountain asceticism that came to be incorporated into a tradition known as Shugendō, was central to this process, developing a sacred landscape in which laypeople and full-time practitioners could worship various deities. Along the way, fierce Sengen Myōjin was transformed into benevolent Great Bodhisattva Sengen (Sengen Daibosatsu), who would eventually be regarded by Fuji worshippers as not only a powerful god but the Original Father and Mother (Moto no Chichihaha) of the world.

Omens, Roads, and Epidemics

The Jōgan eruption was not the first of Fuji's eruptions to appear in the written record. The earliest occurred in 781, when the governor of Suruga Province perfunctorily notified the court that "Fuji is raining down ash, and where it has landed, leaves on trees have withered."[8] The news was more dramatic in 800, when a communication informed the court of activity that had taken place in the spring: "Fuji's summit was on fire, darkening the days with smoke and lighting up the sky at night. It sounded like thunder and the ash fell like rain. The rivers at the base of the mountain all turned red in color."[9] Evidence suggests that this eruption continued until 802. Volcanologist Koyama Masato has theorized that it produced two major lava flows called the Takamarubi and Hinokimarubi no. 2 that traveled far down the northeastern side of Fuji, with the Takamarubi blocking the inlet to Lake Yamanaka and causing the water level to rise to roughly the point seen today (see map 2.1).[10] The dating of these flows is a matter of debate, however.[11]

The Enryaku eruption of 800–802 was not only disastrous for people in Fuji's immediate vicinity but also worrisome for courtiers in the imperial capital of Heian—this despite the fact that they lived far to the west of the volcano in the Kinai region and thus well out of Fuji's destructive reach. According to court records, they had two concerns. One was that "flaming stones from Fuji" had destroyed the road through the Ashigara Pass.[12] The second—and greater—worry was that a divination had revealed the eruption to be an omen of epidemic.[13]

Providing entry to and from the Kantō region and, by extension, the rest of eastern Japan, the Ashigara Pass was an important link in the state's

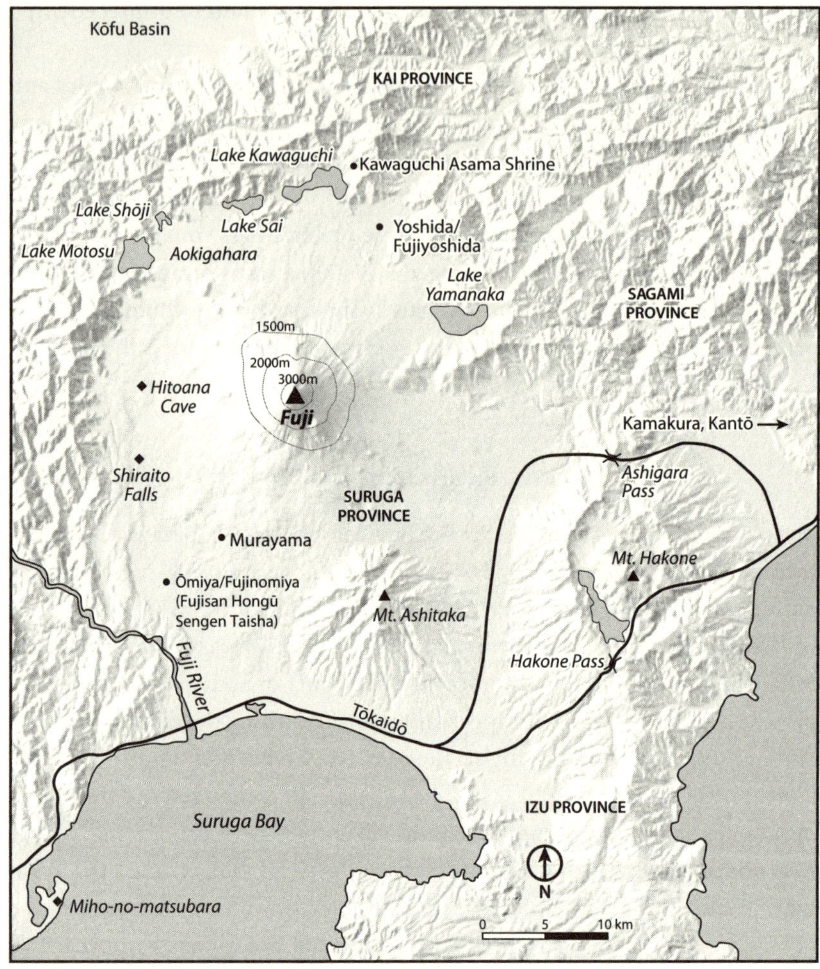

MAP 2.2. Fuji region in ancient and medieval periods.

transportation and communications network (see map 2.2). To extend its authority into the hinterland, in the seventh century the Yamato state had established a system of roads with way stations spaced approximately sixteen kilometers apart.[14] The most important road was the San'yōdō. Dubbed the "great road" (*dairo*), it ran from the capital in the Kinai region to northern Kyushu, with twenty horses stabled at each station. "Medium roads" (*chūro*) typically had only ten horses at each station and "lesser roads" (*shōro*) only five.[15] Fuji overlooked the Tōkaidō, a "medium road" that ran along the Pacific coast from the capital to northeastern Honshu. At the time of the Enryaku eruption, three stations operated along the Tōkaidō just to the south of Fuji—Kanbara, Kashiwahara, and Nagakura. Another station, Yokohashiri, sat on the approach to the Ashigara Pass to the east (the exact locations of the four stations are unclear).[16] Yokohashiri was an exception to the rule for "medium road" stations in that it had a total of twenty horses at the ready for traveling officials and another five post-horses to carry messages, an indication of the strategic importance of the pass.[17] Destruction of the route through the Ashigara Pass was therefore a concern, but at least there was a ready way to deal with it: temporarily use the Hakone Pass just to the south until the Ashigara road could be repaired and reopened a year later.[18]

The news of impending epidemic presented a much greater threat. Disease could spread throughout Japan, killing courtiers just as easily as commoners. In fact, although the entire realm was vulnerable to epidemics—which wiped out an estimated one-quarter to one-third of every generation born from the eighth to twelfth centuries[19]—those living in the well-trafficked capital were especially at risk. Roads, after all, carried not only people and goods but also the microbes that hitched rides with them, explaining why "road feast festivals" (*michiae no matsuri*) were regularly held on approaches to the capital to keep illness at bay.[20] Disease played a central role in the governor of Suruga's decision to ask the court in 864 to close the Tōkaidō's Kashiwahara station south of Fuji. The Jōgan eruption was then in progress, but this petition focused instead on epidemics that had plagued local communities for years, making it difficult for them to support and staff the stations south of Fuji. To alleviate their plight, the governor asked to close the Kashiwahara station and also to move the Kanbara station from west of the Fuji River to the east in order to equalize distances between the remaining stations. The request was granted.[21]

The fear of an eruption portending disease may seem misplaced, but to link volcanic activity with illness was not especially far-fetched. By destroying the agricultural resources of a particular area, an eruption could easily put subsistence-based communities at risk of famine, which could in turn make them more vulnerable to disease. More often than not, earthquakes, floods, and other disasters generate this domino effect, especially in cases where long-distance markets and relief efforts are limited.

Fearful courtiers and provincial officials articulated the relationship between volcanism and disease not in material terms, but according to a metaphysical understanding of the cosmos that had been largely systematized in China by Daoist lineages of belief and practice. In Japan, Daoism never generated the autonomous infrastructure of temples and priests seen in China, but in the seventh century, the Yamato court established a Bureau of Yin and Yang (Onmyōryō) to employ the cosmological knowledge that animated it. Staffed by yin-yang experts mainly of Korean and Chinese descent, the bureau engaged in astrology, timekeeping, prognostication, and the interpretation of omens. These last two functions were especially important, since they not only warned of potential dangers but could also be wielded as weapons to bolster certain people and policies and to delegitimize or outright destroy others, a high-stakes game that Herman Ooms calls "portentous politics."[22] With political considerations in mind, one major form of divination, *kiboku* (plastromancy), was entrusted to sacerdotal lineages in the Council of Divinities rather than to the staff of the Bureau of Yin and Yang.[23]

Official chronicles are replete with perfunctory and seemingly straightforward references to events that take on a whole new meaning when viewed through an omen-seeking lens. An example is the Suruga governor's brief comment about Fuji's eruption in 781.[24] Merely noting that ash fell and leaves withered seems not all that significant, but in the hands of yin-yang masters charged with reading omens, a seemingly simple fact could provide valuable raw material for their portentous work. Although fearsome omens demanded the greatest ritual response, it is also worth noting that this material sometimes pointed to good fortune rather than ill. Consider the governor of Suruga's report in 860 that five-colored clouds had been spotted over Fuji—a propitious omen that turned out to be wildly off the mark in light of the violent Jōgan eruption a few years later.[25]

Because of the political dangers posed by the interpretation of omens, the Bureau of Yin and Yang functioned as much to contain Chinese cosmological knowledge as to employ it. In 820, however, the bureau was shut down—with a number of its functions entrusted to Buddhist clerics[26]—and enterprising yin-yang masters subsequently marketed their services to those willing to pay for them. Teaching that bad fortune was never a mere accident but instead the result of failing to heed portents, avoid unlucky behaviors, and propitiate malevolent forces, these yin-yang masters promoted a culture of fear in which actions were minutely regulated to avoid trouble—down to the level of consulting fortunes before bathing, and cutting fingernails only on days thought to be favorable.[27]

Buddhism contributed to this culture of fear by preaching the law of karma, according to which actions had consequences not just in one life but across multiple ones. To avoid negative outcomes—such as drought, severe illness, or the nasty fate of being reborn as a hungry ghost or hell dweller—meant both regulating one's behavior and employing the ritual knowledge of

Buddhist clerics, who knew how to call upon the help of buddhas, bodhisattvas (those on the path to becoming buddhas), and a host of other divine beings. After all, even the most diligent and ethical person could make mistakes, and who knew what actions performed in a previous life were ultimately to blame for misfortunes in this one?

Looking to protect both its individual members and the realm as a whole, the ritually promiscuous Yamato dynasty appropriated Buddhism along with kami worship and Daoist cosmology. It did this not by setting up a government organ like the Council of Divinities or Bureau of Yin and Yang, but by regulating the ordination of clerics and sponsoring the construction and maintenance of major temples to perform rituals and chant Buddhist scriptures with the explicit aim of protecting the state.

The eruptions of Fuji in 800–802 and 864–866 activated the various mechanisms of divination and ritual appeasement offered by kami worship, Daoist cosmology, and Buddhism. After it was revealed that the Enryaku eruption was a sign of impending epidemic, the court ordered the governors of Suruga and Sagami (the province located just to the east of Fuji) to appease the kami with pacification rituals and the recitation of Buddhist scriptures.[28] In the case of the Jōgan eruption, the court once again ordered a divination and commanded the governors of Suruga and Kai to placate the violent deity.

At the time of the Jōgan eruption, this kami was called Sengen Myōjin. The two characters used to write Sengen (浅間) are appropriate for a volcano deity since they can also be pronounced *asama*, which, when made into an adjective by adding the suffix *shii* (*asamashii*), can mean "terrible" or "fierce."[29] In fact, Asama is the name of a young and famously explosive volcano located about one hundred kilometers north of Fuji. Much of Asama's viewshed overlaps with Fuji's, helping to explain the linguistic association of the two volcanoes.[30] *Myōjin* (明神) was a title awarded to prominent deities recognized by the imperial court. When the Suruga governor notified the capital that Fuji had erupted (his report in fact preceded the one sent by the governor of Kai), he further specified that Sengen/Asama Myōjin was a god of the "senior third rank," a status bestowed by the court in 859.[31] This was a step up from the "junior third rank" awarded in 853. A third promotion would occur in 907, when the kami achieved "junior second rank."[32]

Just about any ritual sponsored by the court and its agents in the hinterland was to some degree political in nature, and politics were certainly at play in responding to Sengen Myōjin. The tortoiseshell divination performed in the summer of 864 placed responsibility for the god's anger on delinquent priests in Suruga and, by doing so, implicitly pointed the finger at Suruga's governor for failing to make sure they were doing their job. Kai's governor turned this to ritual and political advantage by sending a message to the court a year later describing the dramatic possession of one of his top officials, a man named Tomo no Masada, by Sengen Myōjin. Speaking through Masada, whose body reportedly alternated between stretching to eight feet (*shaku*) and shrinking to two, the kami declared that it wanted a new official shrine (*kansha*) dedicated to it in Kai, adding that it had only resorted to an eruption after first trying to attract attention by causing an epidemic, an attempt that unfortunately had been ignored. After the governor had confirmed the authenticity of the kami's demands through a tortoiseshell divination, the court authorized the creation of a new Sengen Shrine in Kai, a move that enhanced the stature of the province and affirmed its authority over the northern foot of Fuji. In a case of possession turned to favoritism, Masada was made into a shrine priest along with his relative Tomo no Akiyoshi.[33]

Laying the blame at the feet of the Suruga priests may have served a more indirect political function as well. In 863 the court sponsored its first *goryō-e*, a high-profile festival designed to appease the angry spirits (*goryō*) of aristocrats and members of the royal family who had either been executed or died in exile and were thus thought to be responsible for the numerous epidemics that had afflicted the Heian capital since the early ninth century.[34] Another epidemic occurring right after the *goryō-e* might call into question the court's ability to mollify angry spirits, but pointing the finger at priests in Suruga who had failed to placate a deity known for spreading disease in addition to causing eruptions would have served the purpose of preemptively deflecting blame. This may not have been a major consideration, but given the high-stakes ritual environment, it would not be surprising if it had.

In sum, the responses to Fuji's eruptions in Heian Japan revealed a volcano entangled with humans and nonhumans across different scales of space and time. The humans included provincial governors, supposedly delinquent priests, commoners killed or displaced from their homes, travelers temporarily unable to cross the Ashigara Pass, and faraway courtiers fearing epidemic. Among the nonhumans were burning vegetation, dying fish and turtles, and microbes that caused disease. Wreaking havoc at the center of it all was the fearsome Sengen Myōjin, source of ruin and death. In the centuries to come, Fuji's volcanic activity would continue to cause suffering, but it would also inspire new relationships—with actors near and far, worldly and divine—distinguished more by creation than destruction.

Immortality and Longing

Instead of encouraging people to keep their distance, the Jōgan eruption had the opposite effect of stimulating the expansion of ritual infrastructure on Fuji's skirt. As the communications surrounding the eruption reveal, a new Sengen Shrine was founded in Kai, with archaeological evidence suggesting it was built close to Lake Kawaguchi on the site where Kawaguchi Asama Shrine now stands.[35] Excavated remains indicate that the precursor to today's primary Sengen Shrine—now known as Fujisan Hongū Sengen Tai-sha but for centuries commonly referred to as Ōmiya Sengen Shrine—was established at the shrine's current location in Fujinomiya (that is, Ōmiya) in the years following the eruption.[36] A ritual site consisting of a few pit dwellings also took shape in the tenth century higher up the skirt from Ōmiya at Murayama.[37]

As the number and size of ritual sites on Fuji expanded, so did a poetic tradition inspired by romantic longing and wonder at Fuji's ability to defy the seasons. The link between Fuji's volcanic activity and forlorn love first estab-lished in the *Collection of Ten Thousand Leaves* is particularly apparent in the *Collection of Poems Ancient and Modern* (*Kokin wakashū* or *Kokinshū*), the re-nowned early tenth-century compilation of poetry that came to function as a manual of metaphorical associations for poets.[38] Here is one example written by the court lady Ki no Menoto (fl. ca. 880s):

> ah burn brightly then
> bitter fires of hopeless love
>   hot as Fuji's flames—
> not even the gods would quench
> these unavailing smoke clouds[39]

All five of the poems that mention Fuji in the *Collection of Poems Ancient and Modern* use its volcanism as a metaphor for burning passion. The consistency among Fuji-themed poems can be explained not only by the editors' commit-ment to a poetic and by extension cosmic order in which certain metaphorical relationships were given precedence over others but also by the linguistic op-portunity inherent in the words *koi* and *omoi*. The first means "love" while the second can be translated as "thought," "feeling," or, in the context of love po-etry, "longing." Because the final syllable of *koi* and *omoi* is homophonous with *hi*, meaning fire, these words offered poets a form of wordplay that readily equated Fuji's volcanic activity with smoldering ardor.[40]

This phonetic sleight of hand is also at work in the following poem by Saigyō (1118–1190), a Buddhist monk celebrated for his poetic talent:

Trailing on the wind,
the smoke of Mount Fuji
fades in the sky,
moving like my thoughts
toward some unknown end[41]

Omoi is translated here not as "longing" but as "thoughts" because Saigyō wrote his widely acclaimed poem when he was already an old man who had taken religious vows. Rather than refer to romantic love, the poem reflects the Buddhist teaching of the transience of all things (*mujō*), a teaching that would apply both to Saigyō's thoughts and his waning life.[42]

Whereas Saigyō emphasized transience, other poets focused on the fact that Fuji transcended the normal rules of time, given that snow could be seen on the soaring mountain even in summer. This theme, like the association of romantic passion with volcanic activity, was first established in the *Collection of Ten Thousand Leaves*. Yamabe Akahito's poem on Fuji marvels that "snow falls on its summit outside the bounds of time."[43] One of the two envoys to Mushimaro's poem on Fuji emphasizes the season-defying power of the mountain as well:

The fallen snow
on Fuji's peak
melts in the sixth month,
    on the fifteenth day,
only to fall again that night.[44]

The fact that snow could be seen on Fuji in summer continued to be a source of wonder in poems composed in the Heian period and beyond. A particularly famous example appears in *The Tales of Ise* (*Ise monogatari*), an anonymous work from the mid-tenth century depicting episodes involving the famous courtier-poet Ariwara no Narihira (825–880) that combines poems with short narratives of a journey to the eastern provinces. This is what the author has to say about Fuji, where snow could be seen falling even in the fifth month:

The peak of Mount Fuji
is oblivious to time.
What season does it take this to be
that the falling snow
should dapple it like a fawn?[45]

"Oblivious to time" (*toki shiranu*) became a favorite turn of phrase for future poets writing about Fuji, appearing, for example, in four out of the ten Fuji-themed poems in a poetry collection compiled by the retired Emperor Gotoba (1180–1239) in the early thirteenth century.[46] It also features in one of the poems about Fuji written by the late thirteenth-century nun and renowned poet Abutsu (1222–1283), and in one composed two centuries later by another Buddhist cleric known for his poetry, the monk Dōkō (1430–1501).[47]

The fact that Fuji was at once an active volcano and a realm "oblivious to time" also inspired the popular belief that it was home to immortals (*shinsen*). This belief takes center stage in *Account of Mount Fuji* (*Fujisanki*), an essay written sometime between 875 and 879 by government official and esteemed scholar Miyako no Yoshika (834–879). Modeled on Chinese "landscape narratives" (*sansuiki* in Japanese), the account relates local stories and practices in addition to describing Fuji's physical features. Yoshika notes that pilgrims traveled long distances to see the sacred peak and that some even chose to climb it, although those who did would turn around midway because of the sliding rocks and ash. He also reports that "jewels" with holes in them were seen falling from Fuji in the Shōwa era (834–848) and that local officials and commoners who gathered for a festival in 875 witnessed two white-robed women dancing on the summit.[48] Yoshika further mentions a cinder cone produced by the Enryaku eruption, but for some reason makes no reference to the more recent Jōgan eruption. From the vantage point of modern science, what appeared to be dancing women was steam rising from the crater and the "jewels" were fragments of volcanic glass; but to Yoshika, an avid student of Daoist teachings that associated sacred mountains with the transcendence of death, these signs indicated the presence of immortals who could, like Fuji itself, defy the passage of time.[49] He reinforced the connection between Fuji and immortals with his fanciful claim that bamboo grew around a steaming pond on the summit, bamboo being associated with the supernatural and longevity.[50]

Fuji was therefore associated with immortality even as the smoke rising from its summit represented worldly passion. These two very different aspects of Fuji came together in one of Japan's most beloved stories, *The Tale of the Bamboo Cutter* (*Taketori monogatari*). The best-known version of the fable that *Tale of Genji* author Murasaki Shikibu (fl. ca. 1000) called "the ancestor of all romances" appears to date to the early tenth century.[51] The tale begins with an old bamboo cutter discovering a tiny girl in a glowing stalk of bamboo. He and his wife raise her as their own, and she quickly grows into a woman so beautiful she is called Kaguya-hime, "Shining Princess." Meanwhile, the bamboo cutter finds gold in the bamboo he harvests and becomes a wealthy man as a result.

Word of Kaguya-hime's beauty spreads far and wide, and five high-ranking aristocrats end up seeking her hand in marriage, but the princess rejects her suitors—all depicted in an extremely unflattering light[52]—when they fail to perform the impossible tasks she demands of them. After she turns down even the emperor, we learn that Kaguya-hime has in fact been sent to this world from the palace of the moon because of an unspecified wrong she committed in a former life. Toward the end of the story, celestial beings from the moon come to retrieve her; and after an attempt by the emperor's soldiers to stop them fails, they instruct her to drink from a jar of the elixir of immortality and don a cloak of feathers that will eliminate all worldly concerns. Before she puts on the cloak, she writes a letter to be delivered to the emperor along with some of the elixir.

This is where Fuji enters the story. The disconsolate emperor decides there is no point in living forever if he cannot have Kaguya-hime, so he orders a group of his men to take both the elixir of immortality and Kaguya-hime's letter to the mountain that reaches closest to heaven—that is, Fuji—and burn them. Accompanied by numerous soldiers, the men follow the emperor's command by climbing the mountain and burning the elixir and letter on the summit. This is why, we are told, smoke continually rises from Fuji and also why the mountain can be written with the characters for "not dying" (不死). The story has also been used over the centuries to explain why the standard way to write "Fuji" is 富士, which can be interpreted as "abundance of retainers."[53]

Like *Account of Mount Fuji*, with its description of the white-robed women dancing on top of Fuji, *The Tale of the Bamboo Cutter* provides a twist on the "white-bird-maiden" motif found in several versions across Asia. Stories with this motif frequently involve one or more people stealing a robe—often made of feathers—from a magical woman (the white-bird-maiden) in order to trap her into generating wealth or bearing children. A fragment of the eighth-century *Ōmi Province Gazetteer* (*Ōmi fudoki*) tells the tale of a young man who makes his dog steal the robe of feathers (*Hagoromo*) belonging to one of eight heavenly maidens who had come to earth in the form of white birds. Stranded without her robe, she marries the man and gives birth to four children before she finally finds the robe and returns to heaven.[54] The subject of a Nō play from the fifteenth century that remains a staple of the Nō repertoire, today's best-known variation of the story is set at Miho-no-matsubara, a pine grove on a spit of land (Miho) jutting into Suruga Bay that has long been famous for the view it affords of Fuji (see the lower left-hand corner of map 2.2). In this version, a fisherman steals the robe of feathers left on a tree branch by a heavenly maiden, who is able to retrieve it only after she agrees to the fisherman's demand that she perform a dance for him.[55]

In a similar fashion, *The Tale of the Bamboo Cutter* tells the story of a beautiful woman from the heavens who descends to earth and eventually returns to the celestial realm after donning a feathered cloak. It is not human theft or deception that temporarily traps Kaguya-hime on earth, however, but her actions in a former life. Toward the end of the tale we find out it was in return for minor good deeds that the bamboo cutter had the privilege of living in Kaguya-hime's presence and growing rich. This is in keeping with the Buddhist message that attachment leads to suffering and that immortality is not to be achieved in this sullied world of ours.

Stories about Kaguya-hime evolved over time, strengthening her bond to Fuji—and Buddhism—in the process. A thirteenth-century travelogue identifies her with the immortals described in *Account of Mount Fuji*, and a fourteenth-century version of *The Tale of the Bamboo Cutter* ends with her on the summit entering a state of eternal meditation together with the emperor.[56] In another version of the Kaguya-hime tale, it is the governor of Suruga, not the emperor, who falls in love and joins her on top of Fuji, where they fuse into the deity Sengen Daibosatsu (Great Bodhisattva Sengen).[57] The identification of Kaguya-hime with Great Bodhisattva Sengen also appears in the Nō drama *Mount Fuji* (*Fujisan*).[58] Referring to the Sengen deity as "Great Bodhisattva" indicates the extent to which Buddhist teachings had gained currency in medieval Japan.

Meanwhile, Fuji's reputation as a realm of immortals traveled to the continent. According to a Chinese work compiled in the tenth century, a Japanese monk who visited China claimed that music could always be heard at Fuji, and that, in addition to fire and smoke rising from the summit, jewels ran down the mountain's slopes during the day and ran back up at night. He also said Fuji was none other than Mount Penglai, which in Chinese legend was one of three mountains of immortals located in the eastern sea.[59]

Of course, Fuji differed from the mythical Penglai in that it was an active volcano. It sometimes erupted violently into the mundane world, and the steam rising from its summit was often associated with fiery passion. Yet its towering, snow-clad peak appeared "oblivious to time," raising it beyond the tumult of worldly affairs. Although the two aspects of Fuji—one volatile and the other eternal—would seem to be at odds, both worked to make the summit into a mysterious realm that appeared off limits to ordinary mortals.

From Climbing to Healing

In the first two centuries following the Jōgan eruption of 864–866, ritual sites sprang up at Fuji's base, but as far as we know, no permanent structures were built at the top of the otherworldly mountain. This changed in 1149, when a man named Matsudai Shōnin (n.d.) is said to have built a chapel dedicated to Dainichi, the "sun" or "cosmic" buddha (Mahāvairocana in Sanskrit). Matsudai taught that the kami of Fuji was Sengen Daibosatsu and Dainichi Buddha at one and the same time.[60]

Legend has it that Matsudai was not the first person to reach the summit. That honor goes to the ascetic En no Gyōja (mid-seventh to early eighth centuries), renowned for possessing the powers of an immortal. The court feared these powers enough that it banished him to an island located off the Izu Peninsula in 699. Supposedly he acquiesced because his mother was held captive, but while he deferred to the court's order by staying on the island during the day, at night he flew to Fuji to practice austerities.[61] According to another legend, even before En no Gyōja was born, Prince Shōtoku (574–622) flew over the mountain on a magical horse en route to China. This episode is depicted in the earliest extant painting that features the mountain. Part of an eleventh-century illustrated biography of the prince, the painting does not represent the mountain in a realistic fashion. Influenced by the portrayal of mountains in Chinese landscape paintings, it instead depicts Fuji with extremely steep slopes, emphasizing the inaccessibility of the ethereal summit.[62] This way of illustrating the mountain persisted for hundreds of years, as we see in a Muromachi period (1336–1573) illustrated biography of Prince Shōtoku (see upper right-hand corner of fig. 2.2.).

Unlike En no Gyōja and Prince Shōtoku, Matsudai had to walk to the top of Fuji rather than fly. This made him the first person in the historical record to have climbed on foot to the summit, although it would not be surprising if there were those who preceded him, their names lost to history. In addition to building the chapel to Dainichi, Matsudai buried Buddhist sutras donated by the retired Emperor Toba (1103–1156), starting a trend of interring sutras on top of Fuji that continued long after his death.[63] Sutra burial on mountaintops was a widespread practice in Matsudai's time. Gripped by the idea that they were living in a degenerate age called *mappō*, "end of the [Buddhist] law," commonly thought to have started in 1052, elites had sutras buried for safekeeping until the advent of Miroku (Maitreya in Sanskrit), the next buddha destined to appear in our world. The act of burying sutras generated individual merit

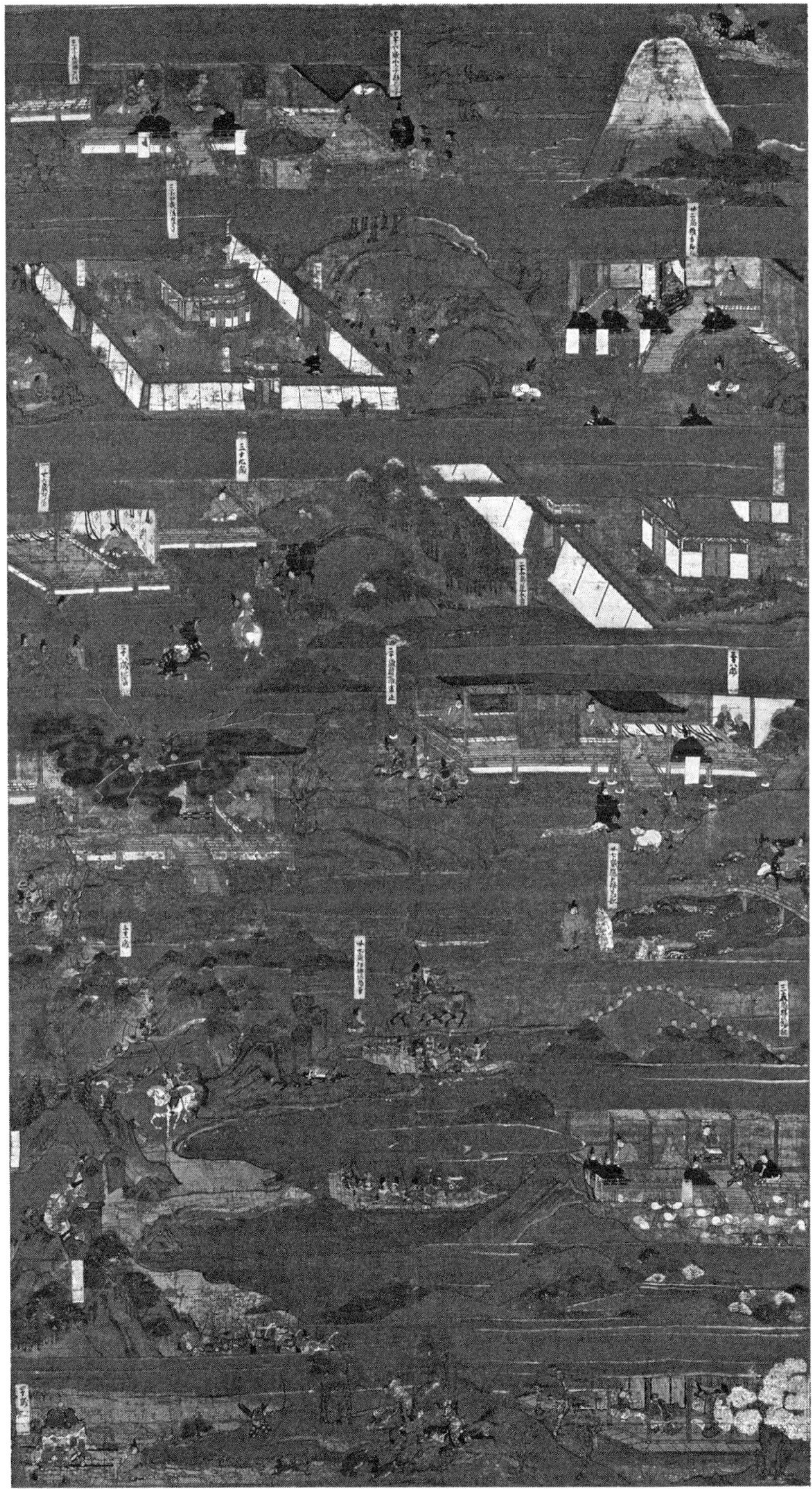

FIGURE 2.2. Prince Shōtoku flies over Fuji. See upper right-hand corner. Third scroll in four-scroll *Illustrated Biography of Prince Shōtoku*. Fifteenth century. Tanzan Shrine, Sakurai. Alamy stock photo.

for both the living and the dead, and was often aimed at achieving rebirth in the Pure Land (*jōdo*) of the merciful Amida Buddha (Amitābha in Sanskrit), where they could reside until the coming of Miroku.[64] Matsudai absorbed Pure Land Buddhist teachings when he was a disciple of Chi'in, a Pure Land priest who also went by the name Amida Shōnin, at Jissōji, a temple close to where the Fuji River empties into Suruga Bay.[65]

Matsudai also established a temple at the southwestern foot of Fuji in Murayama (in today's Fujinomiya City; see map 2.2). Several centuries after his death, Murayama became a thriving center of Shugendō ("the Way to achieve miraculous powers through practice"), a mountain-focused ascetic tradition that traced its mythical origins to En no Gyōja and developed over Japan's medieval and early modern periods by combining local kami cults with esoteric Buddhist rituals and teachings.[66] By the end of the fifteenth century, the ritual complex at Murayama had been incorporated into the Shugendō sect supervised by the Kyoto temple Shōgoin. At this time it was not only a base for full-time ascetics but also a stopping point for pilgrims seeking to climb the mountain. Pilgrims were led by mountain guides (*sendatsu*) who operated under the oversight of the resident temple priests (*shuto*). The priests actively encouraged pilgrims to come to Fuji and managed the lodges where they stayed.[67] It helped that Fuji was relatively subdued between its eruption in 1083—reportedly heard as far away as Heian—and in 1707, which devastated villages and towns to the east of the volcano. Records point to eruptions in the early fifteenth and early sixteenth centuries, but these appear to have been much smaller in scale than those that occurred during the Heian period and in 1707.[68]

As religious institutions and practices evolved at Fuji, samurai (warriors) ruled Japan. Major samurai lineages increased in strength over the course of the Heian period; and by the middle of the twelfth century, one lineage, known as the Taira (or Heike), had come to dominate a fractious imperial court. Yet it was the Taira's greatest rival, the Minamoto (or Genji), that emerged victorious after a war that lasted from 1180 to 1185, which included a famous battle at the Fuji River. The head of the lineage, Minamoto no Yoritomo, established a military government just eighty kilometers to the east of Fuji in Kamakura. Declared shogun in 1192, Yoritomo ostensibly ruled at the pleasure of the imperial court. In reality he was the most powerful man in Japan. The shogunate that he established in Kamakura was overthrown in 1333, but samurai rule would continue in different forms until the middle of the nineteenth century.

Fuji's foothills were used by samurai as hunting grounds, generating popular stories that entwined the sacred mountain with martial exploits. The most famous of these stories, the revenge of the Soga brothers, stemmed from real-life events that took place in 1193, when Yoritomo, to assert his new authority as shogun, led a massive hunt at the base of Fuji. During the hunt, brothers Soga Sukenari (1172–1193) and Tokimune (1174–1193) killed the man responsible for the death of their father and battled other retainers of the shogun. Sukenari died during the fight and Tokimune was executed. Different (and wildly embellished) versions of these events multiplied and grew in popularity over time, with some of them promoting the Soga brothers to the status of gods.[69]

Another popular tale developed around Nitta Tadatsune (also known as Nitta Shirō, 1167–1203), the retainer who killed Soga Sukenari in 1193. During a hunting trip at the base of Fuji, Yoritomo's son Yoriie (1182–1204) is said to have ordered Nitta and other retainers to explore a menacing lava cave at the western foot of the mountain called Hitoana ("person hole"). Multiple versions of the story developed from the thirteenth to seventeenth centuries, all of which claim that Nitta came face to face with the fearsome Sengen deity, often portrayed as a monstrous snake or dragon. In the most popular version, the deity transforms into a human and gives Nitta a subterranean tour of horrific Buddhist hells before showing him Amida's Pure Land.[70] The lesson is straightforward: actions in this life have consequences in the next one.

The Hitoana Cave (see map 2.2) would take on added significance in the sixteenth century as the site of austerities performed by Kakugyō Tōbutsu (1541?–1646), widely regarded as the founder of a popular form of Fuji worship that would eventually challenge the dominance of established institutions like Ōmiya Sengen Shrine and the Shugendō complex at Murayama. But in the meantime, with the blessing of local rulers, these two institutions largely dictated the actions of pilgrims who climbed Fuji.

The relationship between Ōmiya and Murayama involved a certain level of cooperation. For example, they jointly commissioned a new statue of Dainichi that was dedicated in 1478.[71] Yet they were also rivals, both wanting pilgrims to stay at the lodges affiliated with them. Two different artistic portrayals of pilgrimage to Fuji in the sixteenth century reveal this competition. Both are called Fuji pilgrimage mandalas (*Fuji sankei mandara*), "mandala" in this context signifying a sacred site and the deities who inhabit it. Pilgrimage mandalas were popular in medieval Japan, functioning as "both cosmological maps and guide maps to sacred sites, depicting much of what would be encountered on a visit to the illustrated site while also wrapping the landscape with cosmological overtones."[72] Affiliates of a particular site would use these paintings when they traveled to different regions to encourage pilgrimage—and donations.

In the best-known Fuji mandala, attributed to the workshop of famed painter Kanō Motonobu (1476–1559), we see Suruga Bay, with the Rinzai Zen temple Seikenji depicted in the bottom left-hand corner (see plate 1). Located across the Fuji River are the precincts of Ōmiya Sengen Shrine, where pilgrims perform ablutions in Wakutama Pond (fed by water from one of Fuji's lava flows). Further up the slope is the Murayama complex, above which pilgrims make their way up to the summit of Fuji, which consists of three peaks. The origins of the three-peaked depiction of Fuji are unclear; but this way of portraying the mountain—which was already well established when the mandala was painted and became increasingly popular in the centuries that followed— may have been inspired by Chinese representations of mountains (particularly sacred mountains) in literature and art, or by the simple fact that the Chinese character for mountain has three vertical lines (山), or even because Fuji's summit appears to have three peaks when viewed from a particular angle.[73] In the Fuji mandala, these peaks are occupied by three buddhas, who are often interpreted to be Amida, Dainichi, and Yakushi (the Medicine Buddha). The moon and sun, associated with yin (female/negative) and yang (male/positive) energies, flank the mountain, an arrangement typical of Japanese pilgrimage mandalas. A remarkable feature of this particular mandala is the attention it lavishes on the Murayama complex, its halls situated in the middle of the painting and the distances between them exaggerated. "This compositional strategy," as Talia Andrei puts it, "conveys the power of the Murayama institution and its privileged access to the mountain." It is therefore reasonable to presume, as Andrei argues, that the painting was commissioned by Murayama affiliates.[74]

Another Fuji mandala, in contrast, puts Ōmiya Sengen Shrine front and center and reduces Murayama to mere rooftops, suggesting it was commissioned by the shrine (see plate 2).[75] The two mandalas, one highlighting Murayama and the other Ōmiya, appear to reflect the competition between their respective patrons.

The ultimate winner of this rivalry was Ōmiya. Japan during the fifteenth and sixteenth centuries was wracked by near-constant warfare among powerful daimyo (samurai lords). One of these, the Suruga daimyo Imagawa Yoshimoto (1519–1560), backed Murayama, but he was killed in battle against the ascendant Oda Nobunaga (1534–1582). Then, in 1569, the Kai daimyo Takeda Shingen (1521–1573) captured Suruga and threw his support behind Ōmiya. After the founder of Japan's last shogunate, Tokugawa Ieyasu (1542– 1616), won control of Japan in 1600, he, too, chose to patronize Ōmiya Sengen Shrine. To reward the Ōmiya priests for rituals they had performed to help secure Ieyasu's victory, the new Tokugawa bakufu funded the reconstruction of the shrine's main hall of worship, which had burned to the ground in 1582. The shogunate mandated, moreover, that anyone climbing Fuji via Murayama stop at Ōmiya first. Tokugawa Tadanaga (1606–1634), daimyo of both Kai and Suruga, further ruled that the management of activities occurring on the summit fell under the jurisdiction of Ōmiya Sengen Shrine, not the Murayama complex.[76] In the late eighteenth century, the government's Office of Temples and Shrines confirmed the shrine's authority over Fuji's upper reaches.[77] Politically isolated, Murayama was reduced to a shadow of its former self.

Back in the sixteenth century, the ascetic Kakugyō wanted nothing to do with either Ōmiya or Murayama, choosing instead to practice alone in the Hitoana Cave, where, according to legend, he stood tiptoe on a block of wood for one thousand days and doused himself with water from Shiraito Falls six times each day and night.[78] In a departure from earlier views of Fuji as just one place to worship Dainichi, Amida, and other deities, Kakugyō regarded Sengen Daibosatsu (or Sengen Dainichi) as none other than the Original Father and Mother (Moto no Chichihaha), the dual-gendered parent of all things.[79] To him and his followers, Fuji was not just *a* world mountain, but *the* world mountain. They accordingly developed a new Fuji cult that operated outside established institutions even as it incorporated their beliefs and practices.

It would take generations for Kakugyō's view of Fuji to be fully elaborated, and for a comprehensive—and in large part invented—biography of him to be written.[80] He did, however, leave traces by his own hand in the form of "utterances" (*gomonku*) or "spells" (*fusegi*). These were mandala-like arrangements of Chinese characters and images (including Fuji and the sun and moon) that Kakugyō distributed to his adherents.[81] These "image-texts"—known among later Fuji devotees as "body extractions" (*ominuki*)—were also composed by his successors and used as talismans and ritual guides. As their name indicates, the "body extractions" ultimately derived from the physical exertions through which one could, like Kakugyō, commune directly with the Fuji deity and gain insight into the nature of the cosmos.[82]

While courtiers had feared Fuji as a source of epidemics, Kakugyō and his followers turned to the mountain deity to end them. Kakugyō deployed the powers he accrued through his austerities to help his supporters, and he was particularly celebrated as a healer. Leaders of Fuji worship groups (*Fujikō*) in the Tokugawa period (1600–1868) followed Kakugyō's example by distributing talismanic objects, including "body extractions," with the aim of preventing and healing illness. Paper talismans were even ingested or affixed to the body to produce cures.[83] In these and other ways, believers were able to put their bodies into a therapeutic relationship with that of Great Bodhisattva Sengen, the Original Father and Mother.

FUJI WAS a material and cultural shape-shifter that meant different things to different people in premodern Japan. It was a source of life-giving water and devastating eruptions. It was a god of epidemics and a home of immortals. And while the steam rising from its summit suggested romantic attachments, it was a sacred mountain where pilgrims and ascetics looked to transcend mortal cares.

The object of beliefs and practices both imported from the continent and homegrown in Japan, Fuji was also populated by multiple deities who changed in form and status over time. They included the fierce Sengen/Asama Myōjin, Great Bodhisattva Sengen, and the buddhas Dainichi, Amida, and Yakushi. A host of other divinities crowded Fuji as well, such as Kaguya-hime and the dragon goddess Benzaiten, who was said to travel underground on a daily basis between the Hitoana Cave at the base of Fuji to a cave at Enoshima, an island located offshore from Kamakura.[84] Then there was Kakugyō's Original Father and Mother, the cosmic parent who was one and the same with Great Bodhisattva Sengen. Fuji presented a large array of deities for believers to worship.

Even as Fuji gained in popularity as a sacred mountain, it remained a potential source of terrible destruction. People projected all sorts of beliefs onto Fuji, yet this did not change the fact that it acted according to its own geological devices, sometimes to terrible effect. After the Heian period, the volcano went relatively quiet for hundreds of years, but debris and slush flows (*yukishiro*) running down its flanks continued to wipe out everything in their path. Especially devastating slush flows struck Yoshida (see map 2.2) in the middle of the sixteenth century, killing numerous people and horses in addition to destroying homes and fields. To avoid future disasters, the district where pilgrims' lodgings were located (Kamiyoshida) was moved in 1572 to higher ground, where it remains today.[85] No one could prepare for what would occur in 1707, however, when volcanic Fuji reawakened, this time spewing massive amounts of airborne rock and ash. Yoshida was spared, but communities to the east, and downwind, were not so fortunate. It is to the Hōei eruption of 1707 that we now turn.

3

Fallout of the Hōei Eruption

BETWEEN THE END of the twelfth century and the beginning of the eighteenth, Fuji was for the most part well behaved. Fumes sometimes rose from the summit, and minor eruptions seem to have occurred in 1435–1436 and 1511, but the volcano was generally quiet.[1] Then, from December 16, 1707 to January 1, 1708, Fuji erupted on such a massive scale that the sulfur dioxide it thrust into the stratosphere likely contributed to Europe's devastating "Great Winter" of 1708–1709, among the coldest on record for the continent.[2] Closer to the volcano, 1.8 cubic kilometers of fallout rained devastation that was both immediate and long lasting.[3] Fuji buried nearby villages under several meters of debris, and prevailing westerly winds deposited ash up to eight centimeters deep over a hundred kilometers away (see map 3.1).[4] The ash filled rivers and canals, destroying elaborate irrigation schemes that had dramatically increased the output of rice over the preceding century and had, in turn, increased the number of people living in harm's way. The result was widespread famine—and the emptying out of villages, which in some cases would take more than a century to recover.[5] Even today, thick layers of ash prevent the reclamation of formerly cultivated fields.[6]

The Hōei eruption tested the aptitude of a "compound state" in which the shogun's government (*bakufu*) monopolized foreign diplomacy and secured the general peace but collected taxes only from those lands controlled directly by the Tokugawa family.[7] By the time of the eruption, these amounted to about 16 percent of Japan's assessed land. Some of the remaining territory fell under the control of the imperial family, major religious institutions, and the shogunate's immediate vassals, known as *hatamoto* (bannermen); but the bulk of it—almost three-quarters—belonged to domains controlled by approximately 260 daimyo (the number fluctuated over time).[8] The Tokugawa regime asserted its authority by requiring daimyo to reside in the shogun's capital in

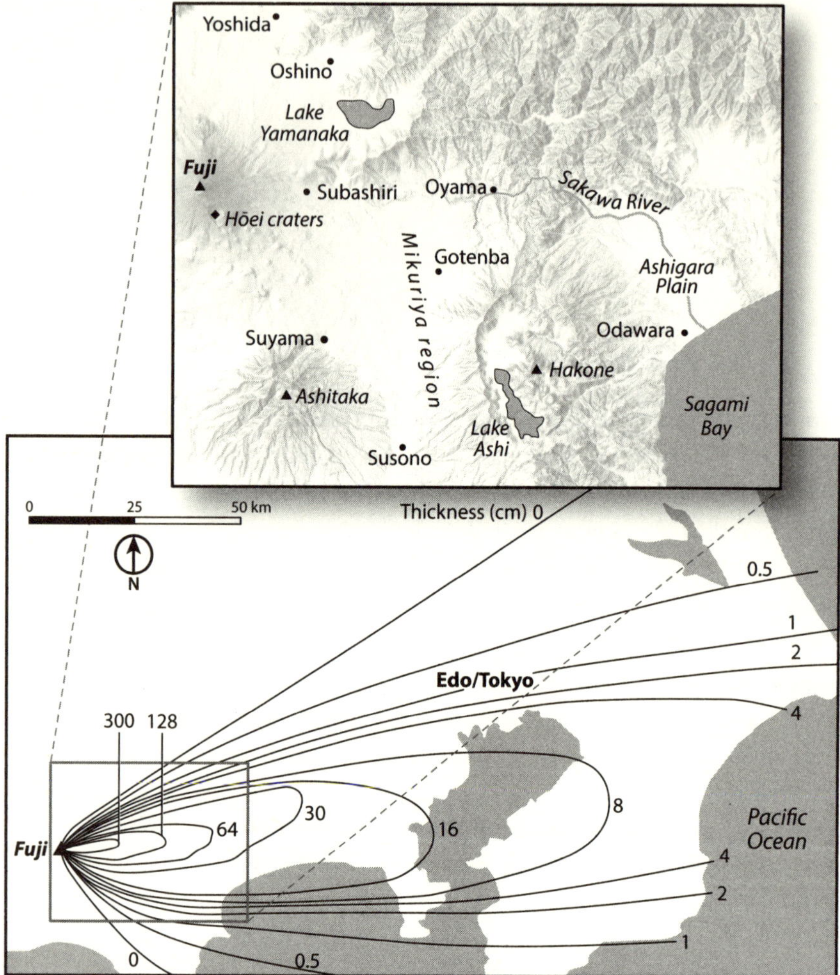

MAP 3.1. Fallout from Hōei eruption. Based on map in Miyaji and Koyama, "Fujisan 1707 nen funka (Hōei funka) ni tsuite no saikin no kenkyū seika," p. 341.

alternate years—and to leave their wives and children behind when they returned to their home domains. Yet in many respects, these lords enjoyed considerable autonomy, which included the all-important right to tax those who lived in their jurisdictions.

This power-sharing arrangement worked tolerably well when times were good (the shogunate lasted for over two and a half centuries), but like virtually all other political systems, it was ill designed to handle a disaster of the nature and scope of the Hōei eruption. In the eyes of farmers who had to sacrifice a portion of their harvest each year, the right to collect taxes brought with it the obligation to govern both responsibly and benevolently. So when crops failed and famine loomed, they sought relief from their lords, appealing to the ideal of *jihi* in written petitions. The term means "compassion" or "benevolence," and although it has Buddhist origins, it resonates with the Confucian ideal of "benevolent governance" (*jinsei*) that circulated widely among rulers, administrators, and intellectuals of the time. Of course, what "benevolent governance" exactly entailed was subject to interpretation and debate.[9]

Farmers resorted to more aggressive protests when appeals to benevolence failed to achieve their aims. To head off or tamp down these protests, domain and bakufu officials frequently (if grudgingly) provided some sort of assistance. The Hōei eruption posed an especially difficult set of challenges, however, because its effects were so devastating, wide reaching, and enduring. It affected multiple—and in some cases, overlapping—jurisdictions, making it unclear who was responsible for providing relief. The hardest-hit domain, Odawara, lacked the resources to meet the demands of villagers faced with both the immediate bodily threat of hunger and the decades-long task ahead of rehabilitating waterways and fields.

Concerned about the escalating protests, the shogunate responded with two measures that were unprecedented—and, it turns out, not to be repeated. First, it expropriated about half of the Odawara domain (measured in terms of agricultural production) so it could assume the responsibility of managing relief and recovery efforts. The bakufu had seized territory from daimyo in the past, but not on this scale and not for the purpose of dealing with a disaster. Second, the shogun's government made the extraordinary decision to levy a tax across all of Japan to pay for those efforts. This highly controversial decision opened it to charges that it was more concerned with filling a depleted treasury than giving assistance to those in need.

In addition to revealing the vulnerability of Japan's agricultural infrastructure, the Hōei eruption thus put the power arrangements of the Tokugawa shogunate to the test. How to judge the regime's success or failure in meeting that test depends on the criteria employed. On one hand, the regime was able to overcome structural limitations to muster a response that helped diminish the death toll and put fields back into production. On the other hand, government assistance fell far short of addressing the needs of devastated farmers who shouldered the burden of recovery largely on their own. Most of the financial aid provided by the government was aimed at rehabilitating the flood control and irrigation infrastructure of the highly productive Ashigara Plain (see map 3.1). Responding to the appeals of devastated villages, officials distributed small sums of money to keep people from starving to death, but they prioritized the reconstruction of Ashigara riparian works with an eye toward reviving the region's substantial tax revenues. Meanwhile, poorer upland villages struggled to get the government's attention as economic concerns trumped humanitarian ones in the distribution of aid. As is often the case when judging relief efforts, one can choose to see the glass as half full or half empty. In the face of institutional constraints, the Tokugawa regime showed the flexibility and will to provide aid—a success by early modern standards. At the same time, due to disparities in wealth and influence, relief was allocated unevenly, causing some groups to suffer far more than others. Unfortunately, this made Tokugawa Japan not the exception in such cases but the rule—both in its time and today.

Fuji Erupts

One of the two largest volcanic events in Fuji's history (the other being the Jōgan eruption of 864–866), the Hōei eruption takes its name from the Hōei era (1704–1711) in which it occurred.[10] Ironically, the era began in 1704 because the reigning shogun, Tokugawa Tsunayoshi (r. 1680–1709), wanted to make a chronological fresh start after the devastating Genroku-era (1688–1704) earthquake of December 31, 1703.[11] The earthquake, with an estimated magnitude of between 7.9 and 8.2 on the Richter scale, laid waste to buildings throughout Edo, the shogunal capital; and horrific tsunamis killed thousands along the Bōsō Peninsula and around Sagami Bay.[12] A delegation of Dutch merchants traveling from Nagasaki to Edo in the spring of 1704 catalogued the ruin they encountered in towns along the Tōkaidō highway. When they reached the Hakone Pass, "everything was upside down, and most houses had burned down," and in Edo, "everywhere one saw areas completely flattened where the houses had collapsed with the earthquake and burned to the ground." Edo castle was in such bad shape that the Dutch heard the shogun had almost called off their audience with him.[13]

Proving that it would take more than a change of era-name to keep things quiet, an even larger, more wide-reaching earthquake struck on October 28, 1707. The Hōei earthquake, with an estimated magnitude of 8.4–8.7, was the most powerful earthquake to strike Japan in the Tokugawa period (1600–1868); and until the Tōhoku earthquake of 2011, it was the strongest in recorded Japanese history.[14] Like the Genroku earthquake, the Hōei earthquake took thousands of lives not only by destroying buildings and starting fires but also by generating tsunamis that slammed into towns and villages along Japan's Pacific coast.[15]

Recovery efforts were still in progress when Fuji erupted forty-nine days later. The mechanisms linking seismic activity and volcanic eruptions are ambiguous at best, but volcanologists believe the Hōei earthquake initiated a sequence of events that culminated in the eruption.[16] It is unclear what happened under Fuji in the first few weeks after the earthquake, but evidence suggests that, from late November or early December, magma originating many kilometers below the volcano began its ascent, triggering swarms of small earthquakes along the way.[17] Movement detected before this point may have been aftershocks of the Hōei earthquake rather than volcanic tremors, but from December 3, according to a contemporary account, people at the eastern foot of the volcano in the village of Suyama (see map 3.1) could hear rumbles originating from Fuji even though the ground stood still. This acoustic evidence indicates that the residents of Suyama had detected not the aftershocks of an earthquake but magma on the move.[18] On December 15 swarms of earthquakes caused by the rising magma intensified, and this time could be felt as far away as Edo, about 100 kilometers to the east, and Nagoya, over 160 kilometers to the west.[19] The next morning Fuji erupted from its southeastern flank, creating the first of three new craters. The newly introduced magma also uplifted older layers of Fuji to form a large protrusion that would come to be called Mount Hōei (see fig. 3.1).[20]

According to an account written under the pen name "bald-headed old man living east of Fuji," on the first day of the eruption, "the ground suddenly trembled, and soon black clouds rose in the west and covered the sky. The sound coming from them was like thousands of thunderbolts. . . . Stones as big as kick balls [lava bombs] rained down and burst into flames when they hit the earth, scorching grass and trees and setting houses on fire."[21] In Edo, prominent scholar and official Arai Hakuseki (1657–1725) wrote in his memoir that "the night before there had been an earthquake, and at noon on this day [December 16], there were rumblings. When I left my house, white ash was falling like snow, a black cloud hung in the sky to the southwest, and there were continuous flashes of lightning."[22]

Thanks to a growing interest in the empirical study of nature in Tokugawa Japan, these are just two of numerous accounts that chart the progress of the two-week-long eruption from a variety of distances and directions. The

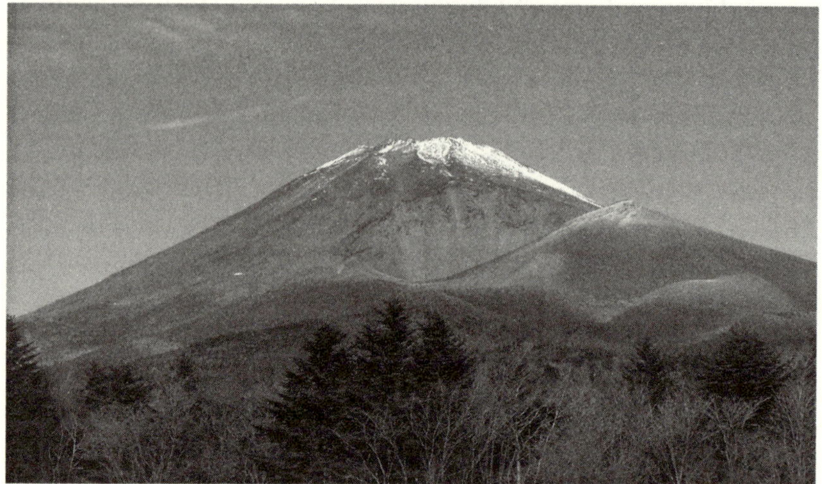

FIGURE 3.1. Hōei eruption crater and Mount Hōei. Photo by author.

empiricism responsible for this rich trove of evidence derived from a mechanistic view of the cosmos that may not have eliminated beliefs in omens and deities, but did provide an alternative for those who sought to make the patterns of the natural world more discernable and thus predictable. Of special concern were earthquakes and eruptions, generally thought to result from the release of positive yang energy trapped by negative yin energy underground. Terashima Ryōan, in his *Illustrated Compendium of Chinese and Japanese Knowledge* (*Wakan sansai zue*, 1712), drew on this theory to explain the relationship between earthquakes and Fuji's eruptions, including the Hōei eruption that took place five years before the publication of his book.[23] The yin-yang framework for understanding the workings of the cosmos provided the means through which to share and make intelligible highly detailed descriptions of a variety of phenomena, and as Gregory Smits notes in his study of earthquakes in Tokugawa Japan, "The precision of the empirical observations of some Tokugawa-period writers was remarkable."[24] In fact, the precision, along with the profusion, of records concerning the Hōei eruption has allowed modern volcanologists to construct a timeline of events down to the half hour. They have used that timeline to create hazard maps to prepare for similar eruptions in the future.[25]

Unlike the Jōgan eruption, in which magma oozed from vents on Fuji's flank and whose devastating effects were limited in reach, the Hōei eruption was Plinian, meaning it produced an eruption column over twenty kilometers in height with impacts that were literally far flung.[26] Volcanologists believe the Hōei eruption was so explosive because of the interaction of multiple magma chambers with different mineral compositions. As explained in chapter 1, Fuji is a remarkable stratovolcano in that it mainly produces basaltic magma. Due to its relatively low silica content, this type of magma is "runny" and therefore usually less explosive than its more viscous, silica-rich counterparts. True to form, the basalt produced by the Jōgan eruption flowed down instead of flying up. In the case of the Hōei eruption, however, volcanologists believe that more silicic magma had formed in relatively shallow chambers, one holding andesitic magma (silica content of 52–63 percent) and the other dacitic magma (silica content of 62–69 percent). When the deeper, basaltic magma ascended through the andesitic magma chamber, it triggered gas bubble formation that raised the pressure and caused the andesitic magma to rise into the dacitic magma chamber. This in turn caused the dacitic magma to rise and erupt. The resulting depressurization then caused the deeper basaltic magma to rise to the surface. The forceful eruption of the basalt was promoted by a high concentration of gases in the magma, perhaps maintained by the lining of the volcanic conduit with magma from earlier stages of the eruption.[27] This is why tephra fell first in the form of silica-rich white or grey ash, as attested by Arai Hakuseki and other observers, but soon turned black and would remain so for the rest of the eruption, which continued intermittently over the next two weeks.[28]

In Edo, the fallout caused respiratory distress—Hakuseki notes that, by the end of the eruption, "there was no one who was not suffering from a cough"[29]—but the accumulation of several centimeters of ash was minimal compared to what people living right by the volcano faced. Hardest hit were villages in the Mikuriya region immediately east of Fuji (in what are now the cities of Susono and Gotenba and the town of Oyama; see map 3.1). By the end of the eruption, they were covered in several meters of debris ranging in size from "grains of barley" to "peaches and plums."[30] A team of inspectors sent by the shogun's government who arrived in Subashiri Village on December 21 reported that the local Sengen Shrine was "buried by hot stones up to the roof" and that burned homes were "buried up to their eaves." The roofs of smaller homes were not even visible. Looking to the future, the inspectors observed that, "because the fields are largely buried, there will be no barley harvest, which will bring hardship to the farmers."[31] Another account notes that thirty people died in Subashiri—remarkably, this is the only surviving record of deaths that occurred during the eruption[32]—but it was the loss of agricultural output, more than anything else, that would devastate Subashiri and many other tephra-covered communities in the years, and even decades, to come.

Precarious Abundance

Fuji's eruptions back in the Paleolithic era could have had a destructive impact on bands of hunter-gatherers living nearby, but the lack of permanent infrastructure and the mobile nature of human populations would have limited that impact—at least by modern standards. At the start of the eighteenth century, in contrast, the regions downwind of the Hōei eruption contained millions of people and were among the most intensively farmed on earth. Agriculture took root in these areas in the Yayoi period (600 BCE–300 CE)[33] and developed in waves over the centuries that followed, with the hundred years prior to the eruption witnessing particularly large-scale and widespread efforts to bring more land into cultivation and boost the output of existing agricultural fields. Transforming the countryside throughout Japan, these efforts involved elaborate irrigation schemes, intensified use of fertilizer, and the spread of both new agricultural products and the practice of double cropping, which enabled farmers to produce two harvests on a piece of land each year. As a result, demographers estimate that the population of the archipelago expanded from about fifteen million at the start of the seventeenth century to over thirty million by the early eighteenth.[34]

The use of fertilizer—whether derived from vegetation, excrement, or, increasingly in the eighteenth century, fish—was particularly important in the Kantō, where Edo was located. Much of the region is covered by the "Kantō loam," a thick layer of volcanic soils (andosols, also known as andisols) formed largely from ash produced by eruptions of Fuji and Asama (located a couple of hundred kilometers north of Fuji). While the Kantō loam is rich in essential nutrients, it also has high levels of aluminum that inhibit the uptake of phosphates by plants on the surface. It therefore requires large amounts of fertilizer to make it suitable for intensive cultivation.[35]

Massive irrigation projects that depended on the collaboration of villages, merchants, and samurai rulers were crucial to boosting output in the Kantō and elsewhere. These projects brought into production alluvial floodplains and other lands that had previously resisted cultivation. The bakufu set an example from on high by rerouting one of Japan's largest waterways, the Tone River, and building canals and draining swamps to allow for the expansion of Edo. In the vicinity of Fuji, levees built along the Fuji River, along with irrigation canals and drainage works, enabled farmers to convert large swaths of marshland located along Suruga Bay (the Ukishimagahara) into new rice paddies.[36]

The Odawara domain, which was located just east of Fuji and bore the brunt of the Hōei eruption, saw a boom in irrigation works in the seventeenth century.[37] Looking to promote the development of new rice paddies and thus expand its tax base, the domain used corvée labor for the construction of flood control and irrigation works along the Sakawa River, which runs through the Ashigara Plain down to Sagami Bay.[38] The domain also supported the tunneling, by hand, of a 1,280-meter-long aqueduct to bring water from Lake Ashi (located in the collapsed Hakone volcano caldera) to the Mikuriya region between Hakone and Fuji (see map 3.1). The taming of the Sakawa River was largely a top-down affair, but the plan to build the Hakone (or Fukuhara) aqueduct was hatched by the headman of Fukuhara Village, who formed a partnership with a group of investors from Edo. The group submitted proposals to both the Odawara domain and a local intendant (*daikan*) for the shogunate, which had direct control of part of the territory to be served by the aqueduct. After striking a deal that included tax incentives and a bakufu loan to help defray costs, the team of investors hired local farmers to dig the aqueduct over the course of four years, from 1666 to 1670. The water from Lake Ashi was then distributed to thirty villages, with both the investors and local farmers using it to develop new rice fields.[39]

Water control schemes were executed on a smaller scale as well. Take the creation of new rice fields at Adanoppara, located on tableland along the Sugawa River in the northern part of the Mikuriya region (now in the town of Oyama). In 1668 the headman of Yubune Village and a merchant in Edo petitioned Odawara officials to develop the new rice fields on the condition that one-tenth of the land would be free from taxation in perpetuity and that the rest would be exempt for seven years. The domain approved the project provided that the partners muster a workforce. They did—and from 1668 to 1671, workers built a holding pond and irrigation system that ended up supporting fifteen hectares (thirty-seven acres) of paddy fields.[40]

Such collaborative projects substantially boosted rice production near Fuji—as they did throughout Japan[41]—but the cost of prosperity was vulnerability, as occasional floods and landslides played havoc with even the best-planned systems. Those systems were certainly no match for the Hōei eruption, which caused damage on such a scale that in some places recovery became a task not for years but for generations.

Benevolence on Demand

While the eruption was in progress, there was not much people could do except to pray to the gods and buddhas and flee to places unaffected (or, at least, less affected) by the falling debris. The "bald-headed old man living east of Fuji" wrote that "men and women, young and old" called on the buddhas and chanted sutras before fleeing their village.[42] In Oshino, which was spared the fallout but offered a terrifying view of what was unfolding to the south, a local account notes that residents imagined they saw the form of the sun deity Nitten amid the rising smoke and began fervently praying to various kami and buddhas.[43] "Thanks to the gods," according to the account, "nothing fell." The common notion that the kami pick favorites, sparing some places while devastating others, also appears in a memoir from nearby Yoshida (see map 3.1), where the diligence of local priests praying for the protection of the Sengen deity seemed to be responsible for driving away the smoke that occasionally approached.[44] Meanwhile, Tokugawa Tsunayoshi did his part by ordering the priests of the Shingon Buddhist temple Goji-in (located in Edo) to perform esoteric rituals. The shogun also sent monetary offerings to Ōmiya Sengen Shrine—the same shrine accused of having failed to perform its duties during the Jōgan eruption—and to several small temples (*bō*) at Murayama.[45]

The divine winds that spared Oshino and Yoshida were the same ones that left villages to the east of Fuji buried in rock and ash. Fallout clogged irrigation works and raised riverbeds, which in turn led to massive flooding, especially in the highly productive Sakawa River Basin in the heart of the Odawara domain (see map 3.1).

Like the eruption itself, the immediate response and long-term recovery are richly documented. Because significant numbers of commoners in the Tokugawa period could read and write—and because those commoners had faith in the written word to get things done—historians in Japan have discovered numerous documents concerning the aftermath of the Hōei eruption produced not only by government officials but also by the villagers whose fields were buried in volcanic debris.[46] This allows a close view of the decisions and actions that took place within and among villages as well as between those villages and samurai authorities.

The outlook for those living immediately east of Fuji was grim, especially in light of the fact that the eruption took place in December, when farmers had already given up a significant portion of their harvest in taxes but were still busy with preparations to sell charcoal and other rural products in order to secure provisions for the winter.[47] Large numbers of them sold off livestock—there was no fodder to feed them anyway—and headed to Edo and other destinations to make ends meet. Others stayed, either by choice or because they were too old, young, or ill to find work elsewhere. Lacking the means to get through the winter, they faced the immediate prospect of starvation. The villagers who remained also had to deal with the long-term challenge of removing massive amounts of debris from their fields. They therefore sought three things from government officials: food aid to stave off famine, funds to pay for the labor (primarily their own) to dig out fields, and both the resources and expertise to rebuild infrastructure for irrigation and flood control.

To whom villagers appealed depended on where they lived. Within the compound-state system, the bakufu directly administered only those territories belonging to the Tokugawa house, while daimyo governed the several hundred domains that made up most of the realm. Tokugawa lands were divided into districts supervised by *daikan* (local intendants) and subfiefs under the control of *hatamoto* (bannermen) who answered directly to the shogunate. Daimyo also parceled some of their domains into subfiefs. Adding to the complexity was the fact that households in a single village could answer to multiple authorities. Some villages in the fallout zone had to negotiate with as many as five different bannermen.[48]

The hardest-hit villages were located in the Mikuriya and Ashigara regions. The majority of them fell under the jurisdiction of Odawara daimyo Ōkubo Tadamasu (1656–1713) or his brother Norihiro (1657–1738), who controlled his own subfief within the Odawara domain. Yet unlike the shogunate, which dispatched its fact-finding team two days into the eruption, the domain was relatively slow to conduct its own investigation, sending an inspector to tour affected villages only after the two-week-long eruption had ceased. Norihiro authorized the release of rice to those at risk of starving in northern Mikuriya villages prior to the domain's inspection, but the delay in sending an inspector was indicative of the sluggish and inadequate response to come.[49]

Three villages in the Ashigara region sent the first appeal while the eruption was still in progress (the exact date is unclear). Addressed to two of the domain's administrators, it catalogued the villagers' distress, reporting that the barley harvest was ruined, their animals had died, and they lacked drinking water. The appeal also noted that it would be difficult for the villagers to rehabilitate their fields on their own and that, because the roads were covered in debris, they could no longer draw income from the packhorse trade or from selling thatch, firewood, and other local goods. Emphasizing the threat of famine—one magnified by the loss of wild plants and game—they requested an official inspection of their villages so the domain could witness the extent of their plight.[50]

The assumption underlying this invitation was clear: if we show you our suffering, you have to do something about it. In *Seeing Like a State*, James Scott emphasizes that governments accrue power by making populations as "legible" as possible, and that those wanting to avoid taxation and other state-imposed burdens often resist efforts to do so.[51] Yet this entreaty and others produced after the Hōei eruption show that people who had "learned to live under the watchful eyes of registrars and see their resources registered by convention"—as Mary Elizabeth Berry describes the villagers of Tokugawa Japan[52]—could also conclude that making themselves legible meant holding the state accountable.

Whether prodded by this specific petition or not is unclear, but after the eruption ended on January 1, 1708, domain official Yanagita Kyūzaemon finally set out to inspect stricken areas.[53] Yanagita's tour gave village leaders an opportunity to make further appeals for emergency rations and help in removing debris. Per his instructions, they quickly submitted written estimates of what it would take to clear fallout from their fields.[54] Cognizant that recovery efforts following the Genroku earthquake had severely depleted domain coffers,[55] he eventually informed supplicants that they would have to clear fields on their own. He also gave them no specifics about food aid, despite having assured them that the benevolent Tadamasu worried about their situation day and night.[56]

Those who had gone through the trouble of submitting estimates were naturally dissatisfied with this response. So a day before Yanagita's return to Edo, where Tadamasu was serving as a senior councilor to the shogunate, the representatives of 104 Ashigara villages handed him a petition to bring to their lord. In it they stressed that both villagers and their animals faced starvation, and appealed to Tadamasu's benevolence (*jihi*)—a term Yanagita himself had used in his assurances to the villagers—to set things right.[57]

Liberal use of the term *jihi* is typical of petitions submitted to the authorities in the months and years following the eruption. In one forwarded to the shogunate in 1709 by thirty-nine Mikuriya villages, it appears no fewer than six times.[58] Samurai rulers invoked the ideal of benevolent governance in order to justify an extractive system ultimately based on military strength and the threat of violence that came with it, but it also allowed the villagers who made their suffering legible both in person and in writing to pressure their overlords to live up to their ideological word. When the lords fell short, the farmers let them know it.

Peasants on the March

The Ashigara villages not only produced written appeals but, helpfully for historians, also kept detailed records of their negotiations with domain officials. We learn from them that, after a number of days had passed with no response to the petition submitted to Yanagita, the increasingly impatient Ashigara village representatives assembled below Odawara castle and gave a domain official written notification that they intended to head to Edo—which could be reached by foot in a couple of days—to appeal directly to the bakufu. They made sure to mention that they lacked fodder to feed the post-horses on which the shogunate's transportation and communications network depended.[59]

Domain officials tried to negotiate with the angry petitioners in Odawara, knowing they would be blamed if villagers jumped the chain of command and made a scene in the shogunal capital. This would have been the case under any shogun, but the man in power at the time, Tokugawa Tsunayoshi, was known for taking an especially hands-on—and in the eyes of samurai critics, high-handed—approach to government.[60] His reputation for autocratic rule reached a peak when he decreed the "laws of compassion," which, among other things, made it a capital offense to kill dogs. Resented in particular by samurai accustomed to using dogs to practice their archery and swordsmanship,[61] this much-maligned policy earned Tsunayoshi the derogatory epithet "Dog Shogun." One (possibly apocryphal) account explicitly links criticism of the law to the Hōei eruption, claiming that the monk Yūten (1637–1718), abbot of the Tokugawa family temple Zōjōji, boldly told Tsunayoshi that debris originating in the earth but falling from the sky was as abnormal as putting people to death for the sake of beasts.[62] Tsunayoshi biographer Beatrice M. Bodart-Bailey warns that we should treat criticisms of the Dog Shogun with caution, considering "the pain he inflicted on the class [the samurai] responsible for the record."[63] She argues that, taken as a whole, the laws of compassion and other measures that had the effect of curbing samurai privileges were designed to benefit the most vulnerable under his rule.[64] So in her telling of the back-and-forth between the Ashigara villagers and officials of the Odawara domain, Tsunayoshi's reputation for keeping a close watch on the performance of samurai officials, combined with his concern for the welfare of commoners, contributed significantly to the calculations of both the villagers and domain officials.[65]

Whether those officials would have been any less motivated to stop the villagers during the reign of another shogun is impossible to know, but they were certainly anxious about the Ashigara petitioners reaching Edo, and they successfully convinced the village leaders to remain in Odawara while the domain made arrangements to provide relief. That was not enough to satisfy the thousands of villagers who had joined their representatives in Odawara, and they decided to march to Edo to appeal to Tadamasu and the shogunate. Panicked, the officials caught up with them and on the spot promised each man five *gō* (almost one liter) of relief rice and each woman two. The villagers found this insufficient, but their representatives convinced them to return home while they continued to Edo to press their case.[66]

When the representatives were within a day's walk of the shogun's capital, a domain official sent from Edo intercepted them with promises to provide twenty thousand sacks of rice and to remove Yanagita Kyūzaemon from any involvement in the relief efforts. In return, he asked them to go back to Odawara. The representatives still wanted aid to dig out their fields, however, so they decided to resume their journey to Edo.[67] They were waylaid again, this time by another official who told them Tadamasu would provide not only the relief rice but also twenty-seven thousand *ryō* (a unit of gold coinage)[68] to pay for the removal of debris. Mollified, they headed back to Odawara, only to be told that the money could not be distributed right away. So they hit the road once more, reaching the southeastern outskirts of Edo before they met with the same official who had given them the figure of twenty-seven thousand ryō. He promised them Tadamasu would indeed raise the money, even if that meant appealing to the shogun, selling his most treasured sword, and forgoing evening meals.[69]

This dramatic assurance convinced the representatives to return to Odawara, but as Nagahara Keiji puts it, they were subsequently "stabbed in the back."[70] First, local officials informed them that only ten thousand sacks of relief rice were immediately available. Second, they conflated the relief rice with the aid to remove debris, saying the rations should be distributed to starving villagers but at the same time presenting an account book that had "Debris Removal Payment" (*Sunahakiryō*) written on the cover. Third, they told the representatives the distribution of rice would be calculated according to the annual tribute of each village as well as the depth of the debris.[71] In response, the villagers submitted yet another petition, in which they took the domain to task for changing the specifics of their agreement. They also complained about the domain's formula for distributing aid, explaining that it took just as much labor to clear fields that produced less tribute as it did the fields that produced more, and that the amount of fallout varied within each village.[72]

Like the villagers, the shogunate expected the lord of Odawara to take action. A couple of weeks earlier it had admonished officials in regions affected by the eruption to press farmers to clear their fields in time for the spring planting and also to provide aid where necessary, taking care not to let people starve.[73] Orchestrated by Finance Magistrate Ogiwara Shigehide (1658–1713), the shogunate's decree put pressure not only on officials but on farmers too by stressing that they should remove debris from their fields as diligently and quickly as possible, even if that meant women and children had to help. It also instructed them to assist neighbors who were shorthanded.[74] In this way, the shogunate made clear that, although officials in the Odawara domain and elsewhere should assist them, farmers would have to do the heavy lifting on their own.

This emphasis on self-help was motivated by financial concerns, not an ideological fear of moral hazard, as was the case, for example, in nineteenth-century British responses to famine in Ireland and India. In his book on the Irish Potato Famine (1845–1852), Cormac Ó Gráda highlights the devastating role played by the "conviction that overgenerous relief would demoralize the Irish poor and merely postpone the reckoning," a conviction responsible for the policy of making hungry people perform physical labor—including building roads and breaking stones in the dead of winter—in exchange for assistance.[75] After the Hōei eruption, in contrast, the push for a public works program to remove fallout came not from above but from below—that is, from farmers seeking financial support for the labor they themselves would provide to rehabilitate their waterways and fields. Although operating within a Confucian moral economy in which rulers were supposed to extend benevolence to those in need, Ogiwara, like the domain officials, simply wanted to bear as little of the cost as possible.

The Shogunate Takes Charge

The shogunate's decree aimed to speed things along, but as the negotiations between the Ashigara villages and the Odawara domain dragged on, it became obvious that exhorting officials and farmers to act was insufficient. So on the third day of the intercalary first month (a lunar calendar "leap month"), just days after the villagers had submitted their complaint about the domain's plan to disburse aid, the shogun approved an extraordinary measure to put the areas hardest hit by the eruption—including over half of the Odawara domain—under temporary bakufu control.[76]

Confiscating domain lands was not unprecedented. During his reign, Tsunayoshi expropriated territory that produced on average about sixty thousand *koku* a year (one *koku*, about 180 liters, was said to be the amount of rice needed to feed an adult for a year).[77] Yet to take over half of a prominent domain that produced over one hundred thousand koku annually, and to do it for the purpose of managing disaster relief, was a first.[78] While affirming the authority of the shogun, this arrangement ultimately benefited Odawara daimyo Tadamasu. Surely thanks to his status as senior councilor, he was assigned new lands in various parts of Japan that collectively produced as many koku as the ones that had been seized, allowing him to dodge the financial burden of the eruption, which was placed on the shogunate instead.[79]

The shogun's government now confronted the task of funding and coordinating relief efforts in the fallout zone. From its earliest days, the bakufu had turned to daimyo for assistance with building projects. Tokugawa Ieyasu ordered them to construct river works; and under subsequent shoguns, they were expected to contribute labor to build and repair castles and religious institutions.[80] Faced with fewer financial resources than his predecessors,[81] Tsunayoshi was particularly reliant on daimyo for help. After the early years of Ieyasu's reign, the shogunate had not called upon daimyo to assist with riparian projects, but Tsunayoshi ordered a number of them to provide aid—mainly in the form of funds distributed to contractors—for work on several rivers.[82] He also ordered groups of daimyo to provide funds to repair the damage caused by the Genroku and Hōei earthquakes.[83]

After the Hōei eruption, the bakufu went a step further. Concluding that it was not enough to call on particular daimyo for help, on the same day it officially announced its expropriation of lands in the fallout zone (the seventh day of the first intercalary month), the shogunate also made the wholly unprecedented move of commanding landholders throughout Japan, in daimyo domains and shogunal territories alike, to pay two gold ryō for every one hundred koku of taxable harvest.[84] In this way, the shogunate nationalized the cost of the disaster, although as Kitahara Itoko points out, by pairing the tax with the takeover of lands in the disaster zone, it stuck with the prevailing logic that the bakufu should provide and coordinate aid only in those territories directly under its administration.[85]

Judging from the total assessed koku in both shogunal and nonshogunal lands at that time, the regime successfully collected all of the tax due, about 488,700 gold ryō.[86] According to an investigation of existing records, the bakufu allocated 54,480 ryō to clear debris and dredge rivers in Musashi and Sagami provinces, 6,225 to disburse as direct aid for villages in Suruga, Musashi, and Sagami provinces, and 1,854 to rehabilitate Subashiri Village. Subashiri was singled out because of its strategic importance as a post station along the highway that connected Suruga to the province of Kai north of Fuji.[87]

It appears, then, that less than 13 percent of the funds raised by the extraordinary tax was spent on relief and recovery efforts associated with the Hōei eruption. Where did the rest of the money go? Arai Hakuseki claimed in his memoir that Ogiwara (Hakuseki's nemesis) never intended for the bulk of the tax to be spent on relief and recovery efforts, using the extraordinary measure to shore up government finances instead.[88] Hakuseki was not the only one who was critical of the tax. After Tsunayoshi died of measles in early 1709, a low-level samurai from the Owari domain, Asahi Shigeaki, impertinently wrote in his journal that the tax had portended the shogun's death.[89] Historians have generally followed Hakuseki's lead in characterizing the tax as a ploy to fill the shogun's treasury, but Bodart-Bailey suggests that the accounting is incomplete, citing Hakuseki's own claim that the shogunate provided one hundred sixty thousand ryō in assistance during the first year following the eruption. If anyone is to blame for misallocating funds, she argues, it is Tsunayoshi's successor Ienobu (r. 1709–1712), who spent the remarkable sum of seven hundred thousand ryō to build a new shogunal residence.[90]

Whatever the case may be, only a small portion of the tax was spent on relief and recovery efforts in the years following the eruption. It is clear that the shogunate's top priority was flood control, particularly along the Sakawa River. This makes sense, considering that villages located in major river basins produced the greatest amount of taxable rice. As noted earlier, Subashiri was an exception to this general rule because of its role as a transportation hub.

Kantō magistrate Ina Tadanobu (?–1712) took charge of the disaster zone, setting up a base of operations in the town of Sakawa, situated at the intersection of the Tōkaidō highway and the road that led to the Mikuriya region.[91] But if farmers thought shogunal oversight would bring with it quick and comprehensive assistance, they were soon disappointed. In a set of instructions sent to villages on the twenty-ninth of the first intercalary month, Ina followed Ogiwara's lead by exhorting residents, including women and children, to be diligent in removing debris from their fields as quickly as they could.[92] He made no mention of funds to support this effort.

Ina did provide details on what to do with the fallout. The "bald-headed old man living east of Fuji," in his account of the eruption and its aftermath, notes that the villagers "employed the power of water" by dumping debris from their fields into nearby streams.[93] Yet this tempting solution contributed to another problem: the raising of riverbeds downstream, which in turn led to flooding and the destruction of the irrigation systems on which farmers—and government coffers—depended. With this in mind, Ina instructed villagers to pile up debris between fields and residences or to dig trenches and bury it in them so that a layer of arable soil was left on top.[94] It is hard to imagine how much work it would have taken to turn over the earth in one field after another, but several excavations in the fallout zone reveal that at least some villagers complied with Ina's instructions.[95]

Although Ina, who worked closely with Ogiwara, was reluctant (or unable) to pay for the labor to remove debris, he provided emergency funds to starving villagers so they could buy food for both themselves and their horses. This aid was especially critical for the hardest-hit villages in the Mikuriya region, where residents had struggled to make ends meet even before the eruption. Buried under meters of debris, their situation became quickly and increasingly desperate, as is clear in village records (created to provide a thorough accounting to authorities) that list the numbers of those in need of emergency relief along with their age, their gender, and the amount of money distributed to them. Records for Tanagashira Village, for example, show that there were 52 aid recipients in the second month of 1708, but that every man, woman, and child in the village, a total of 150 people, received aid in the third, fourth, and fifth months. That number dipped slightly to 146 in the sixth month and was down to 123 by the tenth month, but that was probably because a number of villagers had traveled elsewhere to find work.[96]

Meanwhile, Ina focused most of his attention on dredging the Sakawa River and its tributaries in order to prevent floods that would not only compound the suffering of villages in the Ashigara region but also impede travel along the Tōkaidō. The dredging project, along with the reinforcement of levees, had the added benefit of providing paid work to villagers in the fallout zone.

The shogunate had already levied a national tax to fund relief and recovery efforts after the eruption, but for the Sakawa River project it once again commanded daimyo to provide aid. In this case, five of them were ordered to give assistance. Each domain was assigned a portion of the river system, but Ina coordinated the project as a whole while Ogiwara dictated how to finance it. In the end, Ogiwara contracted the bulk of the work to a stonemason in Edo, with the domains providing funds to defray the cost.[97] This sort of public-private partnership typified irrigation and flood control projects in the seventeenth century. It had also recently been employed to deal with the flooding of the Tone River in 1704 and to repair damage along the Tōkaidō caused by the

Hōei earthquake. In both cases, the shogunate ordered daimyo to give assistance, but private contractors performed the work.[98]

Work on the Sakawa River and its tributaries lasted from the second to fifth months of 1708, but it proved to be insufficient when a typhoon struck soon after, causing the collapse of key levees, which triggered massive flooding.[99] This compounded the misery of Ashigara farmers, and meant more funds had to be spent on an even-larger effort to remedy the situation. But while the floods were disastrous for those directly affected by them, the new riparian project—which lasted from the eighth to tenth months of 1709—was a welcome opportunity for those who needed jobs.[100]

This was especially true for farmers from the devastated Mikuriya villages whose fields were still buried under a meter or two of debris. Since early 1708, Mikuriya farmers had submitted petitions in which they stressed the need for both famine relief and funds to pay for the removal of debris from their fields.[101] One submitted by Yozawa Village in the eighth month of 1708 dramatically asserts that local farmers would have to abandon the area and that it would become a "dead zone" (*bōsho*) if they did not receive aid to clear their fields.[102] The outlook for Mikuriya farmers turned especially bleak when famine relief ended in the second month of 1709. In response, representatives of thirty-nine villages journeyed to Ina's residence in Edo to submit a joint petition (the one in which the word "benevolence" appears six times) to inform Ina and other officials that people were dying of hunger and had turned to begging despite the aid offered so far. The petition also notes that the villages could resume paying tribute if they had the resources to rehabilitate their fields.[103]

Given their desperate situation, it is understandable that large numbers of Mikuriya farmers traveled to the Ashigara region to work on the new Sakawa River project. In the case of Yozawa Village, records show that every ablebodied man, woman, and child over the age of twelve made the journey.[104] Unlike locals, however, they had the added expense of paying for lodging, so they appealed for compensation to make up the difference. Ina had seen their distress firsthand on an inspection tour of Mikuriya villages several months earlier, so he was sympathetic to their request and successfully convinced Ogiwara to boost their compensation.[105]

By the end of 1709, Ina had also secured funds to help pay for the labor to remove fallout—albeit far less than farmers had requested. Tanagashira Village, for instance, had asked for about 3,698 ryō to cover two years' worth of work but received only a little more than 7 ryō.[106] Meager as they were, the payments to Tanagashira and thirty-five other Mikuriya villages (all covered in at least ninety centimeters of debris)[107] at least demonstrated that officials would provide money to clear fallout if farmers pushed hard enough.

Having successfully intervened to boost their pay on the Sakawa River project and to secure funds to remove debris from their fields, Ina won the affection of Mikuriya villagers. According to local legend, he also provided rice from government storehouses without prior authorization and was punished with house arrest. Whether true or not, this story shows that villagers came to see Ina as a sympathetic ally.[108] Generations later, in 1867, a shrine was erected to him in the village of Yoshikubo (now part of the town of Oyama), with another one built in Subashiri soon after. Locals then successfully petitioned the government to award him posthumous court rank in 1915.[109] Memorializing Ina not only expressed enduring gratitude but also functioned as a reminder that state authorities, in whatever form and in whatever age, should look after the welfare of their subjects.

Continuing Struggles

Ina's intervention gave Mikuriya farmers hope, but in the years to come, trying to get government assistance remained a constant struggle. In 1710 representatives for fifty-eight villages submitted another petition, in which they complained that the payments authorized by Ina only went to those villages with at least a meter (three *shaku*) of fallout, disregarding those that had less than that amount but were suffering nonetheless (the situation for twenty-two of the fifty-eight villages). Echoing the joint appeal submitted a year earlier, they emphasized that those with the strength to clear debris had gone elsewhere to find work, children and the elderly had turned to begging, and people were dying every day of starvation.[110] It is difficult to tell from surviving records how much aid villagers received after making this request[111]—perhaps Ina did, in fact, open government storehouses—but judging from the continued desperation of villages in the years that followed, assistance was insufficient, inconsistent, and temporary.

One document cataloging the situation of seven villages in 1712, the year when Ina died and Hakuseki engineered the dismissal of Ogiwara as finance magistrate, shows in numerical terms just how devastated they were.[112] Yozawa Village, for instance, was buried in well over a meter of suffocating fallout, and had lost nearly 65 percent of its population (it is unclear how many residents died of starvation and how many left the village to make ends meet elsewhere) and nearly 90 percent of its horses, which had been sold off by villagers desperate to earn money. The document also reveals that villagers had been able to clear only about 6 percent of their fields.[113] At least this put them ahead of the other six villages, three of which had removed nothing at all.[114]

In 1716 Tokugawa Yoshimune (r. 1716–1745) became shogun, and in that same year his government instituted a new policy of providing stipends of rice (calculated at almost one liter per man and a little over half a liter per woman for each day's work), but farmers only received them based on the labor they had already expended on removing debris.[115] In order to move things along, the shogunate decreed that laborers would acquire rights to the areas they cleared, meaning that if short-handed villages needed outside help, they would have to relinquish fields to newcomers.[116] Tanagashira Village, for one, was unwilling to do this, even though that meant its fields would not be fully rehabilitated until the waning years of the Tokugawa shogunate over a century later.[117]

While villages banded together to seek government aid, each one, like Tanagashira, made sure to look out for its own interests. We see this in a contract from 1716 specifying that, in exchange for helping Adano Shinden (*shinden* means "newly developed rice fields") fix its waterways, Suganuma Village could use a portion of Adano Shinden's water to flush debris from its own fields (despite instructions just after the eruption to bury fallout, it was easier for villages to wash it away using ditches dug for that purpose). The contract also stipulates that Suganuma had to stop using its neighbor's water once those fields were brought back into cultivation.[118]

While some villages reached mutually beneficial agreements, others engaged in bitter, protracted battles as they tried to secure the resources they needed to recover. The right to access common lands (*iriaichi*) was a particularly heated point of conflict, since they provided a wide range of items key to the survival of rural communities, such as fodder for animals, thatch for roofs, fertilizer for agriculture, and brushwood for heating and cooking. Disputes over these communally regulated lands were endemic to Tokugawa Japan, but they were especially pronounced after disasters like the Hōei eruption, which destroyed resources essential to everyday life and therefore intensified competition over those that remained.

A major conflict broke out between groups of villages over access to the Ōnohara, an expansive grassland at the eastern foot of Fuji. Due to prevailing wind patterns during the eruption, the common lands of nine villages to the north of the Ōnohara were especially hard hit by fallout, and they subsequently turned south to acquire the resources they needed to survive.[119] Seven southern villages resisted this incursion, so the northern villages, which included Yozawa, submitted a lawsuit to the bakufu to secure official access to the Ōnohara (which they had, in fact, used to a limited degree before the eruption).[120] The northern villages won the suit, but they had to file another one in 1727, this time against eleven villages that tried to block access to the Ōnohara.[121] In 1729 the shogunate reaffirmed that the nine northern villages—as well as two other villages that had filed their own suits—had the right to access the Ōnohara common lands.[122] The shogunate's decision did not bring conflict over the territory to an immediate end—as demonstrated by the theft of thirty scythes from six villages the following year—but it did provide a legal basis for handling disputes in the decades to come.[123]

Villages were also at odds when it came to rebuilding infrastructure for irrigation and flood control. This was especially the case for the densely populated Sakawa River Basin, home to numerous tax-bearing fields that shogunal officials wanted to rehabilitate as soon as possible. Work focused in particular on the Ōguchi levee, situated where the Sakawa River flowed from a mountain valley into the fertile lowlands of the southern Ashigara Plain. In the spring of 1708, an Edo stonemason won the contract to rebuild the levee and dredge the river. The job was rushed, however, and in the sixth month, just weeks after the work was completed, the levee burst due to torrential rains, and the course of the river quickly shifted west, inundating the villages in its path.[124] The Ōguchi levee was reconstructed once again, this time reinforced by a second levee; but in 1711, during another torrential rainstorm, the twofold defense system failed, and the river shifted to the west once more.[125] The course change not only flooded villages but also stranded irrigation sluices and canals that had been designed to take advantage of the river in its original form; so villages built competing infrastructure to divert water from the new course to their fields, which in turn led to a bitter conflict over water rights.[126] It also led to a debate over a proposal to stop trying to put the river back in its old course and instead reinforce the new one—a proposal made by villagers located on the bank of the new river course who were clearly willing to sacrifice the fields of neighbors directly in its path.[127]

The river was ultimately returned to its original course—thanks in large part to the involvement of Odawara domain. In 1716, as part of a larger effort by the newly installed shogun Yoshimune to reduce expenses, nearly half of the territory expropriated by the bakufu after the eruption was returned to the domain, including a number of villages along the Sakawa River.[128] Six years later, after making one more failed attempt to coax the river back into its old course, and concluding that it would be futile to reconstruct the Ōguchi levee until the fallout that had raised the riverbed had washed out to sea, the shogunate ceded responsibility for the Sakawa River.[129] To its credit, the domain took charge of the situation and quickly drew up a detailed seven-year plan to return the river to its original course and get farmers back on their feet. Rather than immediately rebuild the Ōguchi levee, domain officials focused on dredging the river and strengthening the embankments along its old course in order to reduce the risk of flooding once the levee was repaired. Performed by local villagers employed by the domain, this crucial work was completed in only three years—well ahead of schedule.[130]

At that point the bakufu decided to finish the job by providing the funds and expertise to rebuild the Ōguchi levee.[131] The government put renowned agricultural and water management expert Tanaka Kyūgu (1662–1729) in charge of the endeavor, which lasted from the second to sixth months of 1726. Tanaka employed new engineering techniques to lengthen, widen, and strengthen the main Ōguchi levee as well as a smaller levee located on the opposite side of the river.[132] Pleased with the results, the shogunate awarded Tanaka samurai rank. The promotion in status was apparently well deserved: although portions of the levees were breached in 1734 and again in 1791, they have, for the most part, stood the test of time, and continue to play a key role in managing the Sakawa River today.[133]

The road to recovery for both the Mikuriya and Ashigara regions was difficult, uneven, and long. It took decades of backbreaking labor for Mikuriya villagers to put fields back into production, and intermittent floods attributable to fallout from the Hōei eruption repeatedly plagued the Ashigara Plain until the early 1800s.[134] Along the way, the bakufu returned villages deemed sufficiently rehabilitated to the Odawara domain, with some Mikuriya villages transferred to other jurisdictions.[135] The village of Ōmika remained in such bad shape, however, that the shogunate held onto it until the end of the Tokugawa period.[136] Tax records kept by Odawara domain confirm that there were large disparities in the rate and extent of recovery from one village to another, and that progress was inconsistent wherever it took place.[137]

THE HŌEI ERUPTION tested both the ecological and political limits of Tokugawa Japan. Massive amounts of fallout buried fields and destroyed the elaborate systems of irrigation and flood control on which they depended. Faced with immediate hunger and long-term ruin, tens of thousands of farmers demanded assistance from rulers who worked within a system of quasi-autonomous domains designed to extract resources from farmers, not provide them. Prodded into action, the shogunate ultimately provided aid, but it was grudgingly and unevenly distributed.

The methodical way villagers appealed to both the Odawara domain and the shogunate was typical for the time. Central to the thousands of protest movements that sprang up in Tokugawa Japan were petitions that articulated specific grievances and pointedly reminded rulers of the benevolent treatment that loyal, hardworking subjects could expect from their overlords. By appealing to this widely accepted principle of good governance, petitioners hoped they could press their claims without provoking unnecessary confrontation and punishment.

Words were frequently not enough. Petitioners used the rhetoric of the ideal relationship between ruler and ruled to legitimate their case, but fiscally minded officials were often reluctant to make good on the Confucian principle of reciprocity that they professed. The truth was that "honorable farmers had an obligation to work hard; the rulers' decision regarding whether or how to bestow compassion was always arbitrary," as Anne Walthall puts it.[138] So in many cases, what started as a petition turned into a full-blown protest. This often took the form of a march to a domain's castle or even to Edo—the course of action chosen by villagers devastated by the Hōei eruption. Once again exemplifying a pattern seen throughout the Tokugawa period, both Odawara officials and village representatives engaged in negotiations aimed at reaching an agreement before things escalated too far, with the result that the villagers turned back before reaching Edo.

Here, however, the familiar story of peasant protest took a novel turn. Already burdened with the costs of rebuilding after the Genroku and Hōei earthquakes, Odawara lacked the means to dig out fields and repair the water control infrastructure destroyed by the eruption, so shogunal officials seized control. For the bakufu to support and coordinate relief efforts was not unprecedented—Geoffrey Parker, in his encyclopedic study of disasters in the seventeenth century, extols the proactive and vigorous response of Shogun Iemitsu (r. 1623–1651) to the Kan'ei famine, which peaked in 1641–1642, as an instance of "getting it right"[139]—but when it came to dealing with the aftermath of the Hōei eruption, Tsunayoshi's government made two extraordinary moves. First, it expropriated over half of a prominent domain in order to provide disaster relief. Second, it levied a nationwide tax to pay for that relief. Never before had the response to a disaster been nationalized on this scale.

Intervention by the shogunate surely helped keep down the death toll, even if farmers had to rely mainly on their own resources to bring their fields and villages back to life. Villages emptied out as the able-bodied sought work elsewhere, while many of those who remained turned to begging to survive. Limited as it was, however, bakufu aid helped initiate a trend of increased government relief in the wake of earthquakes, eruptions, and other disasters. By the end of the eighteenth century, as Gregory Smits notes, government assistance "was standard practice and had developed into a predictable pattern that included providing food and shelter for the poorest members of society for one to two months."[140]

But while the shogunate's response to the Hōei eruption helped start a new trend, the decision to temporarily take over Odawara lands in order to provide direct relief kept in place the principle of divided governance that had determined relations between daimyo and the shogunate from the start of the Tokugawa period. Going forward, it would remain primarily the burden of individual domains to see to the welfare of their subjects. There were exceptions to the rule—for example, after fallout from the eruption of Mount Asama in 1783 caused widespread flooding, the bakufu took charge of rebuilding water control infrastructure on which villages in multiple domains relied—but in the case of other disasters, such as the eruption of Mount Unzen in 1792, help from the shogunate was generally limited to loans that had to be repaid by the daimyo who had requested them.[141]

As expectations for government aid in both domains and shogunal lands rose, so did dissatisfaction when those expectations were not met. Across Japan, the number and intensity of popular protests increased over the course of the eighteenth century and into the nineteenth, with anger often directed not only toward officials but also toward well-to-do peasants and merchants accused of inflating prices by hoarding rice.[142]

Uprisings were particularly violent and widespread in the 1780s. Cold, rainy weather caused multiple crop failures, resulting in a nationwide famine exacerbated by the recent growth of commercial over subsistence agriculture.[143] Some domains were able to head off unrest by addressing, even minimally, the needs of their subjects. Ōno domain, for instance, successfully kept the peace by increasing its expenditures on rice gruel kitchens, making no-interest "starvation loans," and ordering well-off residents to provide for those less fortunate.[144] In contrast, famine-stricken villages in the Mikuriya region near Fuji started an uprising in 1783 after the Odawara domain refused to provide tax relief. The tax revolt was ultimately unsuccessful, and the domain punished its leaders with fines, imprisonment, corvée labor, and, in one case, death.[145]

As the sharp difference between events in Ōno and Odawara shows, the shogunate's response to the Hōei eruption may have demonstrated the ability to marshal resources across Japan, but given both political and ecological constraints, it failed to produce a consistent, centralized, and durable system to handle future disasters. In that respect, the response of the Tokugawa regime was not unusual but typical of government actions in the wake of disasters, both then and now.

4

Holy Fuji

AN OIL SALESMAN from Edo called Jikigyō Miroku (1671–1733) starved himself to death on Fuji in 1733. Distressed by the state of the world and acting on the instructions of the Fuji deity Great Bodhisattva Sengen (Sengen Daibosatsu), he aimed to save all living things through his sacrificial fast, which also served as a rebuke to a government that had failed to meet the needs of its people.[1] He chose Fuji because, like others who followed in the footsteps of Kakugyō Tōbutsu (1541?–1646), he considered Fuji not just a sacred mountain but a cosmic one. He and other devotees often referred to Fuji as "the greatest mountain of the Three Lands." The "Three Lands" referred specifically to Japan, China, and India, which in effect meant the entire world. Origin of the moon and sun, Fuji was, argued Jikigyō, none other than Mount Meru (or Sumeru), the axis of the world according to Hindu and Buddhist teachings.[2]

A volcano would seem to make an unstable axis. Fuji had erupted violently in 1707, and villages in the Mikuriya and Ashigara regions were still reeling from its effects at the time of Jikigyō's death. But even though Jikigyō would have witnessed the Hōei eruption—he lived in Edo at the time—he largely disregarded it in his teachings. Instead, Jikigyō propagated Kakugyō's vision of Fuji as a source of stability and life rather than uncertainty and death. In the process he articulated what I call a "cosmic bodyscape" in which Fuji's geological body nourished biological ones. To show gratitude for the sustenance provided by the Fuji deity—Great Bodhisattva Sengen, or the Original Father and Mother (Moto no Chichihaha)—people were supposed to show gratitude by behaving ethically in their daily lives.

How quickly and broadly news of Jikigyō's death traveled is hard to judge, but in the decades after his dramatic sacrifice, confraternities devoted to the worship of Fuji called *Fujikō* proliferated. This was especially the case in Edo, where, on a clear day, residents could view the mountain in the not-too-far-off distance. Members regularly met in each other's houses for prayer meetings coordinated by leaders called *sendatsu*. During the summer months some members also went on pilgrimage to Fuji, where they stayed at lodgings owned by *oshi*, innkeepers who were also ritual specialists. These local figures hosted pilgrims during the summer climbing season and visited their parishioners (*danna*) at other times of the year so they could distribute protective amulets along with ritual instruction. Oshi had existed at the base of Fuji since the Warring States period (1467–1568), but after Jikigyō's death, their livelihoods increasingly depended on parishioners who belonged to Fujikō.

The growth in pilgrimage to Fuji also generated conflict between locals that sometimes turned ugly. Origin of the cosmos, Fuji was at the same time a source of competition for the money spent by pilgrims, and heated debates broke out over women's access to the mountain as certain Fujikō members and allied oshi challenged the long-standing practice of barring women from the upper slopes.

Meanwhile, Fujikō ran into trouble with the shogunate. Operating outside the bounds of formally recognized institutions of belief and practice, these worship groups blurred the line between laypeople and ritual professionals. They also brought together people across the spectrum of Tokugawa society, from samurai to day laborers. This status-mixing worried a regime committed to the ideal of a "container society" in which everyone stayed in their appointed place.[3] It did not help that Fujikō members petitioned the government to embrace their teachings, including one charging that the shogun and emperor should do their part in fully realizing the "Age of Miroku." Initiated by a cosmic upheaval (*furikawari*) in 1688, this new age would bring universal prosperity, but only if people accepted and lived by the tenets propagated by Kakugyō, Jikigyō, and their successors. Far from accepting these tenets, the shogunate repeatedly tried to suppress Fujikō from the late eighteenth to the mid-nineteenth century. The fact that officials had to issue multiple bans shows that this was easier said than done. Fujikō only increased in popularity as time went by, defying government efforts to restrict them. In the end, faith in the saving power of Fuji overcame deference to the Tokugawa status quo.

The vision of Fuji as steady and nurturing rather than volatile and destructive grew all the more powerful as the Hōei eruption faded from memory. Often envisioned by Jikigyō and others in the form of a human body, Fuji sustained and linked the bodies of all living things within a cosmic bodyscape that was cyclical in nature. And despite generating political, economic, and ritual conflicts, popular devotion to Fuji helped make the mountain into a unifying force for the Japanese people. Although few Fujikō remain active, that force is as strong as ever today.

Jikigyō and the Cosmic Bodyscape

Jikigyō's teachings played an outsized role in the development of Fujikō. Kakugyō, the ascetic who communed with the Fuji deity in the Hitoana Cave and achieved a reputation as a healer, was widely acknowledged as the founder of the Fuji cult as it developed during the Tokugawa period (1600–1868). Yet the spread of Fujikō did not take off until Jikigyō and another Fujikō leader, Murakami Kōsei (1682–1759), created their own lineages to spread the faith. Murakami's lineage made its physical mark at the base of Fuji. The wealthy founder provided funds to rebuild the Sengen Shrine in Kamiyoshida, located at the northern foot of the mountain, and the generations that succeeded him erected memorials to sendatsu in an enclosure outside the Hitoana Cave. Memorials to sendatsu in Jikigyō Miroku's line of descent were relegated to the surrounding woods. But while the Murakami lineage dominated the Hitoana Cave, Jikigyō's lineage eventually eclipsed it, thriving into the nineteenth century while the Murakami lineage declined.[4]

Jikigyō performed his ultimate sacrifice after he had already spent decades promoting the gospel of Fuji. Born in Ise Province, he left home as an adolescent and established himself in Edo, where he married and allegedly became a thriving oil merchant with a reputation for being scrupulously frugal and honest—although one of his successors wrote in his diary that he was also a "difficult man" who was "quite unpleasant to be with."[5] While still a young man he devoted himself to Fuji worship and adopted the name Jikigyō, consisting of the characters *jiki* (食) and *gyō* (行). *Jiki* means "eating" or "food," but joined with *gyō*, "practice," signifies ritual fasting, which Jikigyō performed on a regular basis as a way to express his reverence for Fuji—that is, Great Bodhisattva Sengen/the Original Father and Mother—as the source of food and, more specifically, rice.[6]

The Fuji deity instructed Jikigyō to assume a second name: Miroku. This was the Japanese equivalent for Maitreya, the bodhisattva destined to be the next buddha of this world who presided over the Tushita (in Japanese, Tosotsu) Heaven but would eventually reintroduce the true dharma (Buddhist law) here on earth. Belief in the coming of Miroku inspired the practice of burying Buddhist sutras on the summit of Fuji and other mountaintops, a devotional act that reached its high point late in the Heian period (794–1185).[7]

By the eighteenth century Miroku was widely imagined not as the Maitreya of Buddhist scriptures but as a deity who would initiate a golden age of prosperity for all.[8] So when Jikigyō took on the name Miroku, he avoided the characters traditionally used for Miroku-as-Maitreya (弥勒), using 身禄 instead. Depending on the context, the first character (*mi*) can be translated as "body," "oneself," or "one's station in life." The second (*roku*) refers to a samurai stipend but also means "prosperity" or "material benefit" more broadly. This combination of characters appears to reflect the popular view of Miroku as a god of plenty. Yet in his teachings, Jikigyō often writes *roku* not with the Chinese character 禄 but in the Japanese syllabary (*hiragana*) to indicate the word for "proper" or "rectified" rather than "prosperity"; so in choosing the name Miroku, he may have had something closer to "proper/rectified person" in mind.[9]

Jikigyō inherited Kakugyō's belief in Great Bodhisattva Sengen as the Original Father and Mother and source of the sun and moon.[10] He also affirmed the immediate as well as long-term efficacy of talismans, healing water, and the ritual formulas found in the esoteric image-texts known as "body extractions" (*ominuki*) that were produced by Kakugyō, other sendatsu, and himself.[11] Yet Jikigyō was highly critical of those who charged for ritual services, whether Buddhist clergy, shrine priests, Shugendō practitioners, or other Fuji devotees. His critique was in line with a broader merchant ethos that people should focus on their livelihoods rather than spend excessive amounts of money and time on prayer rituals and other activities to win the favor of the kami and buddhas. It also stemmed from his belief that fee-based practices interfered with sincere and singular devotion to Fuji. As Janine Sawada explains, Jikigyō taught that "human beings in their purified state were the dwelling places of the True Father and Mother, and cultivating this unalloyed identity should be everyone's priority. In this view, turning to small outlets of putative spiritual power rather than seeking access to the cosmic source of all life intrinsic to one's own being indicated in the first place the individual's lack of faith in Mount Fuji."[12]

Rather than put resources into "small outlets of putative spiritual power," people should adhere to the virtues prescribed by the Fuji deity: honesty, frugality, hard work, and benevolence. If everyone lived according to these virtues, taught Jikigyō, the world would realize the full potential of the Age of Miroku, an era of peace and plenty that had supposedly begun with a cosmic upheaval in 1688. Unfortunately, all around him Jikigyō saw hardship and misrule. In one of his tracts, the *Addendum* (*Osoegaki no maki*, 1733), he pointedly accuses Japan's rulers of hoarding money and indulging in entertainments at the expense of the people, large numbers of whom were reduced to begging to survive.[13] In another work, the *Unspoken Word* (*Ichiji fusetsu no maki*, 1729), he blames the imperial court's failure to acknowledge and act on the cosmic upheaval of 1688 for a series of natural disasters.[14] Here he mentions the Hōei eruption of 1707, an event otherwise ignored in his teachings. Conditions were especially bad in 1732 and 1733, when Tokugawa Japan experienced one of its worst famines after bad weather and a plague of leafhoppers wiped out crops in the southwest. Intensified tax collection in the years leading up to the famine had left rural areas ill-prepared, and when the shogunate tried to steer rice to the most devastated regions, the cost of the staple rose for everyone else, including people living in and around Edo.[15]

To rectify the evils of the world, Great Bodhisattva Sengen instructed Jikigyō to fast to death on Fuji. This sacrifice would save people through a "Buddhist-type transfer of merit" and also function as a reprimand to Japan's greedy rulers, helping to bring about "tangible reforms that he felt were urgently needed."[16] During his fast Jikigyō sat in a portable shrine by the Eboshi rock. Said to look like a lacquered courtier's hat (*eboshi*), the rock sits around 3,250 meters (over 10,000 feet) above sea level at what is now the lower eighth station of the northern Yoshida climbing route on Fuji.[17] Jikigyō originally wanted to starve to death at the very summit, but locals were convinced that an earlier ritual suicide had caused disastrous weather, so they pressured Ōmiya Sengen Shrine, which had jurisdiction over the summit, to block Jikigyō's original plan.[18] Accompanying Jikigyō up the mountain in the summer of 1733 was his disciple Tanabe Jūrōemon. In addition to supplying him with one cup of snowmelt each day, Tanabe recorded Jikigyō's last teachings, which were then assembled by another follower into the *Thirty-One-Day Transmission* (*Sanjūichinichi no otsutae*). The *Thirty-One-Day Transmission* takes its name from the number of days Jikigyō predicted he would spend starving to death, although it seems to have actually taken a few more.[19] The *Transmission* may not be a verbatim record of Jikigyō's words, and it was likely edited over time, but its contents align with those found in texts written in his own hand.[20] Embraced by Fuji devotees as an authentic expression of Jikigyō's teachings, it had a profound influence on Fujikō beliefs and practices.

Jikigyō's decision to save all living beings by fasting to death on Fuji resonated with the esoteric Buddhist idea that one could "become a buddha in this very body" (*sokushin jōbutsu*) by entering a state of perfect meditation or enlightenment (*jō* in Japanese, *samādhi* in Sanskrit). The most famous Japanese said to have accomplished this goal was Kūkai (774–835), the progenitor of

Japan's Shingon sect of Buddhism. According to legend, he achieved a deathless state on Mount Kōya, where he had founded a monastic center, to await the coming of Miroku. In medieval and Tokugawa Japan, particularly zealous Buddhist monks and practitioners of Shugendō gradually starved themselves to death while seated in meditation. Some of them became mummified icons (*miira*), most famously at Mount Yudono in northeastern Japan.[21]

Likewise, Jikigyō characterized the endpoint of his fast not as death but as the achievement of *jō* through the union of his body with Fuji's.[22] Tanabe dutifully piled up stones around Jikigyō's portable shrine before heading back down the mountain; and although it seems doubtful that Jikigyō intended to become a mummy,[23] when Tanabe visited the tomb a year later, he was reportedly still in a seated position with his hands folded. Tanabe added that his face was smiling and his flesh intact. Yet Jikigyō was unable to sit the test of time. According to several accounts, his remains were removed, with a portion of them interred at his family temple. Meanwhile, the Eboshi rock became a destination for pilgrims and the site of a small shrine that now adjoins a shelter where climbers can spend the night.[24]

Jikigyō adopted a number of Buddhist practices and teachings over the course of his life. He regularly worshipped the Sun and Moon Buddha in addition to the Original Father and Mother.[25] He also identified Fuji with Mount Meru and employed the Buddhist term *jō* to describe his ultimate sacrifice. Jikigyō accepted the Buddhist idea of karmic retribution across multiple lifetimes as well. In both the *Unspoken Word* and the *Transmission* he warns that the wicked will be reborn in one of three "evil paths" described in Buddhist teachings—those of beasts, hungry ghosts, and *asuras* (violent "antigods" or "titans").[26]

Yet Jikigyō's teachings departed significantly from Buddhist ones. His conception of Miroku aligned more with popular religious beliefs than Buddhist doctrine. Similarly, while he referred to the Fuji deity as Great Bodhisattva Sengen, Jikigyō put his own spin on the meaning of "bodhisattva," equating the True Bodhisattva (Makoto no Bosatsu) with the rice consumed by humans.[27] Jikigyō believed that it was through the consumption of rice that Fuji, Great Bodhisattva Sengen, and the human body were unified as "one buddha–one body" (*ichibutsu ittai*).[28]

Jikigyō thought of Fuji/Great Bodhisattva Sengen itself in terms of the human body. The *Transmission* emphasizes the sacred nature of humans by noting that "the two eyes in one's body are the moon and sun," significant given that Great Bodhisattva Sengen is also referred to as "Moon-Sun

Sengen."[29] On the flip side, early in the *Transmission* we read that Fuji takes the form of a person—with the summit as the head, the bulk of the mountain the torso, and the base of the mountain the feet. Later in the *Transmission* Jikigyō reinforces his teaching that Fuji is a body (implicitly, a human body) by equating its rocks with bones, its water with blood, and its earth with flesh.[30]

Jikigyō paid special attention to the female aspects of Fuji, in particular the multiple "womb caves" (*tainai*) at its base, written with the characters for "womb" (胎内) or "inside the body" (体内). The *Unspoken Word* mentions four tainai located in the Kenmarubi lava flow not far from the oshi town of Kamiyoshida. Exactly which caves Jikigyō meant is open to speculation. The one that became the most popular, however, is the Funatsu tainai (see fig. 4.1). It consists of crisscrossing lava tree molds that formed when lava toppled trees that burned slowly enough to leave hollow spaces as the lava cooled and hardened around them.[31] The *Transmission* describes the popular belief that, if a new mother who fails to produce milk drinks the water that seeps into the tainai, her milk will start to flow.[32] We see this belief depicted in a woodblock triptych produced by Utagawa Sadahide (1807– ca. 1878–1879) in 1858 (see plate 3). One of the three prints highlights the "breasts" (lava stalactites) found in the cave's inner recesses. A pilgrim suckles a breast while another captures its "milk" to take home. Another print in the triptych reinforces the association between the tainai and the human body by exaggerating the size of the "ribs" toward the cave's entrance.

In addition to mentioning the tainai, the *Transmission* highlights the life-giving power shared by both Fuji and women by equating the vegetation growing midway up the mountain with a woman's pubic hair.[33] It also emphasizes the maternal rather than the paternal side of the Original Father and Mother by dwelling repeatedly on Fuji as a source of water,[34] which, according to yin-yang thinking, is a yin/female element.

Jikigyō also rejected the belief prevalent in Tokugawa Japan (and in many other times and places) that menses and the blood produced at childbirth were polluting. According to Buddhist teachings as well as taboos surrounding kami worship, blood pollution supposedly rendered women's bodies impure, which was one of the justifications to exclude them from numerous sacred spaces, including the summits of holy mountains like Fuji.[35] Women also faced the horrible prospect of "blood pool hell"—unless, of course, they turned for help to the Buddhist priests who promoted the threat of it.[36] Jikigyō, in contrast, attacked the belief in blood pollution. The *Addendum* argues that women ought to be able to climb Fuji even while menstruating; and the *Transmission*

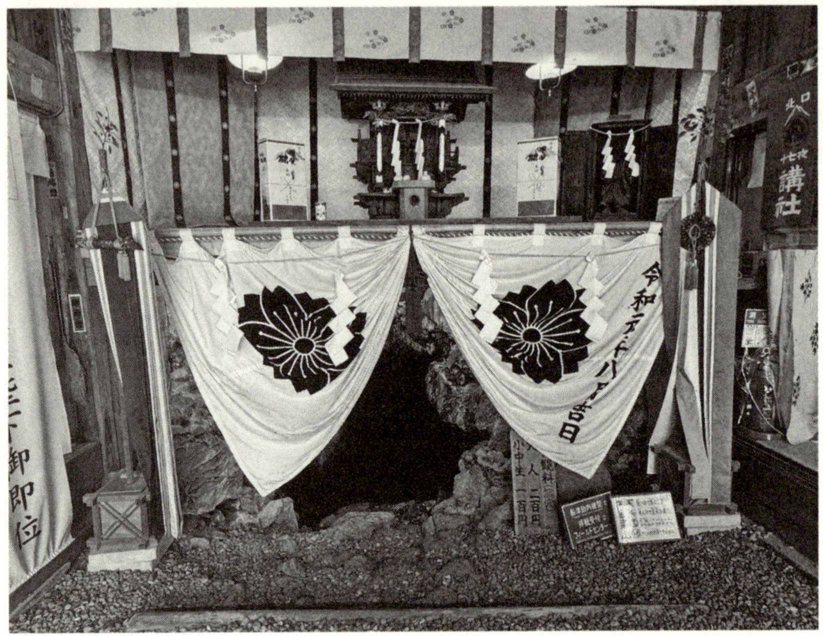

FIGURE 4.1. Entrance to Funatsu womb cave. Photo by author.

says in no uncertain terms that menstrual fluid, or "moon water," is essential for life, so the belief that it is impure is "gravely mistaken."[37] It follows that women are no more sinful than men. As the *Transmission* bluntly puts it, "If a woman does good, she is good; if a man does bad, he is bad," adding that men and women are "equally human."[38] Fujikō members propagated this sanguine view of women and their bodies after Jikigyō's death. His youngest daughter, Ichigyō Hana (1724–1789), became a prominent Fujikō leader.[39] It would nevertheless take nearly a century and a half for women to gain official permission to climb to Fuji's summit because of local opposition (discussed later in this chapter).[40]

Jikigyō's celebration of fertility translated into the promotion of gender equality in the spheres of worship and ethics. In this way, he challenged the conventions that dictated the roles of men and women. Yet his teachings also reinforced the long-standing and widespread Confucian emphasis on procreation and filial piety. Just as we should show gratitude to Great Bodhisattva Sengen, our Original Father and Mother, so too should we be thankful to our human parents for giving us life. The *Transmission* reiterates again and again the importance of filial piety and emphasizes that "being absolutely filial to one's parents is at the core of devotion" to Great Bodhisattva Sengen.[41]

Jikigyō promoted other conventional virtues as well, like generosity, frugality, diligence, and benevolence.[42] Tokugawa Japanese generally accepted that these virtues were supposed to extend from the imperial court down to the lowest levels of the status order. They were used both to justify that order and to press everyone, no matter their standing, to play their assigned roles within it. Commoners appealing for relief from their superiors regularly invoked the ideal of benevolence in their petitions. Similarly, Jikigyō's dismay at the contemporary behavior of the governing class translated to a call not to dismantle the orthodox status order of samurai, farmers, artisans, and merchants but instead to create an idealized version of it. Jikigyō refers to the fourfold hierarchy several times in the *Transmission* and enjoins members of each status group to work hard at their respective duties in maintaining the common good.[43]

The *Transmission* also provides theological justification for the political order by explaining the relationships of different status groups to rice—in Jikigyō's words, the True Bodhisattva. Samurai owe their exalted status to it, which is why Fuji, identified with the True Bodhisattva, is written with the characters for "wealth" (富) and "samurai" (士).[44] Farmers, of course, perform the all-important role of growing rice, while artisans and merchants do their part by acquiring rice and distributing it for the benefit of others.[45] Jikigyō concludes that the four status groups are "originally one body" since all partake of the True Bodhisattva. In this way, he argues for the ultimate unity of all people while at the same time reinforcing the worldly distinctions among them.[46]

There is little room for a sporadically violent volcano in Jikigyō's vision of both a biological and political order nourished by the fertile cycles of life. The production of children, the seasonal cultivation of crops, the monthly discharge of menstrual blood, and the daily act of eating—all are aspects of a dynamic equilibrium with Fuji as its steadily beating heart. Through its continual production of water and rice, Fuji is the source of the mundane bodyscape in which we lead our unremarkable yet cosmically significant lives.

In turn, we must repay this ongoing benevolence with ethical conduct. As Jikigyō declares in the *Transmission*, those who go on pilgrimage to Fuji might attempt to purify themselves with cold water ablutions and by refraining from eating fish and meat, but none of that matters if there are "evil thoughts in one's heart."[47] Certainly Jikigyō believed it was meritorious to come in physical contact with the mountain—he climbed it numerous times and chose to die there—but only for those who honored the True Bodhisattva by behaving virtuously in their everyday lives.

Worshipping and Replicating Fuji

Fujikō multiplied in the years following Jikigyō's dramatic sacrifice, particularly from the late eighteenth century. The boom in groups dedicated to Fuji worship was pronounced in Edo, which, by the mid-nineteenth century, was said to be home to as many as 808 Fujikō—perhaps an exaggeration, but a sign of how widespread they had become.[48] Each of the worship groups was distinguished by a unique insignia (*mon*) that appeared on the stone memorials erected at the Hitoana Cave and other sites around the sacred mountain, as well as on the sedge hats and banners (*maneki*) that they took on pilgrimage.

Virtually all the Fujikō acknowledged Kakugyō as their founder, carefully preserving his "body extractions" (the "image-texts" used as talismans and liturgical guides) while at the same time refashioning his biography and teachings to suit their purposes. This is clearly evident in *The Book of the Great Practice* (*Gotaigyō no maki*). A prominent account of Kakugyō's life and work thought to date to the late eighteenth or even early nineteenth century, it anachronistically aligns his doctrines with Jikigyō's—for example, by using the character for rice to refer to the True Bodhisattva.[49] It also makes the highly dubious claim that Shogun Tokugawa Ieyasu (1543–1616) met Kakugyō on three occasions,[50] a claim apparently designed to bolster the legitimacy of Fujikō at a time when any mass movement outside the bounds of officially sanctioned institutions was immediately suspect.

Fujikō were mainly neighborhood-based groups of men and women who convened each month at the homes of leaders who possessed ritual implements, texts, and expertise. Worship differed somewhat from one group to another, but it generally consisted of chanting "body extractions" (Kakugyō's as well as those devised by later leaders) and performing a fire offering (*takiage*) in front of an altar with a miniature stone Fuji at the center of it. Modeled on the *goma* fire ritual performed by esoteric Buddhist and Shugendō priests, *takiage* involved burning incense sticks piled in the shape of a steep cone (that is, Fuji) and then burning prayer slips that ascended to the ceiling as they turned to ash.[51] While the Hōei eruption seems to have had little to no effect on Jikigyō's theology, in this ritual, at least, there was an implicit acknowledgment that Fuji was not only a cosmic mountain but a volcanic one.

Fujikō members also transported basalt from Fuji to create *Fujizuka* ("Fuji hills" or "mini-Fujis") in Edo and the rest of the Kantō region. Since as far back as the Kamakura period (1185–1333), shrines to the Sengen deity had been built on hills called Fujizuka, but it was only from the late eighteenth century that worshippers constructed the Fujizuka rather than just the shrines on top of them.[52] The idea was initially Jikigyō's. Wanting to make Fuji more accessible—especially to women, who were banned from climbing beyond the mountain's lower slopes—he instructed in the *Addendum* that his followers should build a "replica" (*utsushi*) of the mountain in Edo.[53] So in 1765 one of his prominent disciples, Takada (or Takata) Toshirō, acted on his wish by beginning work on what Edoites dubbed "Takada New Fuji," constructed in what is now part of the Waseda University campus in Tokyo (and dismantled in the 1960s to make room for a new building).[54] Completed in 1779 with the assistance of hundreds of believers, the ten-meter-high Fujizuka incorporated large chunks of basalt from Fuji's summit. True to form, a climbing path with switchbacks led to the top of Takada New Fuji, which was circled by a small-scale version of the Middle Path (Ochūdō) that wound around the middle of the original mountain. Those who visited the Fujizuka also encountered a miniature version of a tainai as well as a shrine at the fifth station and, perhaps most important to worshippers, the Eboshi rock where Jikigyō achieved ultimate union with Fuji.[55]

Through its faithful replication, Takada New Fuji gave women, as well as men who could not climb the original mountain because of their health or financial circumstances, the opportunity to experience a condensed version of Fuji. As Melinda Takeuchi puts it, "In a mini-pilgrimage to a Mini-Fuji, time and space, ritual and religious 'reality' are collapsed,"[56] a maneuver situated in a widespread Asian, and particularly Buddhist, tradition of creating sacred replicas and "shortcut/optimal-gain rituals." These include, for example, the practice of turning a prayer wheel instead of walking around a stupa/pagoda, a ritual prevalent in Tibet and several other Buddhist countries.[57] We also see this dynamic at play in Japan in the miniature pilgrimage circuits that enable believers to accrue the merit of the original ones. Some are condensed to the point that they fit within the grounds of a single temple, allowing people to complete a pilgrimage that might otherwise take weeks or even months within a matter of minutes.[58]

Inspired by the example of Takada Fuji, from the late eighteenth to the early twentieth century, worshippers built numerous Fujizuka ranging in height from one to ten meters throughout the Kantō region (see fig. 4.2, 4.3).

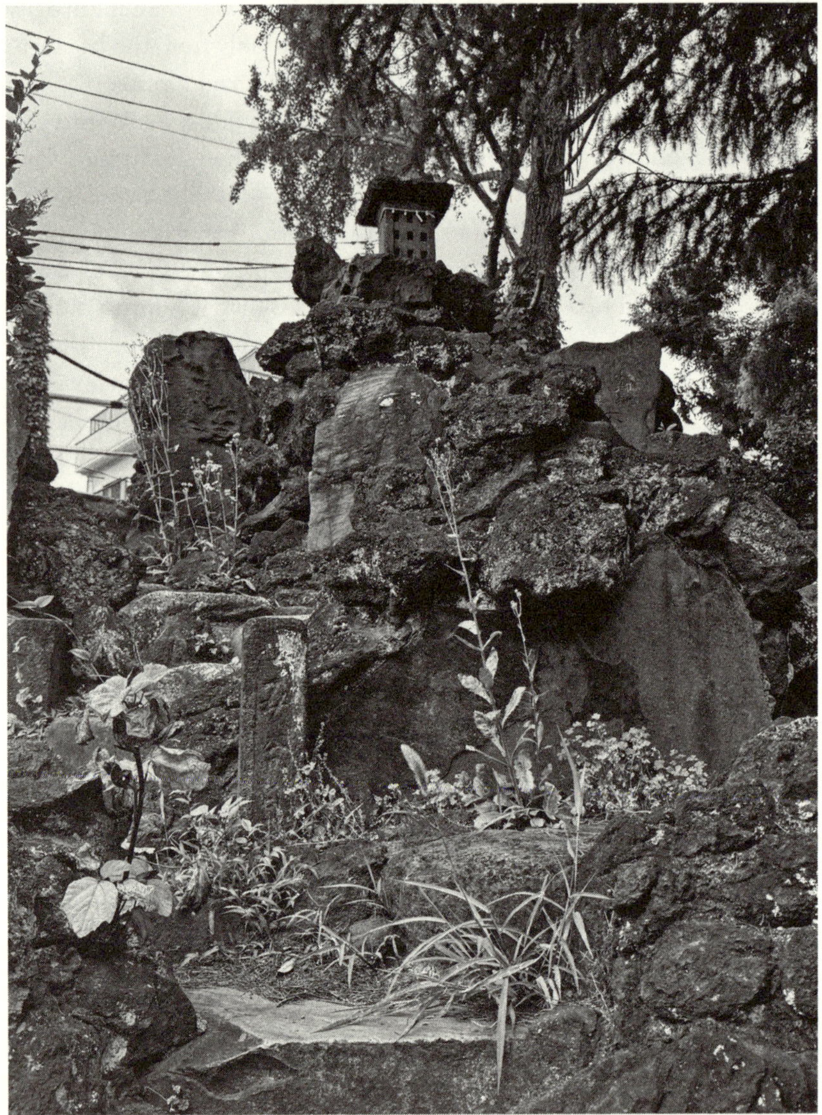

FIGURE 4.2. Fujizuka at Senjūkawada Sengen Shrine, Tokyo. The rectangular stone one-third of the way up the mini-Fuji reads: "Fifth station." Photo by Philip Ellway.

About seventy of them had been constructed in Edo/Tokyo alone by the mid-1930s.[59] During the late Tokugawa period, these handcrafted Fujizuka were incorporated into local festival calendars along with the older Sengen shrines built on preexisting hills, and crowds of people flocked to them on designated days early in the Fuji climbing season.[60] Like pilgrimages to the

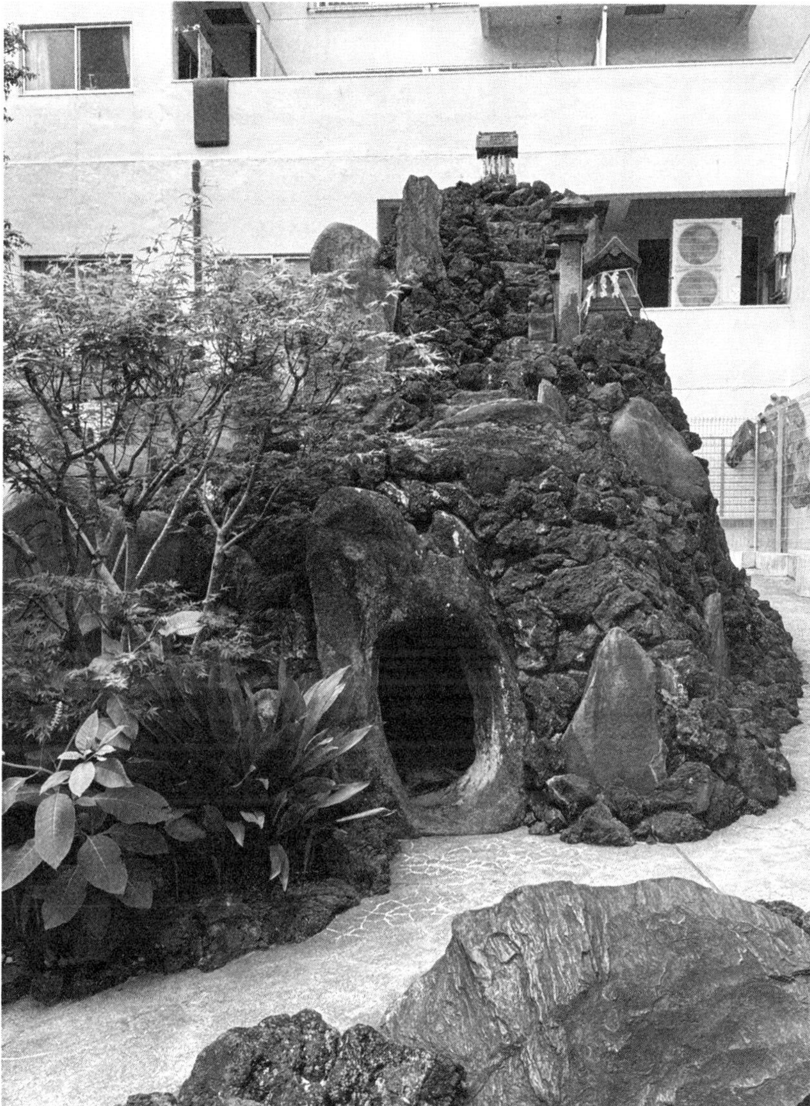

FIGURE 4.3. Fujizuka at Teppōzu Inari Shrine, Tokyo. Note the prominent replica of the Hitoana Cave where Kakugyō performed his austerities. Photo by Philip Ellway.

original Fuji, these festivals mixed pleasure with worship, providing money-making opportunities for vendors who sold food, drink, and souvenirs to "fashionably dressed people of all ages, samurai and commoner alike."[61] Some of these festivals, including the one at Komagome Sengen Shrine, continue to this day.

Destination Fuji

In addition to visiting Fujizuka close at hand, Fujikō members pooled funds so some could go on pilgrimage to the original Fuji during the summer climbing season. In this respect they behaved like the numerous other confraternities devoted to various deities and sacred sites around Japan. It helped that the tight travel restrictions of the early part of the Tokugawa period were increasingly loosened by domain and bakufu authorities from the mid-eighteenth century on. This was especially true regarding pilgrimage, which functioned both as an act of piety and as a way to escape the humdrum of everyday life and enjoy the sites and entertainments of the wider world.[62] It was also a way to mingle with fellow Japanese from various regions and all walks of life.

Resembling Shugendō practitioners, the pilgrims wore white clothes that signified their willingness to confront death. They refrained from the harsh austerities practiced by the full-time mountain ascetics, and only a small number of them died,[63] but their garb was stamped with seals at different sites on and around Fuji as proof of their devotion (see fig. 4.4). They also declared their membership in specific Fujikō with the customized insignia on their hats and on banners.

Pilgrims could choose from a few different climbing routes, but the most popular was the northern one. Those coming from the west along the Tōkaidō generally followed the route that began near the Ōmiya Sengen Shrine (today's Fujisan Hongū Sengen Taisha) and passed through the old Shugendō center of Murayama, but numerous pilgrims from Edo and other parts of eastern Japan followed Jikigyō's example by traveling the more convenient Kōshūkaidō highway to the village of Kawaguchi and especially the oshi district of Yoshida, called Kamiyoshida, or "Upper Yoshida" (both "Upper" and "Lower" Yoshida are now part of the city of Fujiyoshida) (see map 4.1).[64]

Oshi were innkeepers who also served as ritual guides. Each Fujikō had an established relationship with a particular oshi. Like their counterparts at other pilgrimage sites, the oshi at Fuji would host parishioners during the climbing season and travel to their homes at other times of year to distribute amulets and collect donations. The number of parishioners claimed by each oshi ranged anywhere from ten to fifteen thousand. Oshi households jealously guarded their lists of parishioners, which were inherited from one generation to the next, although sometimes oshi sold the rights to certain parishioners to fellow oshi.[65]

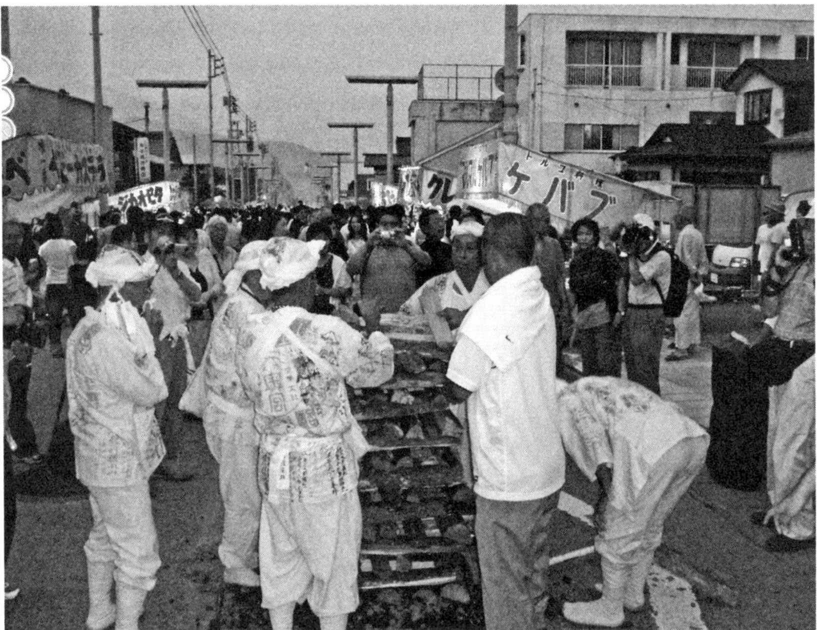

FIGURE 4.4. Fujikō members wearing pilgrims' clothing at Fujiyoshida Fire Festival. Here they are building one of the bonfires lit during the festival. Photo by author.

A Fujikō leader named Nagashima Taigyō produced a guidebook called *True View Illustrations of Mount Fuji* (*Fujisan shinkei no zu*) in 1847 with illustrations of the northern route up Fuji, starting with the town of Kamiyoshida. Depicted in figure 4.5 are pilgrims entering the town through an enormous *torii*, a gate that demarcates sacred space. The main road is lined with inns owned by oshi (see fig. 4.6).[66]

Since the 1960s, the vast majority of climbers have taken cars or buses to the fifth station to make a "bullet climb" (*dangan tozan*) to the summit in a matter of hours to greet the sunrise.[67] During the Tokugawa period, in contrast, Fujikō members experienced the mountain and the surrounding region in its entirety, from the towns, cultivated fields, and expansive grasslands at Fuji's base, to the Middle Path (Ochūdō) that circled the mountain's flanks, and to the cold and barren crater at the summit. Before making the climb, pilgrims entered the sacred body of Fuji by crawling through one of the tainai located in the Kenmarubi lava flow, with many also making the trek to the magnificent Shiraito Falls and sacred Hitoana Cave where Kakugyō performed

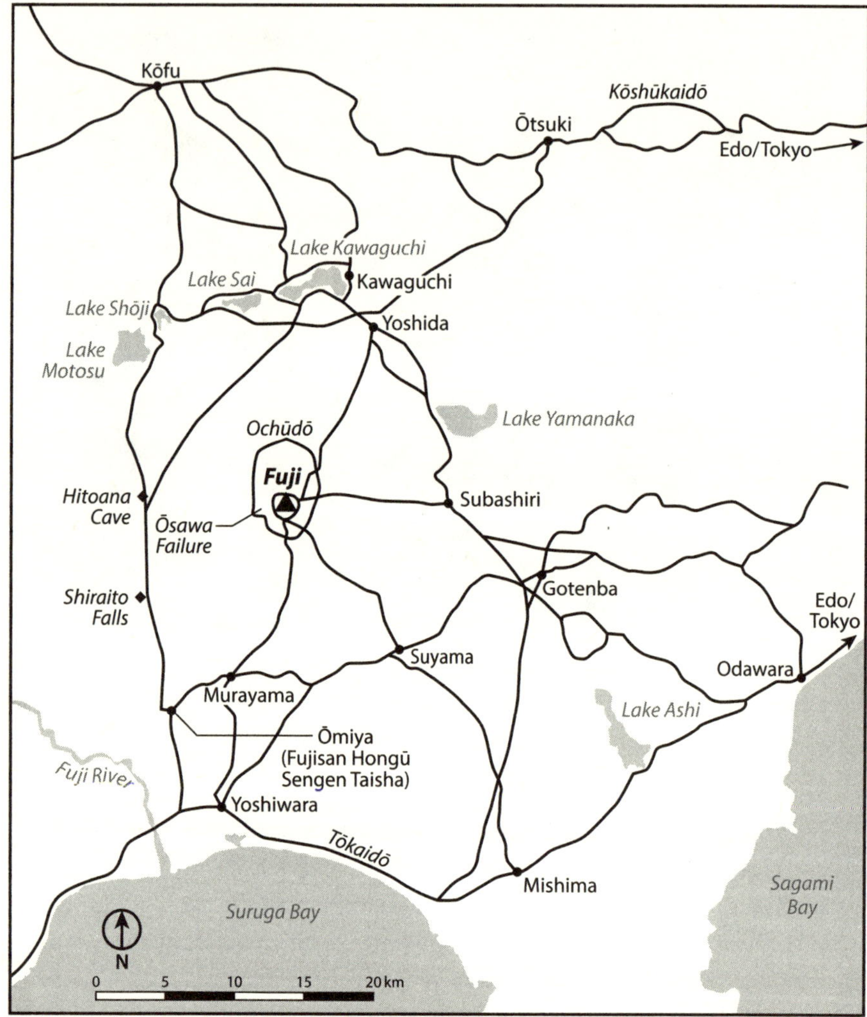

MAP 4.1. Tokugawa-period pilgrimage and transportation routes in Fuji region. Based on map in Sugimoto, "Sankeidō to shite no Kamakura kaidō," p. 13.

his austerities. The more ambitious among them also made Kakugyō's circuit of eight lakes, including the famous "Fuji five lakes" located on the northern side of the mountain. This took them through the rocky, forested landscape of the Aokigahara lava field.[68]

Pilgrims' holistic experience of Fuji from the bottom up—and from the inside out—is reflected in another woodblock print by Utagawa Sadahide (see plate 4). A product of the broader trend to replicate Fuji in different sizes and

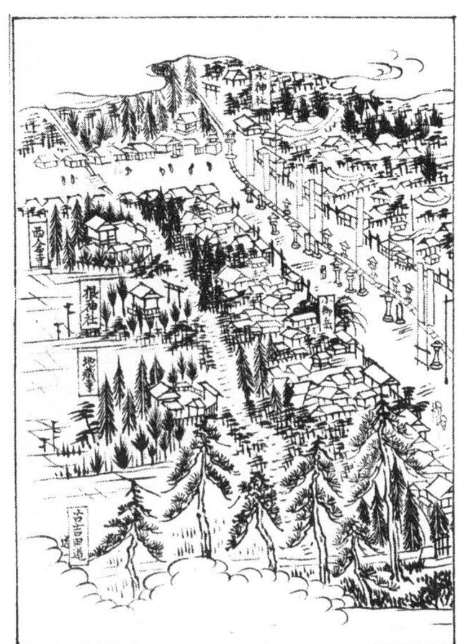
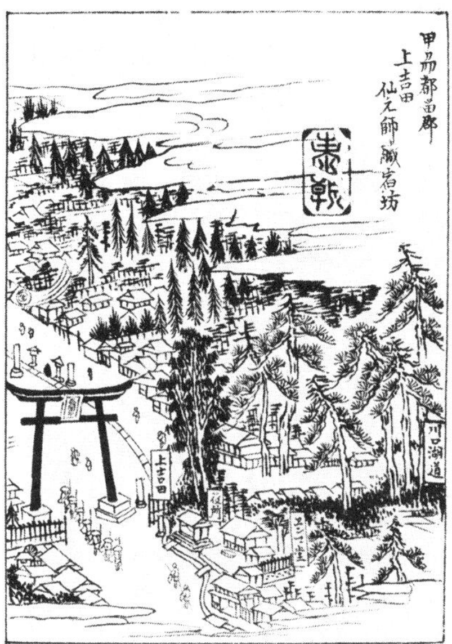

FIGURE 4.5. Illustration of Kamiyoshida. In Nagashima Taigyō, *True View Illustrations of Mount Fuji* (*Fujisan shinkei no zu*). 1847.

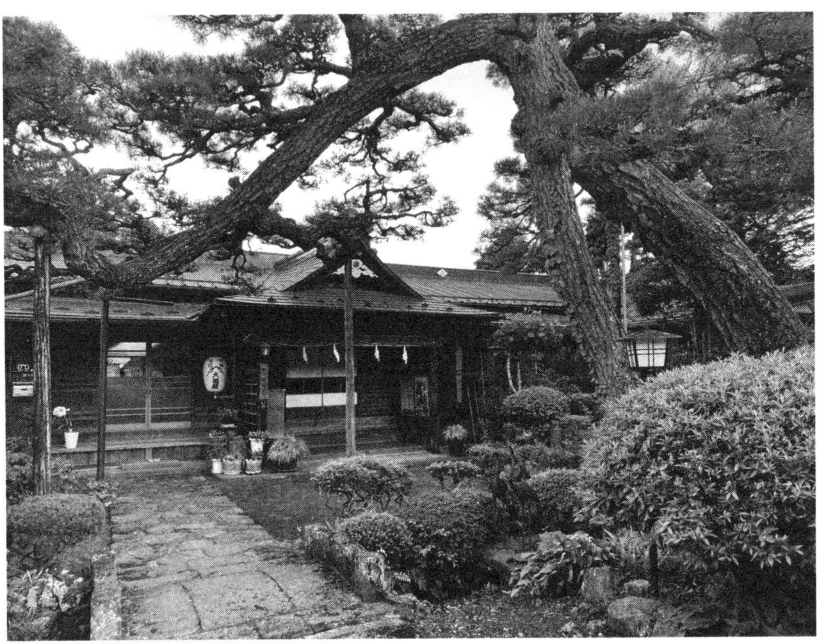

FIGURE 4.6. Entrance to Daikokuya, an oshi inn in Fujiyoshida. Photo by author.

formats, including the Fujizuka and miniature Fujis located on the altars of Fujikō members, the print is a picture-map of the mountain that can be folded into a three-dimensional cone that highlights the Yoshida entrance and its climbing route. It also features poems about Fuji as well as images of pilgrims and a torch used during Yoshida's annual Fire Festival (Himatsuri). Demonstrating that Fuji worship in the Tokugawa period focused not only on the summit but also on the interior of the mountain, it has a flap that, when lifted, reveals the Funatsu tainai. When the entire picture-map is unfolded, we see the holy figures of Kakugyō and Jikigyō seated in caves deep inside. In the hands of Kamiyoshida oshi (who, according to Sadahide, owned the printing blocks), the entertaining print could serve as a highly effective pedagogical and recruitment tool.[69] It is also possible that pilgrims purchased copies of the picture-map to keep at home.

Pilgrims encountered the full range of Fuji's ecological zones as they made their ascent. After passing through the butterfly-filled grasslands, pilgrims climbed through montane and subalpine forests filled with the songs of birds. It was also possible, if not likely, to come across one of the black bears inhabiting the mountain. They would not have seen monkeys, however. Found in mountainous areas throughout Japan, monkeys do not live on Fuji, which is a relatively young, isolated volcano and therefore ecologically "underdeveloped" in comparison to other mountains.[70]

The absence of monkeys is ironic, considering Fuji's strong symbolic association with them. According to legend, it was in a *kōshin* year, one of the years in the sixty-year cycle of the Chinese calendar used throughout East Asia, that Fuji first appeared (with earth scooped out where Lake Biwa is today). The second of the two characters used to write *kōshin* (庚申) can mean "monkey," so the legend explains why monkeys—in particular the monkey god Sarutahiko as well as the three monkeys who see no evil, hear no evil, and speak no evil—came to be associated with Fuji from the seventeenth century.[71] To this day, monkeys appear on a variety of items connected to Fuji, ranging from votive tablets to locally made sweets.

Upon reaching the fifth station, about 2,300 meters (over 7,500 feet) above sea level, pilgrims walked the Middle Path circling the mountain. At one point this required them to clamber through the treacherous Ōsawa Failure (Ōsawa Kuzure), a debris-filled and landslide-prone ravine that has been growing ever deeper and wider over the past one thousand years (see fig. 4.7, 4.8). Because the danger has increased over time, this section of the Middle Path is today off limits.[72]

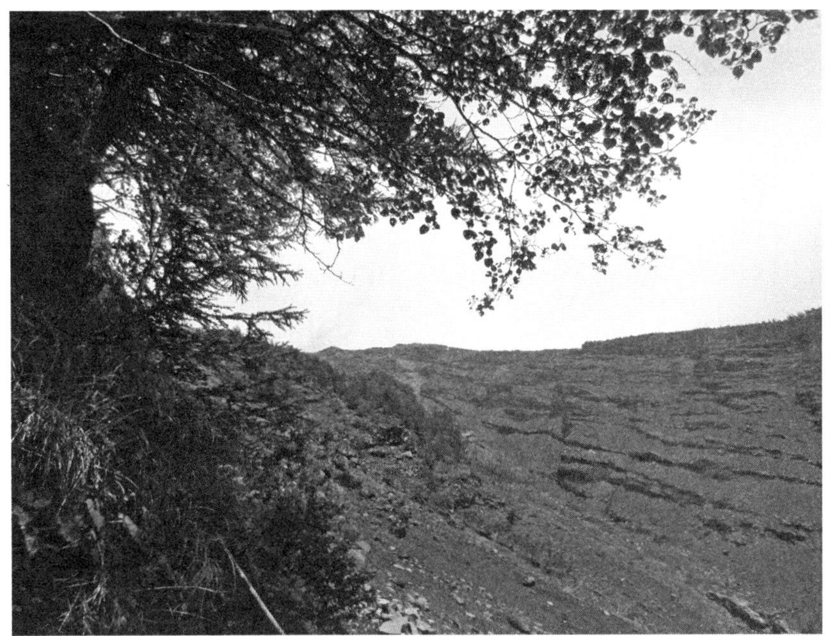

FIGURE 4.7. View of Ōsawa Failure from the Middle Path. Photo by author.

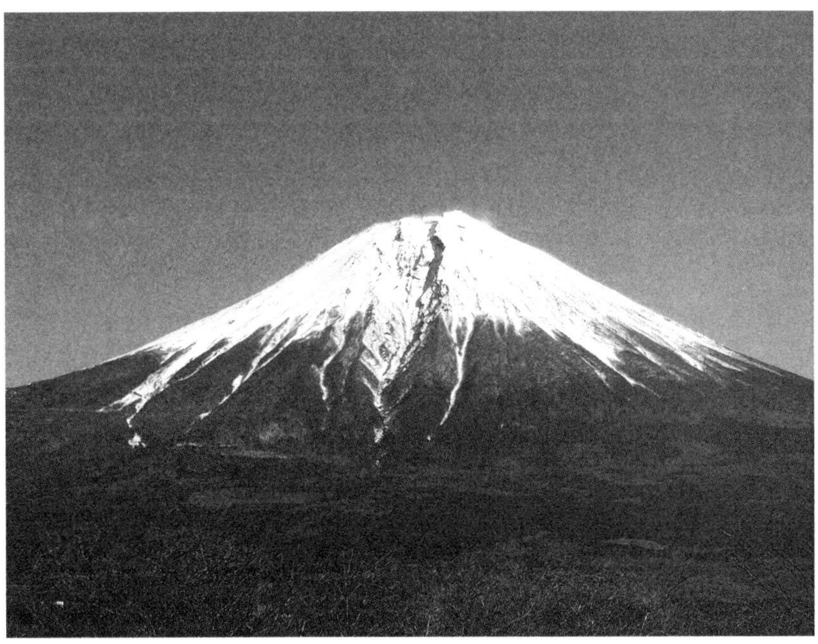

FIGURE 4.8. View of Ōsawa Failure from a distance. Photo by Watanabe Michihito.

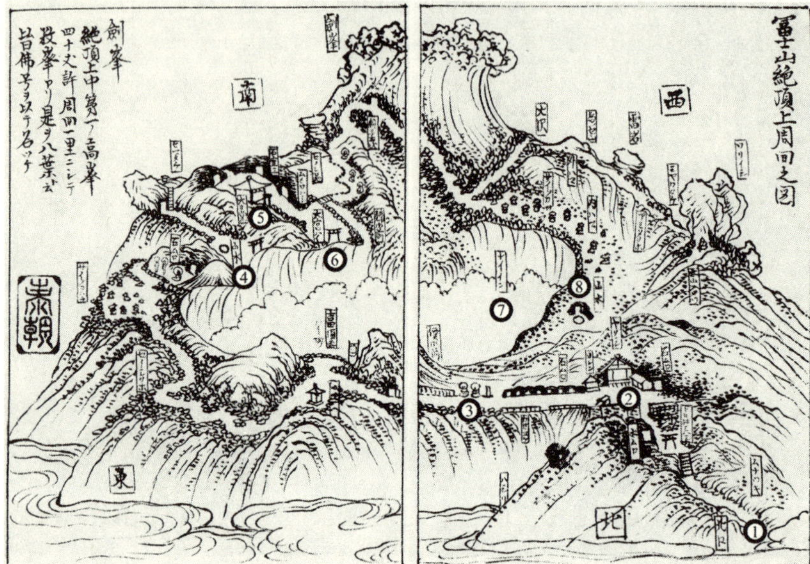

FIGURE 4.9. Illustration of Fuji's summit. The text in the upper left-hand corner notes that all the peaks have Buddhist names. Major sites include: (1) route from Kamiyoshida and Subashiri, (2) Yakushi Hall, (3) Buddhist statues, (4) Ginmeisui (silver light water), (5) Dainichi Hall, (6) Ōmiya Sengen Shrine worship site, (7) Naiin ("inner sanctuary"; i.e., the crater), and (8) Kinmeisui (golden light water). In Nagashima Taigyō, *True View Illustrations of Mount Fuji* (*Fujisan shinkei no zu*). 1847.

The next destination was the site of Jikigyō's fast, the Eboshi rock, located high above the tree line. Beyond this point the thin air and steep, rocky slopes made climbing increasingly difficult as pilgrims wound their way to the summit.

Pilgrims often timed their arrival at the summit, or at least the eighth station, at dawn, so they could witness the *goraikō*. Now used generally to describe a sunrise witnessed from a sacred peak, *goraikō* originated as a Buddhist term for a vision of Amida Buddha and his attendants welcoming a dying person into Amida's Pure Land. Pilgrims who experienced the sunrise on Fuji hoped to see the shadow of Amida Buddha, flanked by the bodhisattvas Kannon and Seishi, cast upon the clouds below.[73]

The entire summit was replete with Buddhist icons, practices, and terminology (see fig. 4.9). Like the summit of Mount Kōya, it was imagined to be an eight-petaled lotus blossom, the symbol of Buddhist enlightenment. Each petal was associated with a Buddhist deity and took the form of one of eight hills surrounding the crater.[74] As pilgrims circumambulated the crater, they encountered statues of the bodhisattvas Jizō and Kannon as well as Dainichi (the Sun Buddha) and Yakushi (the Medicine Buddha). The two buddhas were installed in chapels devoted to them.[75] The summit was also home to two sacred wells where pilgrims could buy "silver light water" (*ginmeisui*) and/or "golden light water" (*kinmeisui*), which can still be purchased today.[76] The volcanic material on Fuji's summit is too permeable to retain water, so it appears that these are not natural wells but instead containers designed to hold snowmelt for sale to those seeking its curative powers.[77]

Locals made money from pilgrims up and down the mountain. Pilgrims reaching the summit tossed coins into the crater and purchased dumplings, adzuki bean mochi, sweet sake, and tea in the shelters managed by local residents during the climbing season.[78] By this point they had already spent considerable sums at the shelters clustered at each station and in the villages at the base of the mountain, where they paid for lodging, meals, and climbing supplies, including the distinctive eight-sided walking staffs that continue to be popular with climbers.[79] Fuji pilgrimage was therefore not just a sacred act but a vehicle for financial gain—and conflict, as we will see below.

Pleasures, Perils, and Profits

The Fuji bodyscape was animated by acts of pleasure as well as piety. Going on pilgrimage in Tokugawa Japan was an opportunity to enjoy new vistas, sample local delicacies, participate in entertainments, and engage in cultural pursuits. For some it was also an opportunity to visit the brothels located along highways and in the towns that grew up around pilgrimage sites.[80] It was apparently common for pilgrims to Fuji to indulge in drinking parties where they were entertained by performers from Edo.[81]

Wanting to make the most of their time on the road, large numbers of pilgrims visited multiple sacred sites, not just one. Pilgrimage to Fuji was incorporated into a popular circuit that included the picturesque island of Enoshima, located just offshore from Kamakura and said to be connected to the Hitoana Cave via a subterranean network of caves traversed by the dragon goddess Benzaiten,[82] and the sacred mountain of Ōyama, located about halfway between Fuji and Edo. So many people combined a visit to Fuji with one to Ōyama that the shogunate, concerned that pilgrims were neglecting their work, tried to ban the practice—though it seems to little avail.[83] There were many travelers who had no desire to climb Fuji but instead stopped at renowned vantage points to enjoy its beauty and cultural associations. A particularly famous spot to view the mountain was Miho-no-matsubara, a pine grove on a spit of land sticking out into Suruga Bay (see the lower left-hand corner of map 2.2) and the setting for the famous tale *Hagoromo*, named after the feathered robe that a fisherman stole from a heavenly maiden in order to make her dance for him.

Travel presented an array of pleasures, but while the roads were safer in Tokugawa Japan than in much of the early modern world, they also presented dangers and discomforts such as illness, con artists, bad food, and bedbugs.[84] Pilgrims who climbed Fuji also had to contend with altitude sickness and sudden, at times fatal, storms; and those who decided to circle Fuji along the Middle Path had to brave the hazardous Ōsawa Failure. To pilgrims who experienced it in person, then, Fuji was not only the life-giving force of a cosmic bodyscape but also a site of immediate perils. Yet realities on the ground did not change the belief in Fuji as an ultimately beneficent force on which the prosperity of the world depended; and, like Jikigyō, pilgrims sought to come into physical contact with this force by entering and climbing the body of the mountain.

What pilgrims generally did not see was the intense conflict that broke out among those looking to profit from them. Pilgrimage was an important source of income for the communities around Fuji, especially those on the northern and eastern sides of the mountain, where cold weather and the fallout produced by the Hōei eruption made large-scale agriculture difficult. One traveler to Fuji in 1803 noted that the village of Subashiri, although surrounded by "infertile land unsuitable for growing the five grains," was a "lively" village full of homes, shops, and inns thanks to the money spent by pilgrims.[85]

Competition for this money was fierce and sometimes nasty, with villagers coming into conflict both with each other and with the powerful Sengen Shrine in Ōmiya. In 1703, for example, three oshi and a shrine priest in Subashiri involved the shogunate's Office of Temples and Shrines (Jisha Bugyō) in a dispute with the Ōmiya Shrine, arguing that it had unlawfully encroached on the right of Subashiri residents to collect a portion of the money offered at the ritual opening of the summit's Yakushi Hall each summer and to gather coins tossed into the crater throughout the climbing season.[86] They also fought the shrine's decision to allow Yoshida villagers to establish two new rest stations at the Yakushi Peak. In this case the office ruled in Subashiri's favor.[87]

Conflict erupted again in the 1770s, now over who was responsible for disposing of pilgrims who had died above the eighth station. Ōmiya Sengen

Shrine, Subashiri, and Yoshida all laid claim to the dead, partly because conducting burial rites was a potential source of income for Buddhist temples at the base of the mountain (provided that relatives of the deceased were willing to pay for them). More important was what removal of the dead signified: managerial authority of Fuji from the eighth station to the peak. The battle over the dead became so intense that, at one point, villagers from Yoshida forced their Subashiri counterparts from the shelters they managed on the summit. Ultimately the Office of Temples and Shrines confirmed Ōmiya Sengen Shrine's authority above the eighth station, which was where the Yoshida and Subashiri climbing routes converged.[88] In doing so, they followed the precedent established by Tokugawa Tadanaga (1603–1634), daimyo of both Kai and Suruga, who had ruled that activities occurring on the peak fell under the jurisdiction of the shrine.[89]

This new ruling did not mean an end to conflict. A case in point was a protest lodged with Ōmiya Sengen Shrine by the headman of Subashiri after Yoshida residents cordoned off one of the sources of water on the summit (which one is unclear) in order to sell the water to pilgrims. In his petition, the headman pointed out that pilgrims who stopped to purchase water sullied the area around it by defecating and urinating. In response, the shrine ordered the Yoshida residents to stop selling the water. A number of years later, however, Subashiri villagers ended up doing exactly the same thing.[90]

So while the boom in pilgrimage to Fuji after Jikigyō's death profited communities at the foot of the mountain, the competition for financial gain often produced acrimony rather than harmony. If Fuji was the Original Father and Mother, the oshi, priests, and villagers appeared to be squabbling children.

Women on the Move

Ōmiya Sengen Shrine banned women from its climbing trail in addition to the summit, and locals afraid of angering the Fuji deity pressured the oshi at the other entrances to keep women from climbing more than a short way up the mountain if at all. They were particularly worried about their presence inviting bad weather that could ruin their harvests.[91]

In 1832 a woman named Tatsu (1814–1876), disguised as a man, defied the ban and climbed to the summit. She was accompanied by a group of men who belonged to the lineage of Fuji worship established by Jikigyō's daughter Ichigyō Hana that would come to be known from 1838 as Fujidō (the Way of Fuji).[92] After Ichigyō's death, leaders of the lineage were all men, but they followed the example set by her and her father not only by stressing the importance of moral rectitude but also by encouraging the full participation of women in Fuji worship. Ichigyō's designated successor, Sangyō Rokuō (1745–1809), provided an intellectual framework for this position by arguing that the problems of the world grew from an imbalance between the cosmic forces of yin and yang. According to Sangyō, the domination of yin (gendered female) by yang (gendered male) had spawned both natural disasters and inequities in the social order, with men excessively valued relative to women and samurai accorded too much power over commoners.[93] The answer, then, was to find ways to put yin and yang back in balance.

Taking seriously (and literally) the need to elevate women, Sangyō's successor, Rokugyō Sanshi (1765–1841), walked the walk by accompanying Tatsu up Fuji. This was the same man who taught that women should take a more dominant role in having sex with their husbands, which was "the only way to propagate human beings who could embody the correct configuration of yin and yang qualities and thereby be capable of building a new society."[94] He further encouraged women to take active roles in the public sphere, even arguing that women should be allowed to enter lines of work, such as sake brewing, that were generally off limits to them because of the impurity supposedly caused by menstruation and childbirth.[95] With his support, women took leadership roles during prayer meetings, participated in both charity campaigns and public works projects, corresponded with believers in different regions, and actively proselytized.[96] It is therefore not surprising that he thought women should be able to climb to the top of Fuji as well. To celebrate Tatsu's accomplishment, he gave her one of Sangyō's works with a postface reading, in part, "Today, in a pilgrimage of thanksgiving, a woman managed to reach the peak and thereby liberate the mountain."[97]

A significant number of oshi on the northern side of the mountain were sympathetic to the idea that women should be allowed to climb Fuji. Miyazaki Fumiko points out that "at least some Yoshida oshi must have secretly aided Tatsu's party. Otherwise it could not have employed luggage carriers or stayed at the huts on the mountain managed by Yoshida oshi."[98] Back in 1800, in fact, oshi from Yoshida had erected signs along major highways and in various other locations announcing that women would be allowed to climb as far as the fourth station, the idea being that the policy would attract not only women but also the men who would accompany them. Given that 1800 was a kōshin year, the year in the sixty-year cycle when Fuji supposedly came into being, the Yoshida oshi already stood to benefit from an increase in the number of pilgrims, but they wanted to boost that number further by allowing women to climb higher than they usually could. Local farmers who feared the wrath of the Fuji deity fiercely opposed the scheme, and the fact that the spring of 1800 was unusually wet strengthened their resolve. When the residents of twenty-eight villages went so far as to threaten a formal complaint with the shogunate, the oshi decided to abandon their plan.[99]

In the years to come, many oshi might not go so far as to encourage women like Tatsu to climb all the way to the summit, but they were more than willing to turn a blind eye to women climbing beyond the second station as long as appearances were kept up. By agreeing to this ruse, the oshi engaged in the "*performance* of obedience" that Luke Roberts argues was a central feature of the Tokugawa order.[100] Of course, one had to be careful about the degree to which "inside agreement(s)" (*naishō*) conflicted with "surface" (*omote*) authority,[101] because opponents could sometimes appeal to that authority to achieve their ends. This is what happened in 1838, when local bakufu officials decided to patrol the northern climbing route in response to the complaints of villagers. Miyazaki speculates that the authorities were likely motivated by the fact that these same villagers had engaged in an uprising to protest high rice prices just a couple of years before.[102] Officials also ordered the Yoshida oshi to create a checkpoint with a gate to keep women from climbing beyond the second station (formerly designated the third station).[103]

Wanting to get an unobstructed view of Fuji, women regularly bypassed the checkpoint by taking an unofficial path that led them to a clearing about one hundred meters higher up the mountain, thus allowing the oshi to "perform obedience" even though they knew full well what was actually going on.[104] They also turned the checkpoint to their financial advantage by using it to collect fees from those climbing Fuji.[105] This, in turn, generated opposition from the Kawaguchi oshi, who had long competed with the Yoshida oshi for pilgrims and the money they brought with them.[106] In 1853 the Kawaguchi oshi filed a suit to stop the practice of collecting fees at the checkpoint. To strengthen their claim, they pointed out that the Yoshida oshi flouted the rules by allowing women to climb beyond the second station, even going so far as to post a sign showing women how to reach the clearing above the checkpoint where they could worship. In the end, the Kawaguchi oshi won their suit over trail fees, and the authorities also ordered the Yoshida oshi to remove the sign as well as a torii that had been built at the clearing.[107]

Just seven years after the Kawaguchi oshi filed their suit, resistance to letting women climb to the summit finally collapsed. As was the case in 1800, when

Yoshida oshi tried to drum up business by allowing women to climb to the fourth station, 1860 was a *kōshin* year. This time the Yoshida oshi secured permission from the Office of Temples and Shrines to allow women to climb the entire route to the eighth station.[108] As a result, Yoshida drew a disproportionate number of pilgrims. In order to hold their own, the oshi and priests who controlled the other climbing routes decided to relax restrictions, with Ōmiya Sengen Shrine ultimately agreeing to allow women to climb to the summit.[109] In the end, it was the allure of increased profits, not a sudden change in beliefs, that allowed women to participate fully in the Fuji bodyscape.

Troublemakers

Tokugawa officials were keenly aware of the potential danger posed by any institution or movement that laid claim to the hearts and minds of its subjects. In the centuries before Ieyasu was declared shogun, the largest Buddhist sects had wielded their economic and military power to fight both daimyo and each other, so the new regime decided from the outset to keep them firmly in hand. It did this by quickly establishing rules aimed at major Buddhist temples, and in 1635 it established the Office of Temples and Shrines to oversee the activities of ritual institutions throughout Japan.[110]

The shogunate was especially intent on suppressing Christianity—introduced by missionaries in the sixteenth century—after Christian peasants in the Shimabara domain in Kyushu staged a rebellion in 1637–1638. In the wake of the failed uprising, the government ordered all Japanese families to register with designated Buddhist temples so priests could monitor parishioners for any signs they might be practicing Christianity in secret. Temples further supported the goals of the state by reporting the deaths of parishioners to domain and/or bakufu officials on a regular basis.[111] In return, the government backed the authority that temples wielded over their parishioners. The formally recognized Buddhist sects, as well as officially sanctioned Shugendō institutions, could also appeal to the Office of Temples and Shrines to settle disputes and maintain their privileged status against any upstart competitors.

The shogunate authorized select lineages of kami worship too. In a set of regulations issued in 1665, it recognized the custom of acquiring sacerdotal ranks through "imperial mediators"—that is, noble households. The most important of these was the Shirakawa, which descended from a long line of court ritualists. The government also threw its weight behind the Yoshida lineage (no relation to the town of Yoshida). In addition to mandating that "shrine personnel without ranks shall wear white robes," it declared that "other apparel may only be worn after obtaining a permit from the Yoshida family."[112] It therefore created a powerful incentive for shrine priests and oshi in the communities around Fuji, like their counterparts throughout Japan, to purchase affiliations with either the Yoshida or the Shirakawa lineages. They did this not only to gain legitimacy in the eyes of the shogunate but to compete for status with one another.

While many oshi and shrine priests around Fuji bought permits to gain official standing, Fujikō lineages and the sendatsu who led them remained independent. This put them at odds with government policy, since laypeople were not supposed to adopt the appearance and behaviors of professional ritualists. Nevertheless, for most of the eighteenth century officials largely turned a blind eye. There was zero tolerance for Christianity, but busy officials often tolerated worship that took place outside the bounds of the formally sanctioned institutions as long as it did not become a public nuisance. So while the Edo magistrates (*machi bugyō*) issued an order in 1742 that banned "Fuji disciples" (*Fuji montei*) from selling Fuji water to cure illnesses, they apparently did so out of a concern for people's health, not because they were alarmed by Fujikō in general. It is also unclear to what degree the ban was actually enforced.[113]

A few decades later, the magistrates targeted a wider range of activities that transgressed the boundaries established between ritual professionals and laypeople, although they did not ban Fujikō per se. In an order issued in 1775, they noted that laypeople from different walks of life inappropriately adopted the paraphernalia of Shugendō ascetics, chanted in front of the gates of people's houses, offered prayers for sick people, and lit bonfires. Most troublesome to officials charged with maintaining public order was the fact that crowds of these rogue laypeople paraded through town performing purification rituals (*harai*) in front of the gates of houses and demanding payment for their services. If the money offered by homeowners was deemed insufficient, arguments would break out. The magistrates therefore determined it was necessary to put an immediate stop to this "extreme outrage."[114]

The 1775 order made no specific reference to Fujikō, but in 1789 the government explicitly directed its attention to Fuji worship. In that year a woman named Nagai Soyo was audacious enough to submit a petition directly to the bakufu's chief councilor, Matsudaira Sadanobu (1759–1829), outside Edo castle. Written by her husband, it called on the shogunate to recognize the Age of Miroku. It also invited officials to inspect Jikigyō's teachings.[115] This petition set off a two-month-long inquiry involving government officials, oshi in

Kamiyoshida, and Fujikō members in Edo.[116] Not surprisingly, the shogunate decided not to endorse Fujikō beliefs, but it also did not bother to prohibit them. As long as Fujikō presented no serious threat to public order, they could believe and worship as they pleased.

The government's hands-off attitude changed six years later, when it prohibited Fujikō by name, though it was apparently more concerned with disorderly bodies than dangerous beliefs. In the 1795 order implementing the Fujikō ban, the Edo magistrates paid no attention to the teachings of Jikigyō, but instead blamed the confraternities for engaging in the practices prohibited by the order issued back in 1775, which they quoted verbatim.[117] Soon after the magistrates issued their Fujikō ban, the Office of Temples and Shrines outlawed lay worship groups called *Shingonkō*, which, like Fujikō, appropriated the rituals and attire of Shugendō ascetics. In this case too, behavior, not doctrine, was the problem.[118] The office threw its weight behind the ban on Fujikō in 1797, when it forwarded to the Edo magistrates a request from Shugendō representatives asking officials to do a more thorough job of enforcing the law.[119] Worried about their livelihoods, the Kamiyoshida oshi sought out and won an assurance from the office that, despite the prohibition, their own activities could continue as before.[120]

Regulating Fujikō was easier said than done. Fujikō remained a concern among officials into the nineteenth century, explaining why they reissued bans on Fujikō activities in 1802, 1814, 1842, and 1849.[121] But while some believers were disciplined,[122] enforcement seems to have been highly selective and did little to stop the growth of Fujikō. "Although Fujikō were trampled, they grew strong and thick as weeds," as Iwashina Koichirō puts it.[123]

Language from the orders in 1814 and 1849 reveals that officials were particularly concerned that Fujikō sometimes attracted samurai along with commoners. It was bad enough that ordinary people engaged in "stupid" activities, but this was especially the case for samurai who, "forgetting their station," got caught up in Fujikō gatherings.[124] In a "container society" where people were supposed to know their place, it was inexcusable for samurai—the ones in charge of maintaining the status order—to forget their station.

Worries about unauthorized gatherings were not limited to Fujikō but applied to a variety of other confraternities no matter their specific teachings.[125] The language of the 1849 prohibition included something new: a direct reference to Fujikō beliefs. Like the earlier order, this one complained about laypeople, including samurai, behaving in ways inappropriate to their status; but it also referred to the teachings of Kakugyō and Jikigyō, now described (unfavorably) as "neither Shinto nor Buddhist."[126] The 1849 order also differed in that the Edo magistrates who issued it did so under the explicit direction of the Office of Temples and Shrines, which had been investigating Fujikō beliefs and practices with a special focus on Fujidō, the lineage descended from Jikigyō's daughter Ichigyō Hana.[127]

Like the one that took place in 1789, this inquiry was set in motion after a believer submitted a petition in 1847 to the government. The author was a peasant and junior Fujidō member in a village not far from Edo named Shōshichi, who asked in his appeal that the teachings of Jikigyō be forwarded to both the shogun and the emperor so they could publicly acknowledge the Age of Miroku and help realize its promise of prosperity for all.[128]

Although Sanshi, the Fujidō leader at the time, exerted outsized influence, he was ultimately not an autocrat but first among equals. As Miyazaki Fumiko puts it, Fujidō consisted of a "combination of networks" that "had neither a hierarchical structure nor a chain of command."[129] This decentralized, horizontal nature made it appealing to large numbers of people from different walks of life, ranging from humble farmers and townspeople to samurai. The flip side of this strength was the inability to keep wayward members in check. This is what happened in 1847, when Shōshichi decided to seek formal recognition of Jikigyō's teachings from both the shogun and the emperor despite the opposition of other members who rightly feared official scrutiny.[130]

To say this move backfired is an understatement. Despite the periodic crackdowns on Fujikō and other lay associations, Fujidō had thrived; and with its promotion of charity, industriousness, and ethical conduct, it was even welcomed by local rulers.[131] When Nagai Soyo submitted her husband's petition back in 1789, the shogunate had shown little concern about Fujikō teachings, and it was only because of worshippers' disorderly conduct that the Edo magistrates periodically took action in the years to come. But general discontent among both commoners and samurai had grown in the ensuing decades, due in large part to social dislocation caused by an expanding commodity-based economy. With that discontent came a quick succession of local "world-renewal" (*yonaoshi*) gods who "served as divine rectifiers of economic problems that plagued various communities."[132] It was in this challenging environment that the popularity of Fujidō and its focus on the coming Age of Miroku drew the concern of central authorities charged with maintaining the status quo. The result was an investigation over two years long that involved the questioning of prominent Fujidō members and various people having anything to do with Fuji worship, including the leaders of other Fujikō, the priests of the Ōmiya, Murayama, and Kamiyoshida Sengen shrines, and representatives of the Kamiyoshida oshi.[133] The judgment issued by the office was just as sweeping. In addition to banning Fujidō and making its leaders renounce their faith, it also mandated the suppression of Fujikō in general, going so far as to order the removal of a shrine to Kakugyō in Kamiyoshida, the memorials erected outside the Hitoana Cave, and the shrine to Jikigyō at Fuji's seventh station.[134]

As was often the case, the shogunate's bark was far worse than its bite. The ban probably explains in part why there was a sharp increase in the number of Kamiyoshida oshi seeking the protection of Shirakawa licenses in the 1850s, but the Office of Temples and Shrines' harsh orders were never fully implemented.[135] After all, why would it be in the interest of local authorities to make a fuss? This was particularly true in communities at the base of Fuji whose financial health depended on pilgrimage. Fujidō's decentralized nature also worked to its advantage. To kill a group organized around a strict hierarchy, you cut off the head. Things are not so simple when power is distributed among a collection of networks. Like a jellyfish, Fujidō could be hacked into pieces but still grow back. It therefore not only survived the 1849 ban but flourished all the more in the 1850s and '60s—the tumultuous final years of the Tokugawa shogunate—attracting thousands of followers who believed they were about to enjoy the fruits of a thoroughly realized Age of Miroku.[136]

INHERITING KAKUGYŌ'S VISION of Fuji as the Original Father and Mother, and relying on the ritual images produced by him and subsequent leaders, Fujikō grew increasingly popular in Tokugawa Japan. This was especially true after the death of Jikigyō, who articulated a Fuji-centered bodyscape that was both cosmic and ethical in scope. It was not enough to show devotion to Fuji through ritual and prayer. People had to live virtuously too. Jikigyō therefore criticized religious professionals who offered their services for profit and distracted people from a genuine commitment to the Fuji deity. He also taught that, as far as ethical behavior and the worship of Fuji were concerned, women were the equals of men.

Just as Fuji looked beautiful from afar but presented dangers up close, the sacred mountain that sustained the universe was also a source of worldly conflict. Economic competition was fierce—take, for example, the battles over who could sell water and collect donations at the summit. And as Fujikō and their oshi allies pushed for women to climb past the lower slopes, they encountered strong resistance from local villages and Ōmiya Sengen Shrine.

Fujikō also ran into trouble with the shogunate by transgressing the boundary between laypeople and religious professionals. Yet Fujikō continued to increase in size and number as the Tokugawa period came to a close. After all, to its devotees, Fuji was not just any sacred mountain—it was the center of the cosmos.

The "greatest mountain in the Three Lands" was also particularly Japanese. According to Jikigyō's *Transmission*, "even those from foreign countries know Fuji, and needless to say, anyone born in Japan, without even being taught about it, knows Fuji."[137] Chapter 5 will show how this special connection between the Japanese people and Fuji was repeatedly taught—not only through the beliefs and practices of Fujikō but also through art, literature, maps, and a variety of consumer items that made the sacred mountain deeply familiar and, by the end of the Tokugawa period, truly national.

5

From Pride of Edo
to Symbol of Japan

FUJI WAS A SOURCE OF stability and life rather than chaos and death to Jikigyō Miroku and his followers. They were not alone. Not even the devastation of the Hōei eruption made a lasting impact on the collective image of Fuji. A few pushed back on this willful amnesia. For instance, the famous artist Katsushika Hokusai (1760–1849) portrayed the Hōei eruption's explosive impact in a woodblock print found in his *One Hundred Views of Mount Fuji*. But for the most part, Fuji the dangerous volcano—which conveniently refrained from erupting after 1707—quickly receded from view in favor of Fuji as a beautiful, benevolent, and ultimately stable point of reference amid the shifting currents of everyday life.

Fuji was especially celebrated by the people of Edo. The shogunal capital was an urban upstart lacking the historical cachet of Kyoto and the rest of the Kinai region. What it did have were views of the most impressive mountain in the realm. Its residents, known as *Edokko* (Edo's children), embraced Fuji as an emblem of hometown pride and were its ardent boosters.

Those who could not see Fuji in person came to know the mountain through literature, art, games, and maps. Its iconic shape also appeared on clothing, fans, dishware, and other mundane items, making it part of everyday life. A sacred site thus became an object of popular consumption, and as "Edo popular culture became, in large measure, national popular culture," the pride of Edo turned into the symbol of Japan.[1]

Status Symbol, Familiar Sight

When Tokugawa Ieyasu (1542–1616) made it his headquarters in the 1590s, Edo consisted of a dilapidated fort on a coastal bluff and a small village along one of the creeks at its base—a far cry from centuries-old Kyoto, with its imperial court and illustrious past. But thanks to massive investments in projects that included filling in swamps, rerouting waterways, and constructing bridges, temples, and numerous other structures, by 1610 Edo had become a proper city with 150,000 residents. By the early eighteenth century, with a population of about one million, it ranked among the largest cities in the world.[2]

Ieyasu and his successors, as well as the daimyo and numerous other samurai who lived in Edo, sought cultural legitimacy for their military authority. Higher-ranking samurai—and eventually, even middling ones—appropriated the titles, trappings, and pursuits of aristocrats in Kyoto as they looked to elevate both themselves and the city in which they lived. Ieyasu set an example from on high by having the emperor appoint him shogun in 1603. The bakufu wanted not only to share in the authority of the imperial court but to supersede it. One way it did this early in the seventeenth century was to establish mile markers along the major highways of Japan that counted distance not from Kyoto but from Nihonbashi, the "bridge of Japan," located in the heart of Edo. With this move, it made an early and important contribution to the "Edocentrism" that would intensify as the Tokugawa period wore on.[3]

Also instrumental to the regime's ideological project was nearby Fuji, visible from the shogun's capital and durably ensconced in the cultural heritage of Japan, having drawn the admiration of courtiers as far back as the eighth century. The combination of physical proximity and cultural provenance made Fuji ideal for a government looking to assert its newfound status. When Minamoto no Yoritomo (1147–1199) established his own military government in eastern Japan, he implicitly partook in Fuji's charisma by leading magnificent hunts at its base. Centuries later, the Tokugawa regime appropriated Fuji through art, particularly through the work of Kanō Tan'yū (1602–1674), the official bakufu painter who created multiple illustrations of Fuji that elevated the shogunate's prestige.

Tan'yū's main inspiration was a renowned ink painting attributed to the Zen monk Sesshū Tōyō (1420–1506), one of the most admired and imitated painters in Japanese history, that features Fuji, the Miho Peninsula with its famous pine grove (Miho-no-matsubara), and Seikenji Temple. It was widely accepted in Tan'yū's time that Sesshū created his masterwork during his stay

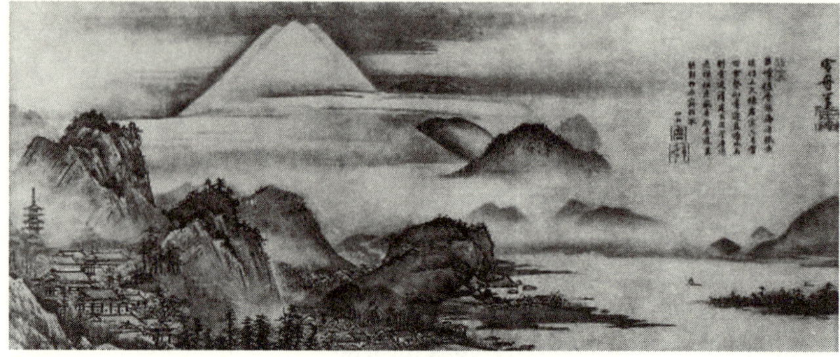

FIGURE 5.1. Sesshū Tōyō. *Mount Fuji, Miho, and Seikenji Temple.* Sixteenth century. Eisei Bunko Museum, Tokyo. Alamy stock photo.

in China (1467–1469), although art historians no longer accept that claim. The earliest extant copy of the painting, now located at the Eisei Bunko Museum in Tokyo, seems to have been made soon after Sesshū died (see fig. 5.1). Nevertheless, the painting displays "the monk-painter's brush habits, especially his penchant for highly simplified axe-cut strokes and vigorous outline," as Yukio Lippit notes, so it "was understood to transmit the essence of his artistry throughout most of its afterlife." What distinguishes the painting is an "unusual combination of continental pictorial norms with a hieratic presentation of the mountain often found in Japanese narrative painting," in this case accentuated by "the negative cast of Fuji's summit, the tripartite crown depicted in reserve."[4] So while the painting employs a "continental mode" that indicates, as Timon Screech observes, "Fuji's international reach," it at the same time displays a particularly Japanese understanding of the mountain with deep roots.[5]

Like other members of the Kanō school, Tan'yū embraced this combination of homegrown conventions and continental influences in his work. Take, for example, his series of three scrolls in which three-peaked, all-white Fuji towers at the center and China's renowned Mount Yuwang and Jinshan temples stand in attendance to the left and right of the lofty summit (see fig. 5.2). These temples were major destinations for Japanese monks traveling to China—including Sesshū—but here they are dominated by Japan's peerless mountain. "The creation of this set of three hanging scrolls might be interpreted as the Chinese cultural tradition, as well as the Chinese emperor, attending the new 'Chinese emperor' of the Tokugawa shogun in a symbolic gesture," as Matsushima Jin puts it.[6]

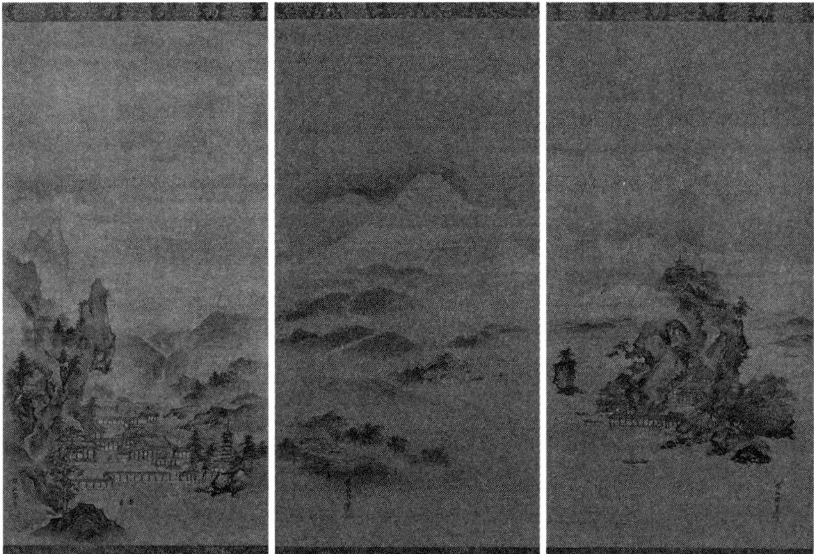

FIGURE 5.2. Kanō Tan'yū. *Mount Fuji, Mount Yuwang, and Jinshan Temple.* Seventeenth century. Private collection.

Perhaps the most ideologically significant of Tan'yū's depictions of Fuji was a rendition of the Mount Fuji, Miho-no-matsubara, and Seikenji Temple motif that covered the sliding doors in the upper section of the shogun's retiring room at the heart of Edo castle. The lower section of the room displayed famous places (*meisho*) in western Japan while the upper section showcased those in the east. Pride of place went to Fuji, which was "placed in the centre as the 'king of *meisho*'" such that "people who made obeisance to the shogun would be entranced by the authority of the Tokugawa shogun's body unified with Mount Fuji," in Matsushima's words.[7]

Although the bakufu made it a symbolic ally, Fuji was not, for the most part, viewed as the shogun's mountain. Those who worshipped Fuji instead saw it as the center of the cosmos and source of salvation. And thanks to Edo's growing influence on popular culture, millions of Japanese across the shogun's realm came to see Fuji as a familiar image. It was the mix of grandeur and intimacy, the sacred and the everyday, that made (and continues to make) Fuji such a powerful presence in the collective imagination.

Many Japanese could encounter Fuji with their own eyes. Fuji's viewshed is enormous, covering most of the densely settled Kantō region to the east, where the lack of high mountains allows for plenty of unobstructed views from up to two hundred kilometers away (weather permitting). Depending on the local topography, Fuji can also be seen from pockets of central Japan over one hundred kilometers to the north and even from certain locations on the Kii Peninsula several hundred kilometers to the west (see map 1.1).[8] This explains the numerous place names incorporating the characters *Fujimi* (富士見), or "see Fuji," in eastern Japan and especially in Edo (now Tokyo), although today tall buildings obstruct views of the mountain from many of them.[9]

Place names incorporating "Fujimi" or just "Fuji" are not limited to the mountain's viewshed. Geographer Tanaka Kei has calculated that, of the 467 places in Japan that include "Fuji" in their name (these include natural features plus residential areas such as neighborhoods and towns), about half are in regions where it is impossible to see the mountain.[10] This is largely thanks to the estimated 370 "hometown" (*furusato*) Fujis that are located throughout Japan and resemble their namesake to greater or lesser degrees.[11] Because Fuji is considered *the* mountain par excellence, naming local mountains "Fuji" indicates their importance to those living nearby. Gazetteers and travel diaries show that the practice was widespread by the middle of the Tokugawa period.[12] When Japanese traveled abroad in the decades following the end of the shogunate in 1868, they also applied the name "Fuji" to mountains located in the Japanese empire and elsewhere around the world. Two examples in the US Pacific Northwest are "Tacoma Fuji" (Mount Rainier) and "Oregon Fuji" (Mount Hood).[13]

Poetic Fuji

Fuji also traveled through literature, especially poetry. Thanks to the sharp growth in printing and literacy from the seventeenth to the nineteenth centuries, by the late Tokugawa period, many Japanese were familiar with ancient poems once confined to a small group of elites.

Particularly famous was a poem about Fuji found in *The Tales of Ise* (*Ise monogatari*), an anonymous work dating to the middle of the Heian period (794–1185) that describes the romances and travels of a nobleman typically identified as Ariwara no Narihira (825–880):

The peak of Mount Fuji
is oblivious to time.
What season does it take this to be
that the falling snow
should dapple it like a fawn?[14]

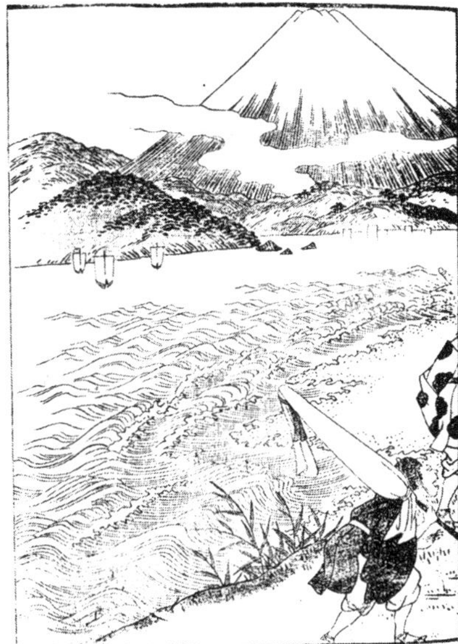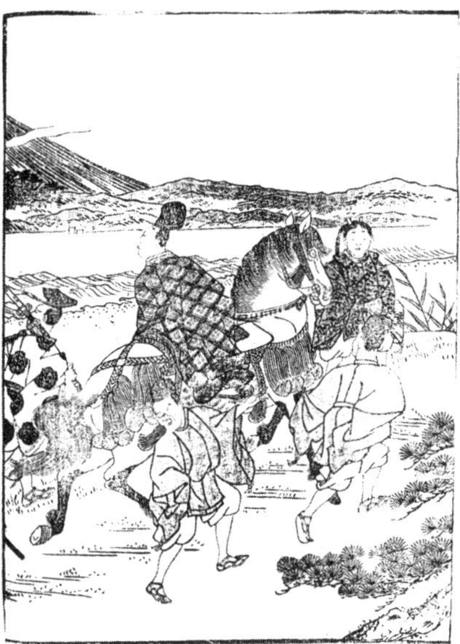

FIGURE 5.3. Ariwara no Narihira passes Fuji. In *Yamato Library: Teaching One Hundred Poems by One Hundred Poets* (*Oshie hyakunin isshu Yamato bunko*). 1829.

Often accompanied by an iconic image of its author riding past Fuji, this poem appeared in numerous school primers, including those specifically aimed at women.[15] Laura Nenzi argues that identifying with Narihira, through both his poetry and artistic portrayals of him, was one way in which Tokugawa Japanese could "create an instant bond with literary tradition" and temporarily reinvent themselves as participants in that tradition.[16] Figure 5.3 shows a depiction of Narihira and his retinue passing by Fuji in a book of poems designed to educate women.

The episode appears on the lacquered *inrō* (a container hung from a kimono's waist sash) in plate 5 as well. On one side Fuji is depicted along with two of Narihira's attendants, while on the other side Narihira looks back toward the mountain.

Widespread knowledge of the literary canon opened the door to parody. We see this dynamic in a nineteenth-century paper board game (*sugoroku*) about Narihira and *The Tales of Ise* that "transforms a literary icon of the past into a veritable Edo period traveler" who praises Fuji at one point by exclaiming, "Is this a great scene or what?"[17] We also see it in Jippensha Ikku's (1765–1831) popular *Shank's Mare* (*Tōkaidōchū hizakurige*), an early nineteenth-century serialized tale about the adventures—and misadventures—of two friends as they journey along the Tōkaidō. In one scene, some postboys sing a bawdy verse that satirizes the time-honored association between Fuji's rising smoke and ardent longing:

> The smoke goes up into the sky
> From Fuji's crest. I wonder why.
> The girls at Mishima should know;
> They light the fires of love below.[18]

It had been over a century since Fuji last erupted, but its metaphorical smoke continued to rise thanks to the appropriation of classical poetry by popular culture.

As Japanese schooled themselves in the classics, innovative poets broke free of traditional proprieties and formulaic expressions to create works that captured the immediacy of the world in which they lived. The most famous of these poets was Matsuo Bashō (1644–1694), who wrote haiku in vernacular Japanese and dealt with lowly topics—such as a man blowing his nose and a dog urinating[19]—but did so with exquisite care.

Like poets before him, Bashō appreciated Fuji's distant majesty, as we see in a haiku that compares the mountain to a towering cedar:

> Rooted in clouds
> Fuji soars,
> The shape of a cedar.[20]

In other haiku, however, he brings Fuji close, linking the magnificent to what is near at hand. Here Bashō draws on a nursery rhyme about a flea carrying a Fuji-shaped tea grinder, a homely tool within easy reach:

> Fuji's shape: a tea grinder
> covered with a cloth
> on the back of a flea.[21]

Bashō links the transcendent and the intimate in this haiku as well:

> The wind of Fuji
> carried on this fan
> —here,
> a gift from Edo.[22]

In place of a conventional gift, Bashō offers the boundless wind (or spirit) of Fuji, made accessible through an everyday object carried on one's body and manipulated by hand. The poem also underscores the close association between Fuji and the nearby city of Edo.

A century later, renowned haiku master Kobayashi Issa (1763–1828) expressed an easy familiarity with the mountain in poems like this one:

Snail—
slowly,
slowly
make your way
up
Mount Fuji.[23]

Here the poet brings Fuji into contact with something humble and ordinary, a snail, to make the point that, with persistence, anyone can surmount a challenge as large as Fuji. In another haiku, Issa links Fuji with the everyday by describing farmers tossing bundles of rice seedlings to plant in a paddy field:

In the morning light
they toss rice seedlings
at the head
of Fuji.[24]

Issa incorporates Fuji into a commonplace activity not to diminish the mountain, but instead, like Bashō before him, to elevate the mundane.

Thanks to the rise of large-scale commercial printing and literacy in the Tokugawa period, the same audience that relished the classic portrayal of Fuji in *The Tales of Ise* as a mountain "oblivious to time" also enjoyed the proliferation of inventive depictions like Bashō's and Issa's that situated Fuji in *their own* time. Just as the aristocrat Ariwara no Narihira was repackaged and commodified for the growing reading public, so too was Fuji. The mountain remained a lofty object of awe and worship but was also linked to less exalted subject matter like liaisons with sex workers in Mishima. This dynamic also appeared in the visual arts, which met the demand for novel depictions of Fuji by highlighting its distant grandeur on the one hand and tying it to everyday activities on the other. Fuji proved to be as versatile as it was enduring.

Picturing Fuji

If written descriptions made Fuji familiar in Tokugawa Japan, the visual arts made it ubiquitous. Artists took advantage of the popular obsession with Fuji to portray it on a variety of objects in new and striking ways, even while long-standing conventions, like showing the mountain with three peaks, persisted. Fuji was a popular motif on mirrors, pipes, ceramics, boxes, and lacquerware (see plate 6). Images of Fuji also appeared prominently on kimonos, combs, and hairpins, making it possible to stay in bodily contact with the mountain all day long.[25]

Seeing Fuji, whether in dreams or reality, was an omen of good fortune. Since it could be written with characters meaning "deathless" (不死), it was especially popular to hang scrolls depicting the mountain at the start of the new year.[26] In a print by Suzuki Harunobu (ca. 1725–1770), *Lucky Dream for the New Year: Mt. Fuji, Falcon, and Eggplants* (*Hatsuyume mitate ichi Fuji ni taka san nasubi*), Fuji is joined by two other lucky objects: a falcon and an eggplant (see fig. 5.4). Why they became popular as auspicious objects in the Tokugawa period is not entirely clear, although a homonym of falcon (or hawk), *taka*, means "tall" or "lofty," and one for eggplant, *nasu* (also called *nasubi*), means "to accomplish." It is also said that they represent "the three highest things in Suruga Province: Mt. Fuji, Mt. Ashi*taka* (sounds like *taka*, hawk) and the exorbitant price of the earliest aubergines of the season."[27]

Fuji also starred in the landscape print genre that developed in the early nineteenth century in response to demand for images of famous places (*meisho*).[28] Popular in their day as well as ours, Katsushika Hokusai (1760–1849) and Utagawa Hiroshige (1797–1858) were masters of this genre; and their depictions of Fuji, particularly Hokusai's, have achieved lasting fame. Fuji appeared in works produced by Hokusai throughout his career, but in the early 1830s, he created the drawings for an entire series of printed sheets dedicated to the mountain called *Thirty-Six Views of Mount Fuji* (although forty-six prints were eventually produced). Hokusai then devoted himself to the production of *One Hundred Views of Mount Fuji*, a series of black-and-grey prints (there are actually 102 of them) compiled into three books, with the first volume issued in 1834 and the last perhaps as late as the mid- to late 1840s.[29]

Appearing at the same time as Hokusai's *Thirty-Six Views of Mount Fuji* was Hiroshige's *Fifty-Three Stations of the Tōkaidō*, a series of fifty-five prints of which six focused on Fuji.[30] After Hokusai's death, Hiroshige created his own *Thirty-Six Views of Mount Fuji* (1852), followed by *One Hundred Famous Views of Edo* (1856–1859), which included nineteen views of Fuji as well as two of Edo's Fujizuka ("mini-Fujis").[31] He then decided to produce drawings for another *Thirty-Six Views of Mount Fuji* and sketches for *A Collection of One Hundred Views of Fuji*, with the former printed in the year of his death (1858) and the first volume of the second work appearing the year after.[32]

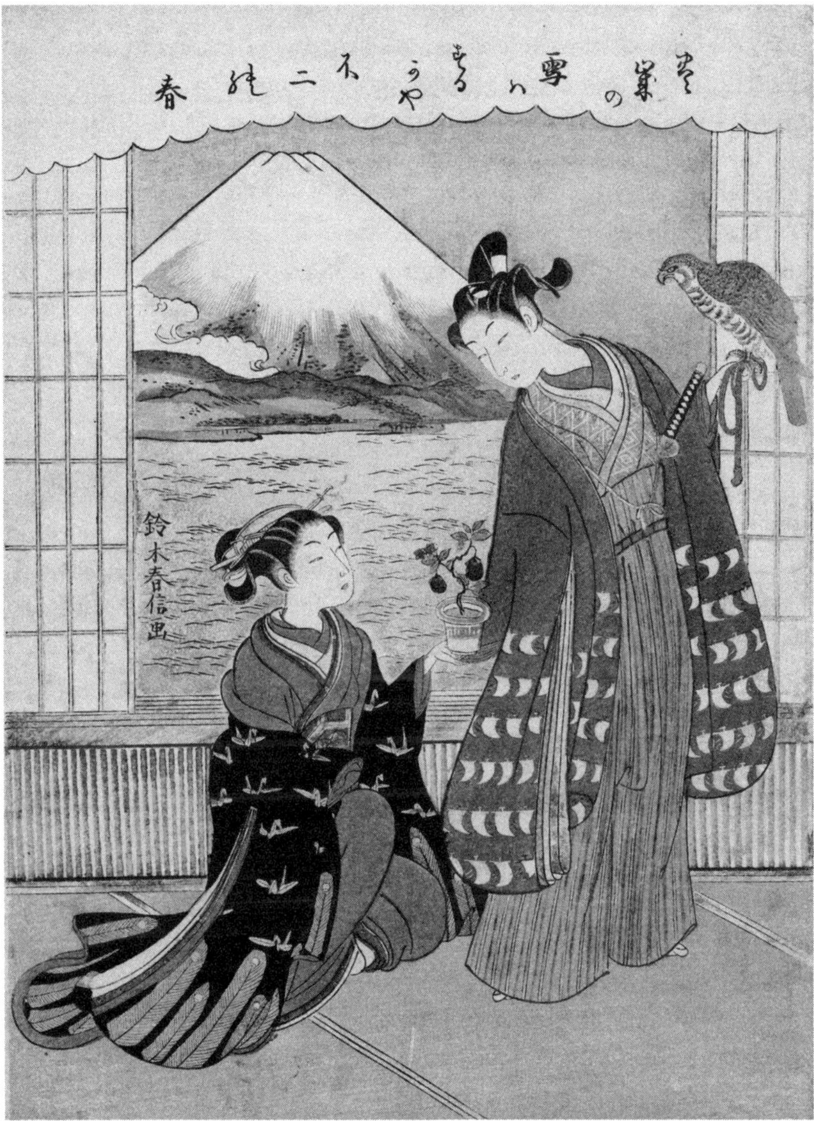

FIGURE 5.4. Suzuki Harunobu. *Lucky Dream for the New Year: Mt. Fuji, Falcon, and Eggplants.* 1768–1769. Los Angeles County Museum of Art.

Hokusai and Hiroshige were not the first to portray Fuji from multiple angles. Poetry collections comprising one hundred poems about Fuji dated back to the seventeenth century, and several artists followed their example by creating their own one hundred views of Fuji. The first of these artists that we know of was Kawamura Minsetsu (dates unknown), whose *One Hundred Fujis* appeared as a four-volume book at the end of the 1760s or the beginning of the 1770s, with each image accompanied by a poem.[33] Soon after, Kitao Masayoshi (1764–1824) began work on his *One Hundred Fujis* (it seems only seven were completed), and renowned painter Ike Taiga (1723–1776) reportedly tried his hand at creating one hundred views of Fuji as well.[34]

Another artist who produced multiple views of Fuji was Shiba Kōkan (1747–1818), an ardent promoter of Western science and art. The shogunate strictly controlled trade with the outside world, and the only Europeans allowed to bring goods to and from Japan were Dutch merchants who were largely confined to a small island in Nagasaki Harbor. But while direct contact with Westerners was limited, a growing number of scholars, including Kōkan, learned about European ideas, institutions, and technologies through the pursuit of "Dutch Studies" (*rangaku*). A much larger cross section of Japanese grew acquainted with the "Western gaze" via telescopes, magic lanterns, anatomical dolls, and other inventions imported either directly from the Dutch or indirectly through the Chinese.[35] An evangelist of the Western gaze, Kōkan was highly critical of those who "draw mountains and water in whatever way strikes them as interesting, giving them free play to their brush." He specifically noted that of all Japanese artists, "none of them knows how to draw Mt. Fuji."[36] Kōkan therefore made a point of painting Fuji numerous times with linear perspective in the European style. He donated one painting that depicts Fuji low on the horizon, *Seven League Beach, Kamakura, Sagami Province* (*Sōshū Kamakura Shichiri-ga-hama zu*), to Edo's Atagoyama Shrine so it could be displayed in its votive painting hall (see fig. 5.5).[37] Along with other paintings donated by Kōkan to shrines around Japan, it was aimed at familiarizing Japanese with what he argued was a superior mode of representing reality.[38]

Other Japanese artists joined Kōkan in adopting European techniques such as linear perspective, but Hokusai was especially successful at combining them with Japanese ones to create dynamic, highly original compositions that would achieve fame in Japan and around the world.[39] This is especially true of *Thirty-Six Views of Mount Fuji*, which reinforced Hokusai's status as a pathbreaking artist eager to depart from convention. Hokusai's publisher advertised the prints prior to their release in a notice that appeared in a book of stories written by popular author Ryūtei Tanehiko (1783–1842). The advertisement highlighted the innovative nature of the forthcoming series as well as its value as a model for aspiring landscape artists: "These pictures show how the form of Fuji differs depending on the place, such as the shape seen from Shichirigahama, or the view observed from Tsukudajima: he has drawn them all so that

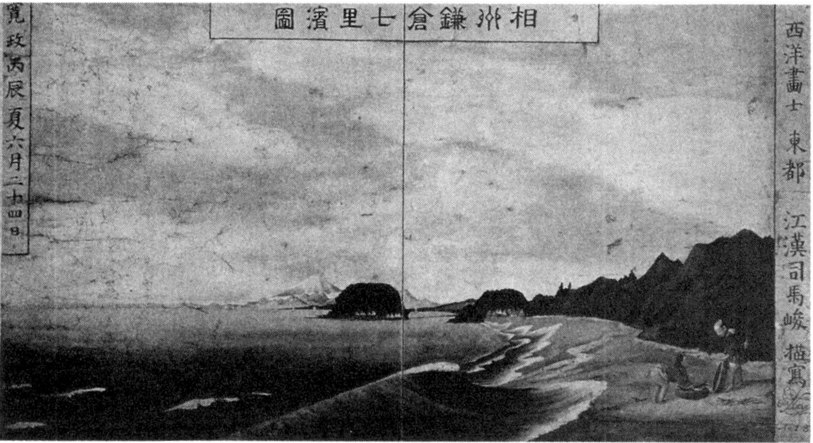

FIGURE 5.5. Shiba Kōkan. *Seven League Beach, Kamakura, Sagami Province.* 1796. Kobe City Museum. Fuji sits low on the horizon to the left of Enoshima Island.

none are the same. These should be useful for those who are learning the art of landscape."[40] As Christine Guth points out, "Few artists in the print medium before Hokusai took note of different angles of vision or the atmospheric and seasonal changes that could inform the appearance of the sacred mountain, nor had they incorporated these features into an implicit narrative of travel."[41] The advertisement also noted that the series would consist of *aizuri-e,* pictures that featured a chemical dye invented in Europe known as Berlin (or Prussian) blue (*bero*). First imported into Japan in the late eighteenth century, the dye allowed artists to create richer and more varied shades of blue than was possible with existing indigo pigments. Initially quite expensive and therefore rare, Berlin blue did not become more widely available at lower cost until the 1820s, so when Nishimura published Hokusai's *Thirty-Six Views of Mount Fuji,* it was still exotic enough to highlight in the advertisement.[42]

Hokusai's willingness to break with conventional portrayals of Fuji is underscored in the preface to the first volume of *One Hundred Views of Mount Fuji.* After noting that Hokusai "has adored the peak of Fuji for many years," the preface reads:

> He must have thought that it was too trite to look up at Fuji from Tago Bay or across from Miho, as obvious as full blossoms under the cloudless moon. So he has walked with cane in hand as far as Fujimigahara and has parked his palanquin at Shiomizaka. He has gazed up at the lofty peak through

trailing willows and through trembling sheaves of rice. We see the open sea with waves beating against the rocks, winding roads buried in valleys white with mist, the ascent of precipitous hills and descent along perilous slopes. Since he has thus shown the true face of nature, his spirit resides in this volume.[43]

Here we see an implicit recognition of the popular demand for views of, and from, real places. But while twenty-six of the prints are labeled with specific place names, most are not, and for even those that do include a name, the connection to specific vantage points is vague at best.[44] Hokusai's prints convey a *sense* of reality, but the views are ultimately a product of his imagination.

In designing the *Thirty-Six Views* and *One Hundred Views*, Hokusai played with different geometrical forms and framing devices to create never-before-seen (and in many cases, impossible-to-see) views of Fuji. One of the most well-known of the *Thirty-Six Views*, called *South Wind, Clear Sky* (*Gaifū kaisei*), depicts Fuji in isolated grandeur (see plate 7), as does *Rainstorm under the Summit* (*Sanka no haku'u*). In the rest of the series, however, Fuji is juxtaposed with foregrounds that echo the mountain and/or draw attention to it. We see this dynamic spectacularly at work in the famous print *Under the Wave off Kanagawa* (*Kanagawa oki nami ura*), commonly known as *The Great Wave*. The print shows a monstrous wave threatening fishermen huddled in their boats, while it situates Fuji as a stable focal point on the horizon. The curves of both the boats and the waves mimic Fuji's, and the whitecaps of the smaller wave under the great wave capture the mountain in its entirety (see plate 8).

Hokusai made a point of depicting Fuji with ordinary human activities in the foreground. Many of the prints in the *Thirty-Six Views* show people traveling or performing some sort of labor. Take, for instance, *Mitsui Shop at Surugachō in Edo* (*Edo Surugachō Mitsui mise ryaku zu*), which portrays roofers working on top of the famous Mitsui store in Edo's Surugachō shopping district (see plate 9). Akin to the farmers in Issa's poem who toss rice seedlings in the direction of Fuji, one of the roofers throws a bundle to someone standing below with outstretched arms. The arms of the man throwing it parallel the roof, which in turn parallels Fuji's left flank, while the arms of the man catching it echo the right flank.

In this print as in others, Fuji's shape is found not only in the things created by humans—such as the kite strings and roof—but in the human bodies that do the creating. Jikigyō developed a theology of a Fuji-centered bodyscape in which the bodies of humans and all other living things are linked through their

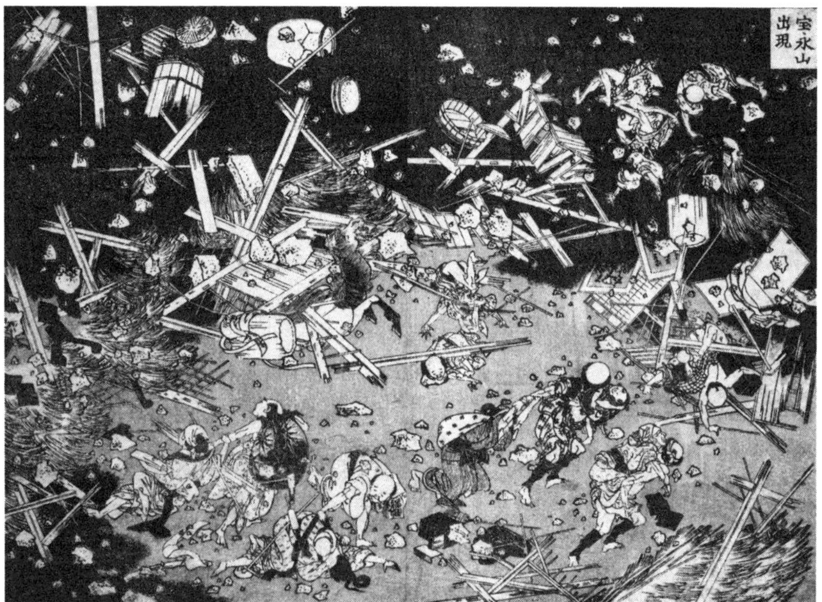

FIGURE 5.6. Katsushika Hokusai. *The Appearance of Mount Hōei*. In *One Hundred Views of Mount Fuji*. 1835. New York Public Library.

dependence on the water and rice produced by the body of Sengen Daibo-satsu. In Hokusai's prints we see a bodyscape in which people unconsciously mimic Fuji in their bodies as they go about their daily work. Fuji is always there, whether noticed or not. It is not just with you but of you.

Hokusai's *One Hundred Views*, like the *Thirty-Six Views*, treats Fuji as a steady, familiar presence amid the comings and goings of people's lives. One of the prints, *The Appearance of Mount Hōei* (*Hōeizan shutsugen*), portrays the havoc caused by the Hōei eruption, but this is the only one in the series to highlight Fuji's destructive power (see fig. 5.6). And while we see the chaos and damage caused by the eruption, we do not see the volcano responsible for it. It is only in an indirect fashion that the print breaks the dominant pattern of depicting Fuji as a fundamentally stable, enduring force—a pattern in accord with the centuries-old association between Fuji and immortality. This association held special meaning to Hokusai, whose cherished goal was to live to one hundred years of age.[45] In Henry D. Smith's words, "By showing life itself in all its shifting forms against the unchanging form of Fuji, with the vitality and wit that inform every page of the book, he sought not only to prolong his own life but in the end to gain admission to the realm of the Immortals."[46]

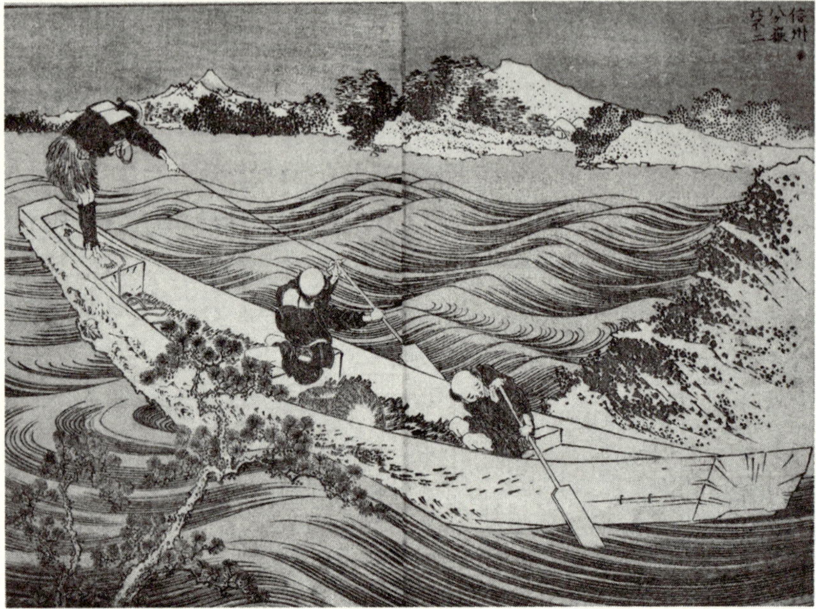

FIGURE 5.7. Katsushika Hokusai. *Fuji and Yatsugatake in Shinshū*. In *One Hundred Views of Mount Fuji*. 1835. New York Public Library.

In "showing life itself in all its shifting forms," Hokusai paid special attention to the human body engaged in various mundane activities. Consider *Fuji and Yatsugatake in Shinshū* (*Shinshū Yatsugatake no Fuji*), the thirty-third print of the *One Hundred Views*. Here the form of the distant mountain is reflected in the three fishermen's bodies, most obviously in the arm of the man to the upper left, which, together with the taut fishing line, echoes Fuji's slope (see fig. 5.7). The bodily incorporation of Fuji repeats itself throughout the *One Hundred Views* "in the arch of spine, in the closure of limbs, and on down to gestures of the hand," thereby expressing the mountain's "life-giving power," as Smith puts it.[47]

Human figures populate the works of Hiroshige as well, but without the attention to individual features and musculature that distinguishes Hokusai's. Like those seen in Song-dynasty landscape paintings, these figures are generic in quality; and thanks largely to Hiroshige's skillful depiction of varying intensities of shadow and light, they contribute to scenes more lyrical and atmospheric than those created by Hokusai.[48] Yet like Hokusai, Hiroshige repeatedly juxtaposes Fuji with the activities of everyday life. For example, *New*

Fuji, Meguro (*Meguro, Shin Fuji*), one of the prints in the *One Hundred Famous Views of Edo*, shows leisure seekers enjoying the cherry blossoms at the base of one of Edo's Fujizuka (the so-called "New Fuji" in the Meguro district), while those who have climbed to the top of the Fujizuka take in the sight of the original Fuji in the distance (see plate 10).

Fuji also appears prominently in *Surugachō*, another print in the *One Hundred Famous Views of Edo* (see plate 11). Here the main thoroughfare in Hiroshige's print points like an arrowhead at Fuji, drawing the eye in the direction of the mountain, which rises above the clouds in transcendent glory. Yet the figures in this print appear to go about their business without paying it any mind. Fuji is both singular and familiar, magnificent and mundane. It is as much a fixture of everyday life as it is an object of extraordinary power and beauty.

Mapping Fuji

The mountain also circulated through guidebooks and maps. Some of these could be used by oshi when they visited their parishioners to encourage them to come to Fuji (and stay at their inns) during the climbing season. But while oshi looked to reinforce bonds with parishioners, the proliferation of guidebooks and maps in the late Tokugawa period reflected an increase in the number of travelers unaffiliated with a specific worship group or leader.[49] Meeting the needs and desires of travelers on the road as well as vicarious travelers who stayed at home, these publications situated Fuji within the broader political and cultural geography of Japan.[50] One of the most famous is Ishikawa Ryūsen's *Map of the Seas, Mountains, and Lands of Japan* (*Nihon kaisan chōrikuzu*), originally published in 1689 and reissued with updates for well over one hundred years. Ryūsen's map depicts the classical provinces, highways, and cities along with the more recent castle towns controlled by daimyo, thus positioning (and legitimizing) the contemporary political order within the enduring frame of imperial Japan (see fig. 5.8). The map also includes famous places that speak to a heritage presumably understood by all Japanese. To distinguish Fuji, one of the most famous of famous places, the map leaves out the green shading used for nearby hills and mountains and relies on (and in turn, perpetuates) the convention of depicting it with three peaks. Other prominent mountains on the map, such as Mount Tsukuba (the mountain to which Fuji is compared unfavorably in the eighth-century *Hitachi Province Gazetteer*), are clearly labeled, but not Fuji. By the end of the seventeenth century, the three-peaked Fuji was so iconic and the mountain's location so well known that labeling it

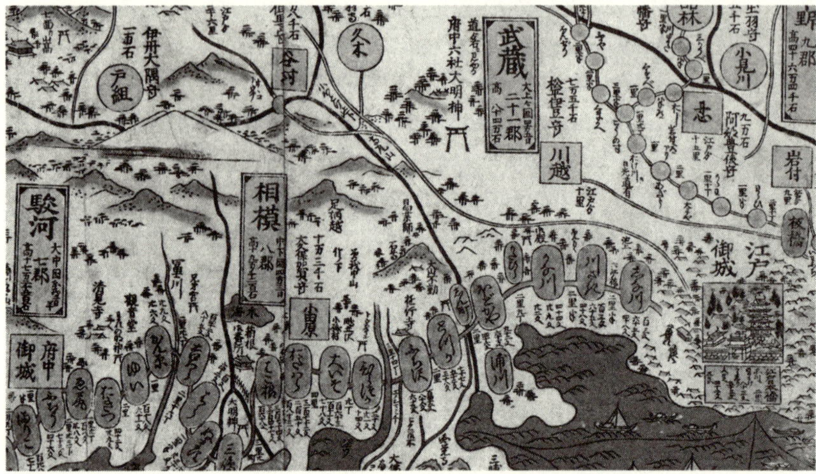

FIGURE 5.8. Ishikawa Ryūsen. Portion of *Map of the Seas, Mountains, and Lands of Japan*. 1691. University of California, Berkeley Library. Fuji is depicted with three peaks in the upper left-hand corner. Edo castle appears in the lower right-hand corner.

was unnecessary. For the remainder of the Tokugawa period, many mapmakers continued to use three peaks to indicate Fuji, with some choosing to add a label, but others, like Ishikawa, doing without.

Other maps produced in the Tokugawa period, particularly in the nineteenth century, highlight Fuji's stature in more dramatic ways. Hokusai's 1818 panoramic rendering of the Tōkaidō highway, *Famous Places of the Tōkaidō at a Glance* (*Tōkaidō meisho ichiran*), depicts Kyoto at the top right, Edo at the bottom right, and a massively out-of-scale Fuji at the upper left (see fig. 5.9).

Fuji's special place in Japan's geography is also reflected in Akiyama Einen's *Map of the Thirteen Provinces from which to View Fuji* (*Fujimi jūsanshū yochi no zenzu*, 1843), which treats Fuji as not just a destination but the main focal point (see plate 12). The map proved to be highly popular, going through multiple printings in the ensuing decades.[51] Attesting to its fame, the renowned Japanologist and Tokyo Imperial University professor Basil Hall Chamberlain (1850–1935) went so far as to call it "the best of the old-fashioned Japanese maps."[52]

Akiyama's map provides a detailed political geography in line with cartographic norms based, for the most part, on *Map of the Seas, Mountains, and Lands of Japan*. Like Ryūsen's map, this one fuses past and present. The title of the map refers to the classical provinces, which are subdivided into districts.

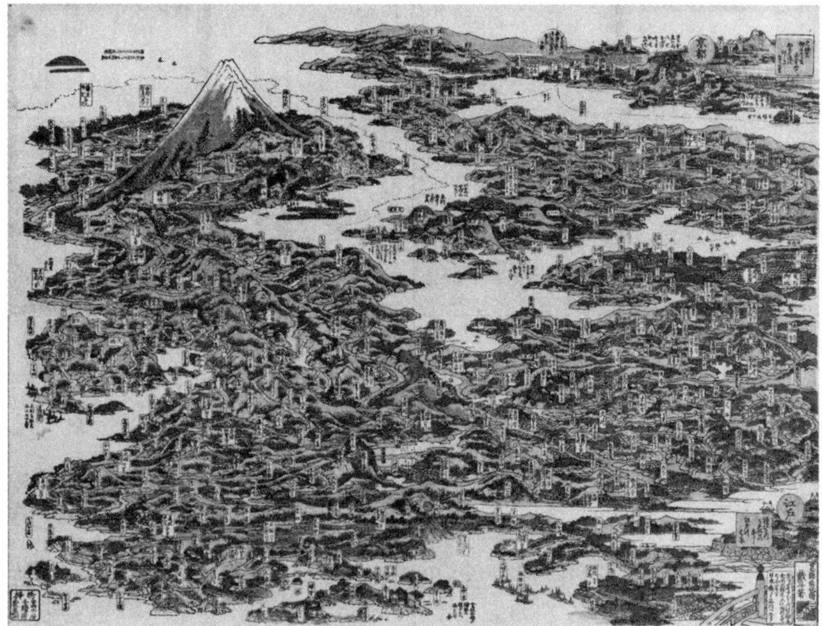

FIGURE 5.9. Katsushika Hokusai. *Famous Places of the Tōkaidō at a Glance.* 1818. University of British Columbia Library.

It shows rivers, mountains, roads, contemporary castle towns, the offices of local intendants and minor daimyo, and post stations. Akiyama also chose to include the names of main and branch villages, ancient castles, old battle sites, barrier gates, shrines, Buddhist temples, assorted famous sites, and hot springs. Aimed at travelers who might want to worship at shrines and temples, visit famous sites, and soak in hot springs while journeying from one town to another, this map presents a Japan unified not only by the geographic order imposed by ancient and modern rulers but also by the cultural and bodily pleasures available to any Japanese with the means to travel. As the title of the map indicates, the pleasure that links all the others is to gaze upon Fuji, displayed prominently in yellow with hachures in the lower left-hand quadrant. Unlike the other mountains, Fuji is portrayed from directly overhead. Combined with the hachures—which appear to signify the numerous ravines running down the mountain[53]—this perspective heightens the centrality of Fuji, whose influence radiates outward in all directions. It also provides a synoptic view of the main climbing trails and other significant features, including the sites of the "silver light water" and "golden light water" and the names of the hills that circle the crater. Edo, indicated by the red blotch to the east, is puny in contrast. The message conveyed by Akiyama is clear: what ultimately unifies the thirteen provinces is not the shogun's capital but Fuji, which has endured from ancient times to the present.

Akiyama could take for granted Fuji's physical and cultural endurance, which was constantly reinforced by time-honored ways of depicting the mountain even as writers and artists showed off their ability to represent such a familiar subject in novel ways. Generations of Japanese memorized the famous poem about Fuji in *The Tales of Ise* and immediately recognized the image of Ariwara no Narihira riding with his entourage past the snow-topped mountain. Meanwhile, mapmakers, painters, and woodblock artists carried on the tradition of depicting the mountain with three peaks. It was this very continuity that enabled poets like Bashō and Issa and artists like Hokusai and Hiroshige to display their inventiveness as they juxtaposed "timeless" Fuji with the vibrant activities of everyday life. In the process, Fuji the destructive volcano faded from the memory of a public that increasingly claimed Fuji the eternal mountain through poetry, artwork, advertisements, housewares, clothing, maps, and even dreams.

Foreign Encounters

The circulation of Fuji in diverse formats made the mountain a familiar presence throughout Japan. Along with other elements of popular culture, this familiarity contributed to a sense of shared landscape and heritage among the country's disparate status groups and regions. In other words, it bolstered a feeling among Japanese that they belonged not only to a realm but to a nation, however different their local customs and allegiances might be.[54] As representations of Fuji helped constitute a sense of "our country" from within, many writers and artists juxtaposed the mountain with the foreign in ways that bolstered Japanese self-esteem relative to the rest of the world. Fuji's role as an emblem of Japanese identity and pride became especially salient in the nineteenth century in the face of growing pressure from Western powers.

By the start of the Tokugawa period, there was already a centuries-old practice of elevating Fuji's stature in the eyes of non-Japanese (both real and imagined). One way to do this was to claim that Fuji was none other than Mount Hōrai, the mountain of immortals in Chinese legend. A Chinese work from the mid-tenth century notes that a Japanese monk visiting China had identified Fuji with Hōrai, claiming that a retainer sent by the Qin Emperor to find the elixir of immortality had settled at the mountain, producing descendants who remained in Japan.[55] The story of a Chinese retainer's journey to the immortal realm of Fuji/Hōrai also appears in the fifteenth-century Nō play *Mount Fuji* (*Fujisan*). In this version of the story he returns to China.[56]

Yet it was in the Tokugawa period, and especially from the eighteenth century, that nationalistic depictions of Fuji versus the foreign proliferated on a mass scale. These nationalistic depictions of Fuji took a variety of forms in both writing and art. As Ronald Toby puts it, "Mt. Fuji was made to speak to the Foreign, to emphasize its Otherness, while reducing the distance separating Mt. Fuji (Japan) from the Foreign, bringing the Foreign into the orbit of Mt. Fuji (Japan), and subordinating the Foreign to the mysterious, universal powers of the sacred mountain."[57] Jikigyō taught that Fuji was "the greatest mountain in the Three Lands" and that it attracted the admiration of foreigners. But while Jikigyō, like many Japanese, took "the Three Lands" (India, China, and Japan) to stand for the world in general, others were more geographically and historically specific in asserting the supremacy of Japan's great mountain.

Artists in the Tokugawa period produced a significant number of woodblock prints showing Korean embassies passing Fuji on their way to Edo to pay respects to the shogun—a motif that remained popular long after the final embassy from Korea arrived in Edo in 1764.[58] *Fuji and Foreign Embassy* (*Raichō no Fuji*), one of the prints in Hokusai's *One Hundred Views of Fuji*, portrays a retinue of Koreans gazing in the direction of Fuji towering overhead. Their banners, displaying the word *raichō* ("coming to the court"), show they are paying tribute to the shogun (see fig. 5.10).[59]

Also capturing the popular imagination was the conceit that Fuji could be seen not just in Japan but from abroad. There was a particularly widespread and tenacious belief that the celebrated general Katō Kiyomasa (1562–1611) could see Fuji all the way from the continent. Kiyomasa led one of the armies dispatched by the hegemon Toyotomi Hideyoshi (1537–1598) in the 1590s to conquer Korea with the ultimate goal of taking over China. Hideyoshi's successes in Korea were short lived, but stories of Kiyomasa's prowess as a military leader carried on. So did the false claim, originating in a biography written soon after his death, that he was able to view Fuji from what was known to Japanese as Orankai, a region of northeastern China just over the border with Korea. Although there were those who debunked the tall tale, its persistence over the years in both writing and art shows a readiness to imagine Fuji as both the world's greatest mountain and a projection of Japanese power to foreign shores.[60]

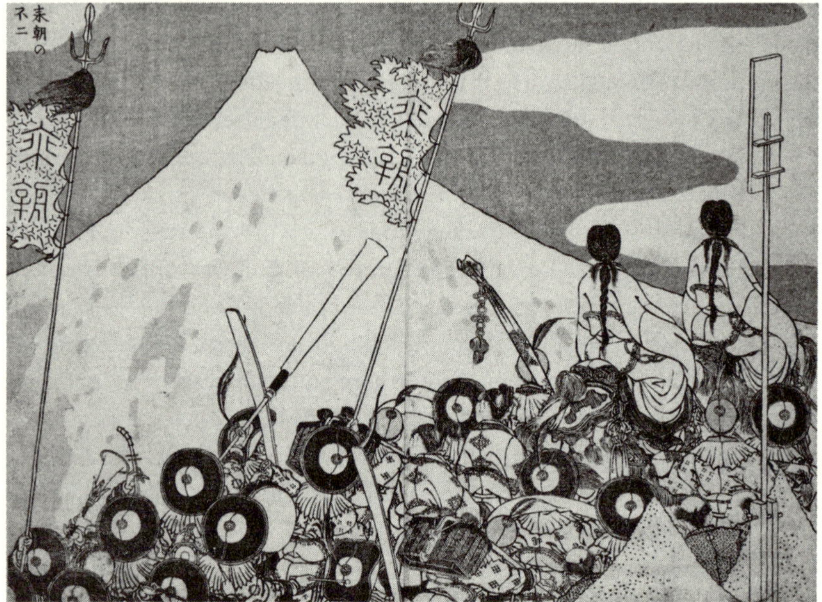

FIGURE 5.10. Katsushika Hokusai. *Fuji and Foreign Embassy*. In *One Hundred Views of Mount Fuji*. 1835. New York Public Library.

If Japanese who had traveled beyond the sea could view Fuji, it followed that non-Japanese could too. One of the prints in Hokusai's *One Hundred Views* depicts two unidentified people in foreign costume in Orankai gazing upon Fuji in the distance.[61] Hokusai also extended Fuji's reach to the south in his series *Eight Views of the Ryūkyūs* (*Ryūkyū hakkei*, 1832), with Fuji visible on the horizon in three of the prints. During the Tokugawa period, the Ryūkyū Islands (today's Okinawa Prefecture) occupied a liminal status: an independent kingdom, it was at the same time in a tributary relationship with both China and Japan and had been largely subjugated by the Satsuma domain in southern Kyushu.[62] By inserting Fuji in the prints, Hokusai implicitly brought the Ryūkyū kingdom into Japan's orbit.

An earlier example of pairing Fuji with the foreign appears in *The Tale of Dashing Shidōken* (*Furyū Shidōken*), a satire published in 1763 by the prolific author, artist, and inventor Hiraga Gennai (1728–1780). The rambunctious tale relates the adventures of Shidōken (aka Fukai Asanoshin), a Gulliver-like Japanese man who travels among foreign and fantastical lands.[63] In China he meets the Qianlong Emperor (r. 1735–1799), and in the course of their conversation, informs him that peerless Fuji is the tallest mountain in the world—taller, even, than the famous Five Sacred Mountains of China. The emperor responds that he had once seen Sesshū's painting of Fuji but had assumed it could not be a true depiction. Learning that Fuji is indeed the grandest mountain in the world, and wanting a copy of his own, the emperor dispatches Asanoshin along with a fleet of ships to create a papier-mâché cast of the mountain to bring back to China. The plan does not turn out well. Alarmed by the attempt to replicate Fuji, the kami of Japan collectively decide to destroy the fleet, entrusting the task to the wind and rain deities who had stopped the Mongols from invading Japan in the thirteenth century.[64] Marcia Yonemoto suggests the ultimate point may be that Fuji-as-symbol is "a meaning in search of a reality." It is, she observes, "a seemingly impressive entity when viewed from afar, but upon closer inspection proves to be an empty paper shell."[65] And yet, Gennai's satire relies on (and to some extent reinforces) the widespread understanding of sacred Fuji, the greatest mountain in the world, as both coveted by foreigners and inextricable physically and symbolically from Japan.

Mapmakers expressed this understanding in cartographic terms as well. Years before Akiyama produced the first version of *Map of the Thirteen Provinces from which to View Fuji*, Matsumoto Yasuoki (also known as Gengendō) created his *Detailed Copperplate Map of Japan* (*Dōsen Nihon yochi saizu*, 1835). Like Akiyama's, this map gives pride of place to Fuji—though in this case through a prominent illustration of travelers viewing the mountain from the Tōkaidō (see fig. 5.11). While Akiyama's Fuji is a focal point for the thirteen provinces, Matsumoto's looms over *all* Japan. The map also shows a keen awareness of the encroaching Western powers. Created with copperplate technology invented in the West, and employing the Western convention of compass roses, it shows sailing ships well enough off the coast that they are presumably foreign. Matsumoto also chose to use roman letters to approximate his name, which reads "gen II 'MATSMOT' JASQOKj" in small print at the bottom of the sheet. It is this emphasis on the foreign that strengthens Fuji's role as a representative—and even a protector—of Japan. Concern about foreigners also motivated the shogunate to sponsor the work of cartographer Inō Tadataka (1745–1818) and his team of surveyors, who produced the first maps of Japan to accurately represent the archipelago's coastlines. Here, too, Fuji came to the nation's aid—in this case as a conspicuous reference point to measure long distances.[66]

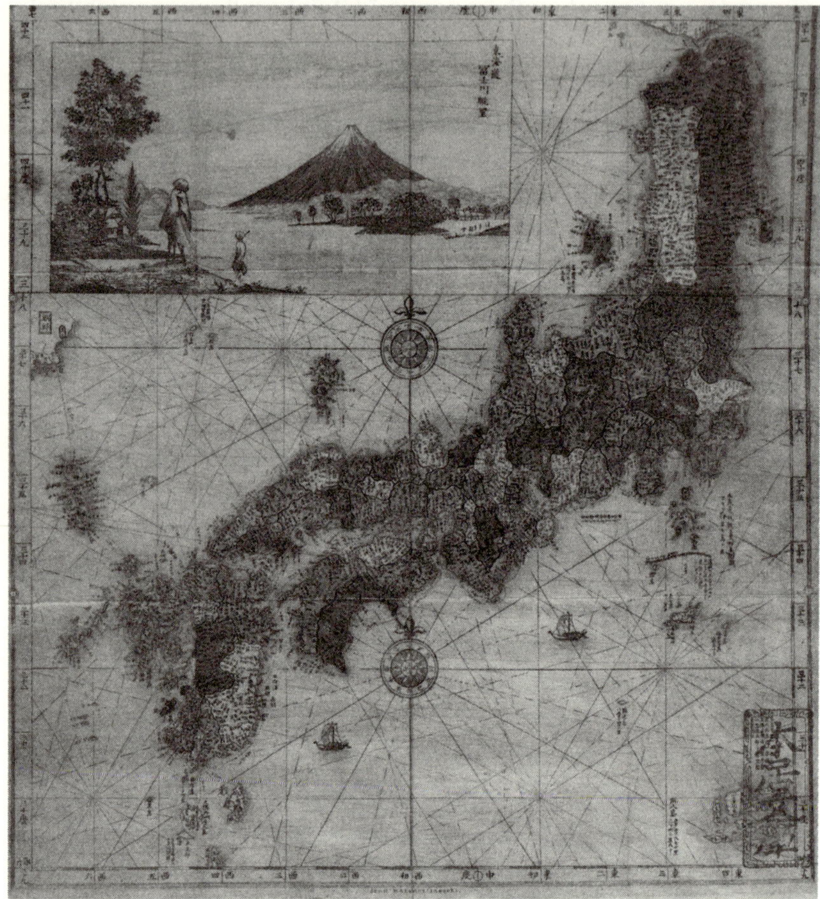

FIGURE 5.11. Matsumoto Yasuoki (Gengendō). *Detailed Copperplate Map of Japan*. 1835. University of California, Berkeley Library.

Another map from the late Tokugawa period that depicts Fuji with an eye to the global context is *Map of the World and its Peoples* (*Bankoku jinbutsu zue*, date and creator unknown). In this fanciful vision of the world, Fuji, and therefore Japan, is located at the center, with various countries represented by people dressed in local costume arranged around it (see plate 13). As Toby notes, the "people in nearby lands labeled 'Ezo' and 'Korea' seem to stare across the intervening space, gazing worshipfully from afar at the sacred mountain that sustains the world."[67] On one level, this map reflects the longstanding view of Fuji as "the greatest mountain of the Three Lands." But although it makes no pretense to accuracy, it also highlights the specific nations that constituted the world at that time. Jikigyō's Fuji sustained a cosmic order. This Fuji anchors an international one.

IN THE FOURTH MONTH of 1834, the same year the first volume of Hokusai's *One Hundred Views of Mount Fuji* appeared, heavy rains loosened ice and snow on Fuji's upper slopes, triggering massive slush flows (*yukishiro*) that devastated communities at the foot of the mountain. Among them was Shimoyoshida (Lower Yoshida), the community that adjoined the oshi-dominated district of Kamiyoshida (Upper Yoshida). Much of Japan was already in the grip of the Tenpō famine that had begun in 1833 and lasted into the late 1830s due to crop failures caused by unusually cold and rainy weather. In Shimoyoshida, a bad harvest and poor silkworm production had, by the spring of 1834, reduced residents to eating leaves and grass. Then came the slush flows, which compounded the misery by destroying fields, forty-eight homes, and most of the village's mulberry trees (which provided food for silkworms). As a result, between 1832 and 1838 the number of households in the village dropped by about 30 percent, from 507 to 352.[68]

News of the disaster spread—for example, via a broadsheet (*kawaraban*) that used a variation of Hokusai's famous *Great Wave* to depict the raging slush flows[69]—but like the Hōei eruption (and unlike Hokusai's prints), it ultimately made little impact on the popular image of Fuji as a timeless source of continuity, beauty, and life itself. Fuji's familiar presence called to mind a long and seemingly unbroken history of literary and artistic associations—a history that, in addition to Fuji's immediate splendor, offered the Tokugawa shogunate an important cultural resource to bolster its authority. The people of Edo and its environs also laid claim to the mountain. Fujikō members worshipped Fuji as the sacred center and origin of the entire cosmos while "the children of Edo" appropriated the mountain as a site of local identity and pride. We see this in Bashō's poem about offering the wind of Fuji on a fan as a souvenir from Edo and in woodblock prints by Hokusai and Hiroshige that show Fuji fully incorporated into the lives—and very bodies—of those lucky enough to view the mountain on a regular basis.

Those not so lucky could take advantage of well-maintained highways to view and even climb Fuji. Circulated through fiction, poetry, art, maps, and a variety of consumer goods, Fuji also entered the lives of people who never saw the actual mountain. Fuji had been famous for centuries, but it was in the Tokugawa period, with the rise in consumerism and print culture—and Edo's growing influence on that culture—that depictions of the exalted mountain saturated the entirety of Japan. By the middle of the nineteenth century, it is probably safe to say that all Japanese, from farmers to daimyo, had encountered Fuji in some form. Popular culture had transformed the sacred origin of the cosmos into an intimate part of the nation.

This widespread intimacy, combined with singular glory, made Fuji the perfect representative not only of Edo but all of Japan. Fuji's role as emblem of Japan grew in importance toward the end of the Tokugawa period, when increasing contact with the Western powers heightened awareness of the country as more than a confederation of domains: it was also a nation in a world of competing nations. The shogunate used Fuji as a means to bolster its prestige, but the mountain was ultimately the property of the people, not the officials who governed them. As faith in the bakufu weakened, Fuji remained a steady and unifying force, linking past to present through its cultural history and physical presence. It was part of the land that stood for all the land. The shogun was only human—he could go.

6

Native Mountain, Foreign Threat

IN 1854 a new edition of Akiyama Einen's popular *Map of the Thirteen Provinces from which to View Fuji* was issued. Unlike the original, it portrayed newly constructed gun batteries located just off the coast of Edo.[1] These were hastily built after Commodore Matthew Perry (1794–1858) arrived at the entrance to Edo Bay with four steam-powered gunboats a year earlier to press the shogunate to open Japan to trade with the United States. Belching smoke and bristling with cannon, the "black ships" (*kurofune*) signaled the industrial imperialism that was reshaping the globe. Despite constructing the batteries, the bakufu knew Japan was woefully unprepared to go to war, so it decided there was little choice but to make deals with the Americans and the other Westerners who followed in their wake.

The foreign interlopers quickly set their sights on Fuji, determined to summit the mountain as an act of both personal achievement and imperial conquest. The shogunate was wary of letting them travel into the interior of Japan, but after putting up some resistance, it agreed to allow foreign diplomats and their retinues to make the ascent.

Permitting foreigners to climb sacred Fuji outraged nativists, many of whom were part of the National Learning (*kokugaku*) movement that took hold in the eighteenth century and appealed to increasing numbers of Japanese in the 1850s and 1860s. Drawing heavily on the eighth-century imperial myth-histories, the *Record of Ancient Matters* and *Chronicles of Japan*, National Learning scholars constructed a fantasy of a long-ago Japan uncontaminated by foreign influences. Japanese in this imagined world engaged in a pure form of kami worship and showed complete loyalty to the imperial household, which was idolized as a divine lineage connecting Japan's past to its present. The idea that foreigners would climb Fuji, which, like the emperor, linked Japanese across the ages, was repugnant to those who reveled in dreams of a purely native realm.

These dreamers also despised "foreign" Buddhism—even though it had existed in Japan for over a millennium—and promoted what they envisaged to be a completely native Shinto (Way of the gods) in its place. Until the bakufu was overthrown in 1868, however, the impact of Shintoization on worship at Fuji, as elsewhere in Japan, was limited. It was true that over the course of the eighteenth and early nineteenth centuries increasing numbers of oshi around Fuji secured licenses from the Yoshida and Shirakawa lineages of Shinto. And during that time the mountain goddess Kaguya-hime, the "shining princess" in *The Tale of the Bamboo Cutter* who came to be revered as a manifestation of Great Bodhisattva Sengen, was steadily displaced by Konohanasakuya-hime, a goddess found in the *Record of Ancient Matters* and *Chronicles of Japan*. Yet generations of Fujikō members continued to embrace teachings, rituals, and terminology derived from Buddhism, while pilgrims to Fuji walked through a terrain filled with Buddhist icons and structures. These ranged from the guardian deities who guarded the entrance to Kamiyoshida's Sengen Shrine to the halls dedicated to Dainichi (the Sun Buddha) and Yakushi (the Medicine Buddha) on the summit. It is also unclear to what extent oshi and shrine priests decided to affiliate with the Yoshida and Shirakawa lineages out of a purely pragmatic concern to gain an advantage in local battles for institutional and ritual control or out of a genuine desire to embrace and propagate Shinto teachings.

Everything changed when an alliance of domains defeated the Tokugawa in a series of battles in 1867–1868. Dissatisfaction with the political order had been growing for decades, but it was the shogunate's decision to bow to foreign pressure and allow trade with the United States and then other Western powers that sparked the revolt. In place of the bakufu, the rebels created an emperor-centered national government that could stand up to the foreigners.

The Hōei eruption had wreaked havoc on the physical landscape on and around Fuji. Policies implemented after the fall of the bakufu upended the religious one. Officials in the new regime who were sympathetic to the nativist cause declared the "unity of rites and rule" (*saisei itchi*). That meant ridding the imperial household of Buddhist practices, reviving the ancient Council of Divinities (Jingikan), and putting major Shinto shrines under state control. They also ordered the "separation of kami and buddhas" (*shinbutsu bunri*) throughout the realm. This last measure revolutionized the worship of Fuji. As elsewhere in Japan, radical nativists took the government policy of separating gods and buddhas to mean they could freely attack Buddhist icons, institutions, and rituals in campaigns to "destroy the buddhas, annihilate Shākyamuni" (*haibutsu kishaku*). The Sengen shrines around Fuji were purged of everything that smacked of Buddhism, as was the mountain itself. Government-appointed Shinto priests asserted their authority over Fujikō, although many of the faithful continued to hold onto beliefs and practices deemed Buddhist in origin.

Persecution of Buddhism at Fuji and elsewhere was harsh but short lived. Over the course of the 1870s, pragmatists in the new government showed less interest in Shintoizing Japan than in modernizing it. Shinto continued to serve state interests to the extent that it encouraged devotion to the emperor and obedience to the law. Realizing, however, that most Japanese were unwilling to give up Buddhist beliefs and practices, officials decided it was better to work with Buddhist institutions than attempt to destroy them.

The arrival of Westerners in Japan confirmed and strengthened Fuji's status as a symbol of Japan, in the eyes of both foreigners and the Japanese people. The foreign threat also boosted nativist efforts to remake Fuji, like the rest of the nation, into a purely Shinto space. While the persecution of Buddhism did not last long, the Shintoization of Fuji as both a site and object of worship endured. To the chagrin of those who wanted to revive a purely native Japan, the foreigners were there to stay as well.

Shinto Fuji

Shinto first made inroads into Fuji worship in response to the Tokugawa bakufu's efforts to control religious institutions and practices. As discussed in chapter 4, soon after its founding, the shogunate created regulations for major Buddhist temples, and in 1635 established its Office of Temples and Shrines to supervise religious activities throughout Japan.[2] After suppressing the Shimabara Rebellion of 1637–1638, the uprising in Kyushu that prominently displayed Christian images and slogans, it was particularly intent on stamping out Christianity. To do that it mandated that households officially register with Buddhist temples, creating a surveillance system that lasted until the end of the Tokugawa period. It also issued a series of regulations in 1665 aimed specifically at shrines devoted to the veneration of kami. One regulation specified that those who worked at shrines without rank could not wear vestments other than white robes unless they had been licensed by the Yoshida lineage.[3] Another recognized the legitimacy of ranks acquired via other "imperial mediators"—in other words, noble households such as the Shirakawa—but singling out the Yoshida worked in the favor of the lineage, which found itself meeting increased demand for permits from professional ritualists around Japan, including those around Fuji.[4]

Descended from a clan of augurs who performed tortoiseshell divinations in the Council of Divinities, such as the ones carried out in response to Fuji's eruptions in the Heian period (794–1185), the Yoshida lineage took a decisive turn during the Ashikaga shogunate (1336–1573) under the leadership of Yoshida Kanetomo (1435–1511). At the time, kami worship was generally incorporated into temple-shrine complexes dominated by Buddhist priests, a prominent exception being the shrine at Ise devoted to Amaterasu, the sun goddess and ancestor of the imperial household. Buddhist priests taught that kami were local manifestations of buddhas and bodhisattvas, but Kanetomo rejected this long-established idea. According to his "One-and-Only Shinto" (Yuiitsu Shintō), Buddhist deities were the avatars of kami; and Shinto, not Buddhism, was the "founding principle of the universe."[5] It followed that Japan, the land of the kami, was superior to China and India.[6] As politically astute as he was theologically and ritually inventive, Kanetomo ingratiated himself with major figures in the Ashikaga bakufu and imperial court, and managed to bring the Council of Divinities, which until then had been directed by the Shirakawa sacerdotal lineage, largely under his control.[7] He also took over the imperial prerogative of awarding higher ranks to kami in response to requests from shrines throughout Japan. This practice continued under his successors, who extended their influence (and made money) by "offering the esoteric rituals devised by Kanetomo, modification of local shrine practices, bequeathing priestly titles, and granting permission to shrine priests to wear aristocratic vestments for shrine rites."[8] After the Tokugawa took control, the Yoshida lineage made sure to guard its privileges by currying favor with powerful daimyo. As a result, the shogunate formally recognized its standing in the shrine regulations of 1665.

Shrine priests and oshi around Fuji, like counterparts around Japan, purchased Yoshida licenses to gain a form of official legitimacy that could be deployed in rivalries for local status and power. Competition came to a head in the early eighteenth century in the pilgrimage town of Kamiyoshida when the Yoshida-certified priest of the town's Sengen Shrine, with the support of allied oshi, fought with other oshi over control of the shrine, its rituals, and even the name of its god. The conflict originated in the eleventh month of 1707, when the priest at the time barred oshi from entering the main sanctuary without permission, despite the fact that Kamiyoshida's oshi had, for generations, jointly managed the shrine with priests who were oshi as well.[9] After the priest once more refused entry in early 1708, oshi representatives appealed to the local intendant (*daikan*), arguing that they were unable to offer prayers on behalf of their parishioners. The result was a compromise: the intendant recognized the authority of the priest over the main sanctuary but allowed the oshi to use an older sanctuary built by the daimyo Takeda Shingen (1521–1573).[10]

Conflict erupted again a few years later after two oshi, without consulting their peers, raised money to be licensed as Shinto officiants (*shinkan*, or "kami officials") by the Yoshida lineage.[11] They also secured the highest kami rank (*shōichi'i*) for the Sengen deity—now called Great Kami Sengen (Sengen Ōkami) instead of Great Bodhisattva Sengen (Sengen Daibosatsu)—and displayed banners both in the shrine precincts and in town that declared the kami's new name and status. In addition, they established new ritual procedures at the shrine, including sacred songs and dances performed for the deity. A group of oshi who opposed the changes appealed to the Office of Temples and Shrines, complaining that the renegade oshi and their supporters (including the shrine's priest) had no right to alter existing practices without consultation.[12] In their defense, the targeted oshi argued that other shrines, including the Sengen Shrine in Ōmiya, referred to the Fuji deity as Great Kami Sengen, not Great Bodhisattva Sengen, and that it was necessary to put up the banners to inform people of the god's true name and rank. They also argued that it was only proper for those who had received certification from the Yoshida lineage to follow its strictures, criticizing fellow oshi who had obtained Yoshida licenses but continued to chant sutras and use Buddhist rosaries in front of the deity.[13] To settle the dispute, the Office of Temples and Shrines ruled that the banners should be removed and both the plaintiffs and the defendants worship as they pleased.[14]

Oshi continued to obtain licenses in the decades that followed, but from the mid-eighteenth century an increasing number turned for ritual instruction and support to the Shirakawa, not the Yoshida, sacerdotal lineage. The Shirakawa, the noble family in charge of court ritual prior to the rise of the Yoshida, had recently reasserted its authority over the imperial court's Council of Divinities and stepped up efforts to issue its own licenses to officiants of different ranks throughout Japan, overtaking the Yoshida as time passed.[15] The shift to the Shirakawa among Fuji oshi therefore reflected a nationwide trend but, like the earlier movement to buy Yoshida licenses, was fueled by parochial competition.

Local rivalry explains why, in 1759, the oshi who served as the priest for the Suwa Shrine that adjoined Kamiyoshida's Sengen Shrine traveled to Kyoto to get certification from the Shirakawa.[16] Although thought to be much older, the Suwa Shrine—one of many shrines dedicated to the Suwa deity across Japan—had by this point been overshadowed by the neighboring Sengen Shrine, which had grown more powerful over recent decades thanks to the patronage of Fujikō.[17] Because the Yoshida-certified head priest (*kannushi*) of the Sengen Shrine treated the uncertified Suwa priest as a subordinate, the decision to acquire a license from another sacerdotal lineage with ancient roots was a way for the Suwa priest to assert his independence. In response, the Sengen Shrine priest filed a lawsuit in 1760 claiming that the Suwa priest was unauthorized to receive a license from the Shirakawa lineage.

The Sengen priest also protested the Suwa priest's actions during Kamiyoshida's annual Fire Festival (Himatsuri), which dates as far back as the late fifteenth century (perhaps earlier) and is renowned as one of the most spectacular festivals in Japan thanks to the torches and bonfires that light up the streets at night.[18] Today locals generally associate the festival with Konohanasakuya-hime, a kami in the imperial chronicles who proved her fidelity to her husband by emerging with her children unscathed from a hut she had set on fire. Until the early twentieth century, however, the festival was widely understood to be dedicated to Suwa Myōjin, a kami worshipped in various regions of Japan.[19] So although a Fuji-shaped *mikoshi* (portable shrine) has been paraded during the festival since the late sixteenth century (see plate 14), it has always been joined by a *mikoshi* for the Suwa deity. In fact, although they are now carried together, for most of the festival's history, it was the Suwa *mikoshi* that took precedence, with the one shaped like Fuji following behind.[20] The Sengen priest nevertheless complained in his lawsuit that the Suwa priest had illicitly performed rituals during the Fire Festival reserved for Yoshida-certified *kannushi* like himself. The Suwa priest was subsequently ordered to give up his Shirakawa license; and in an agreement reached in 1762, the supremacy of the Sengen priest in ritual matters was confirmed.[21]

A similar conflict unfolded in the nearby pilgrimage town of Kawaguchi, although this one pitted nearly all of the town's oshi against the priest of the town's Sengen Shrine and a prominent oshi who supported him. Like the Suwa priest in Kamiyoshida, most of the oshi in Kawaguchi applied for Shirakawa licenses in 1759 in order to challenge the authority of their local Sengen priest, who, after being licensed by the Yoshida lineage as *kannushi* a year earlier, had begun treating the town's oshi (many of whom had lower-ranking Yoshida

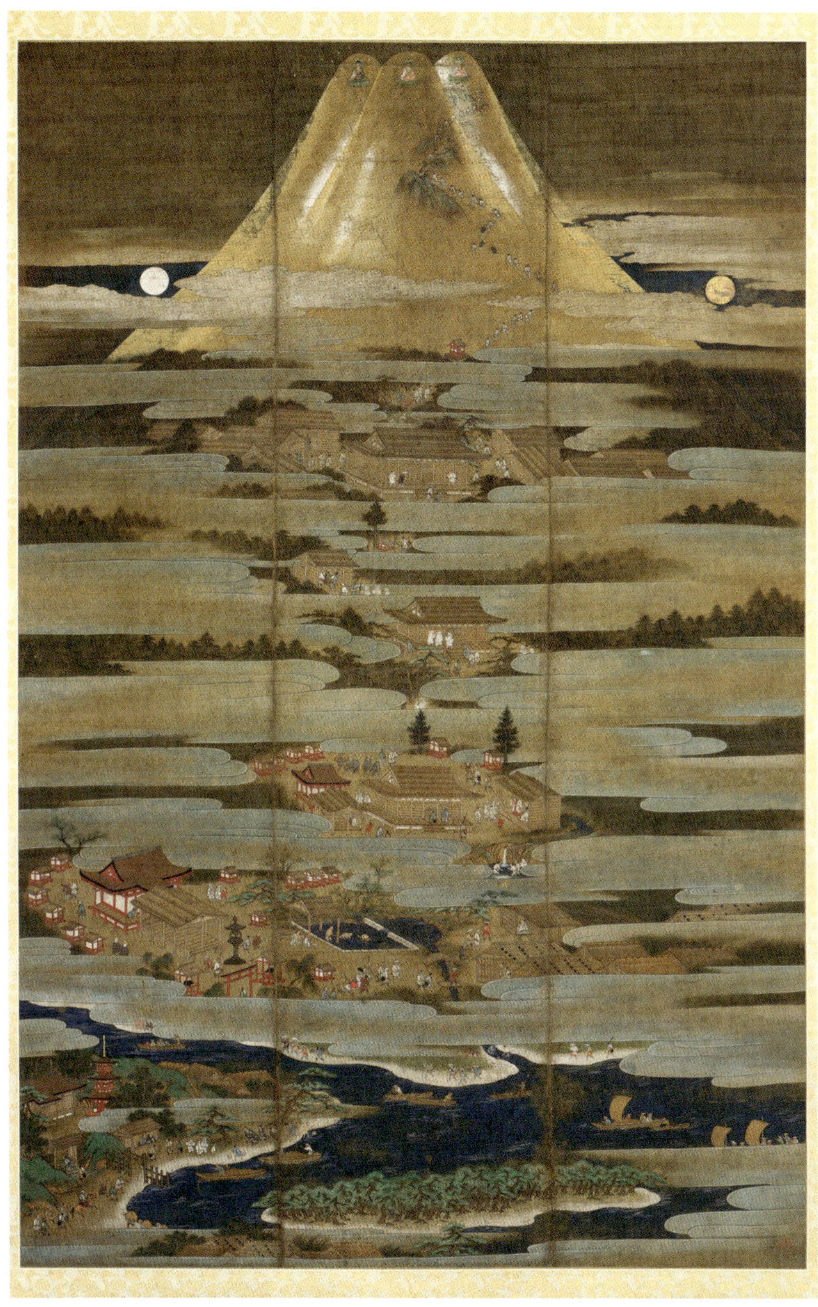

PLATE 1. Kanō Motonobu. *Fuji Pilgrimage Mandala*. Sixteenth century. Fujisan Hongū Sengen Taisha, Fujinomiya.

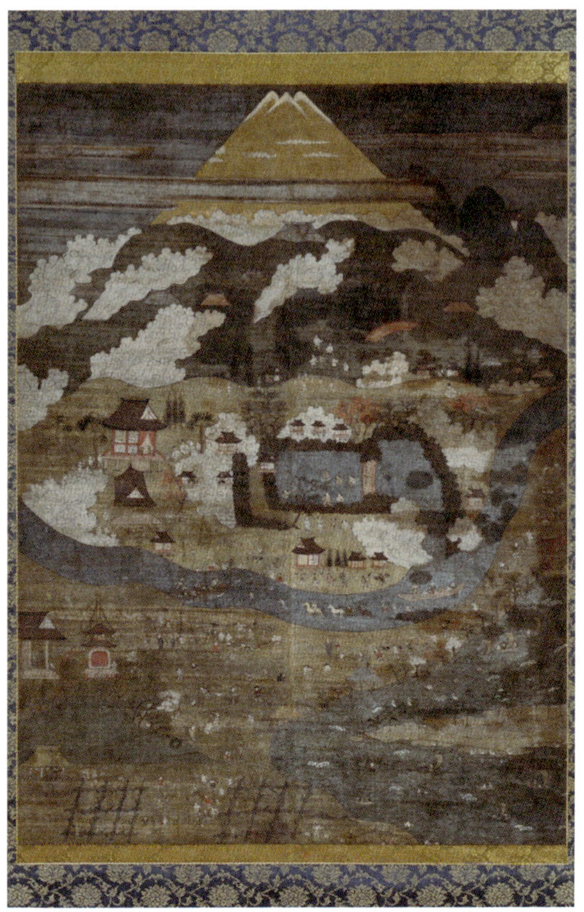

PLATE 2. *Fuji Pilgrimage Mandala*. Sixteenth century. Fujisan Hongū Sengen Taisha, Fujinomiya.

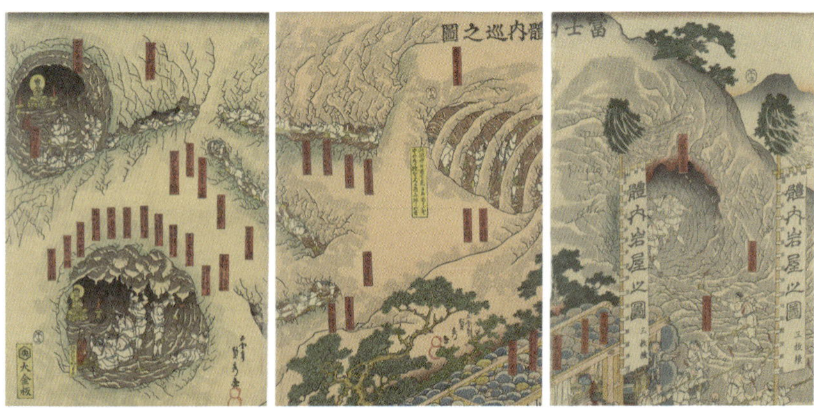

PLATE 3. Utagawa Sadahide. *Mount Fuji Womb Cave Pilgrimage*. 1858. Private collection of D. Max Moerman.

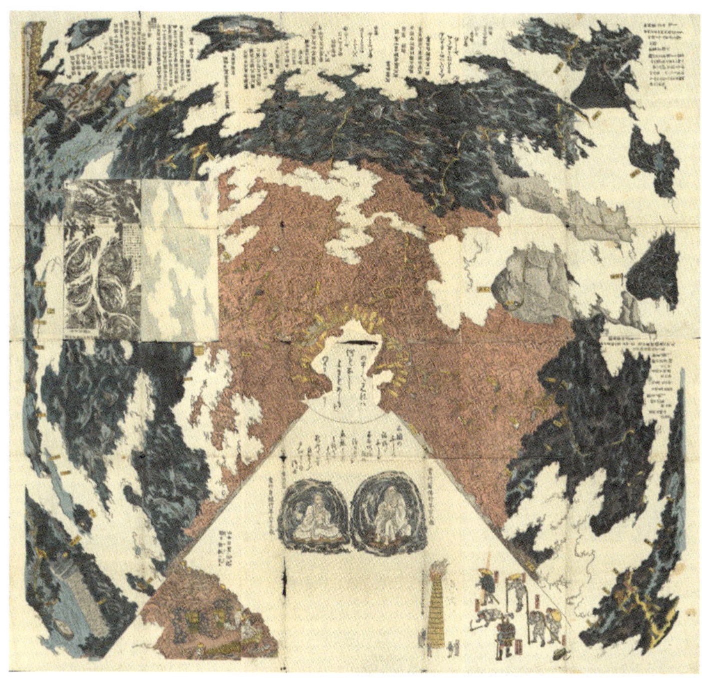

PLATE 4. Utagawa Sadahide. *Portrayal of Mount Fuji*. 1848. University of British Columbia Library. This "picture-map" of Fuji can be folded into a three-dimensional cone. Kakugyō Tōbutsu and Jikigyō Miroku are seated in caves inside the mountain.

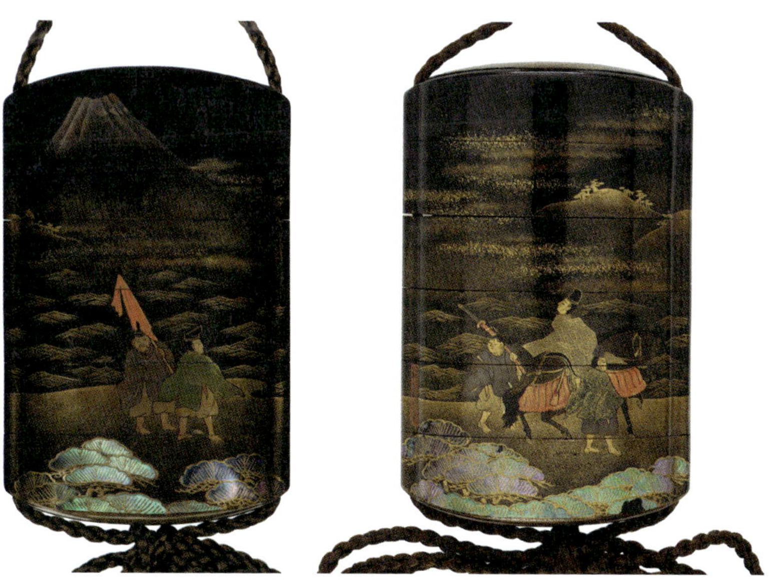

PLATE 5. School of Shiomi Masanari. *Ariwara no Narihara at Mt. Fuji*. Lacquered inrō (portable container) from inrō/ojime/netsuke ensemble. Early nineteenth century. Los Angeles County Museum of Art.

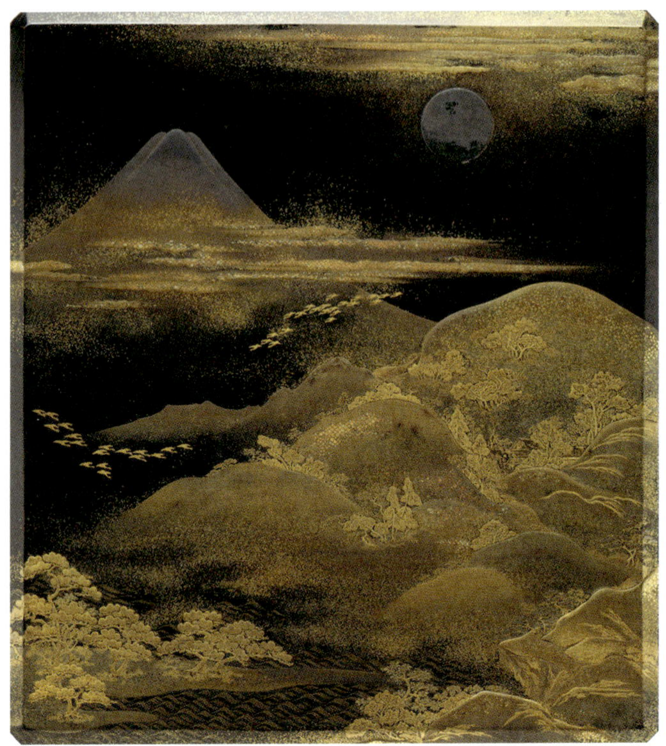

PLATE 6. Box for Inkstone and Writing Implements. Early to mid-nineteenth century. Metropolitan Museum of Art, New York. The moon brings to mind Kaguya-hime, the "shining princess" of *The Tale of the Bamboo Cutter* who served time on earth before rejoining her fellow immortals on the moon.

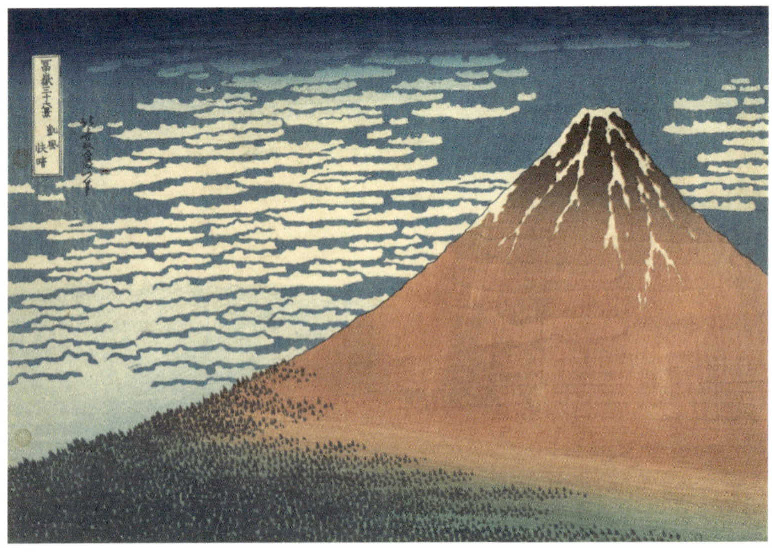

PLATE 7. Katsushika Hokusai. *South Wind, Clear Sky.* In *Thirty-Six Views of Mount Fuji.* 1830–1832. Metropolitan Museum of Art, New York.

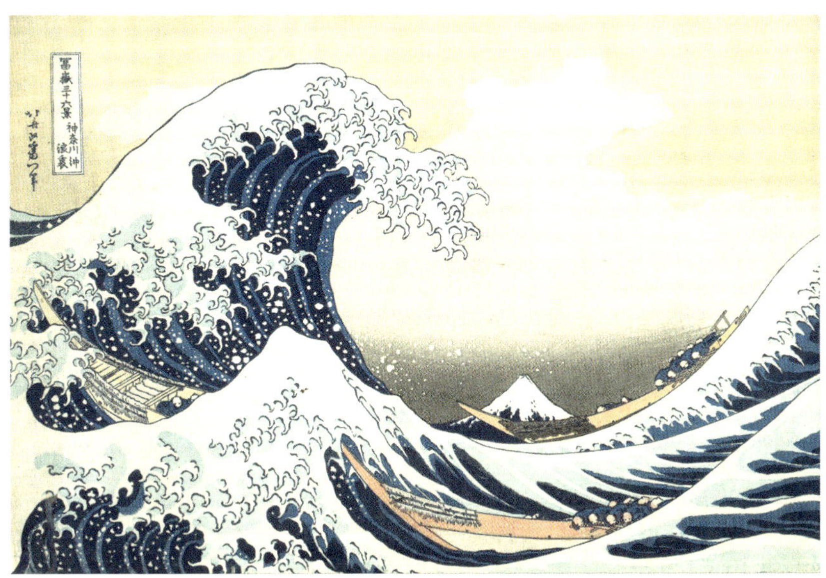

PLATE 8. Katsushika Hokusai. *Under the Wave off Kanagawa* (*The Great Wave*). In *Thirty-Six Views of Mount Fuji*. 1830–1832. Metropolitan Museum of Art, New York.

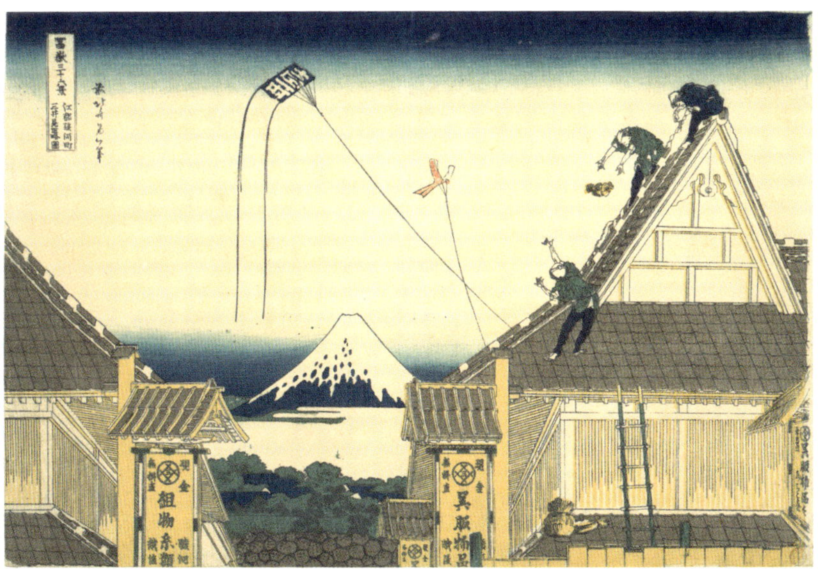

PLATE 9. Katsushika Hokusai. *Mitsui Shop at Surugachō in Edo*. In *Thirty-Six Views of Mount Fuji*. 1830–1832. Metropolitan Museum of Art, New York.

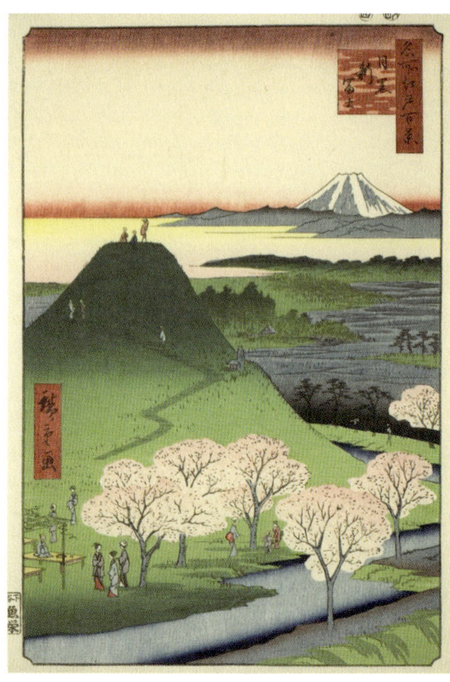

PLATE 10. Utagawa Hiroshige. *New Fuji, Meguro*. In *One Hundred Famous Views of Edo*. 1857. Art Institute of Chicago.

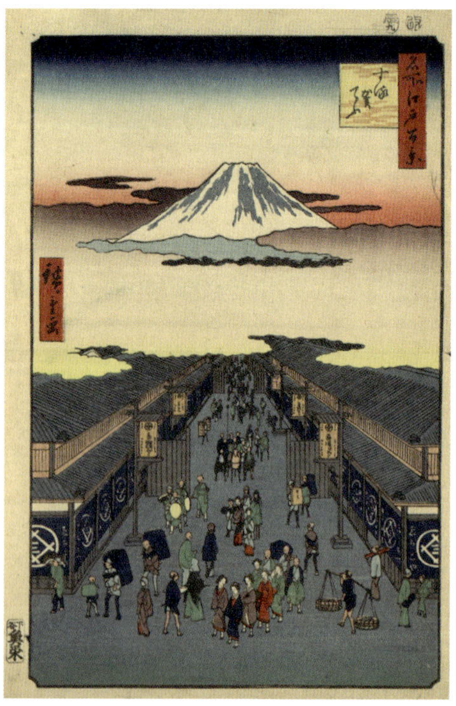

PLATE 11. Utagawa Hiroshige. *Surugachō*. In *One Hundred Famous Views of Edo*. 1856. Library of Congress, Washington, DC.

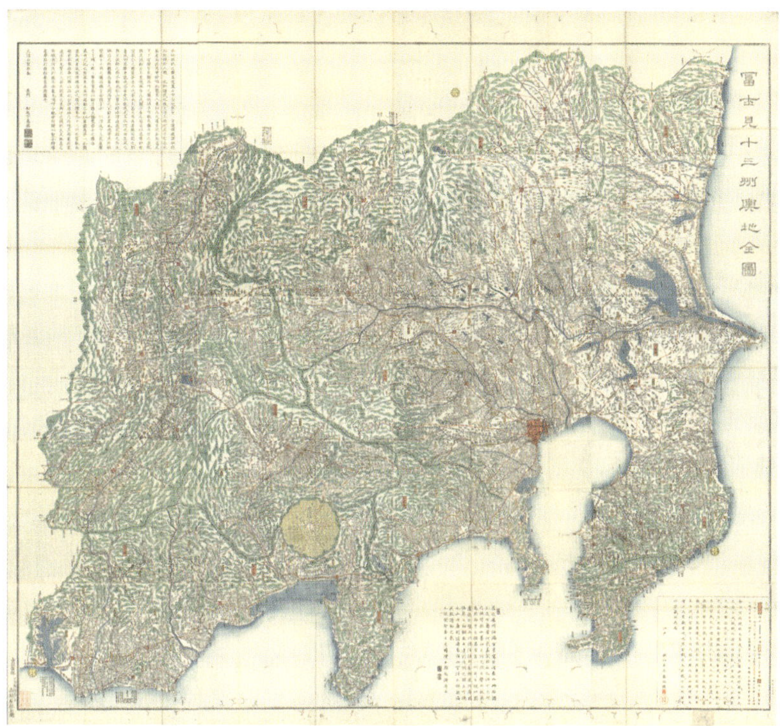

PLATE 12. Akiyama Einen. *Map of the Thirteen Provinces from which to View Fuji.* 1843. University of British Columbia Library. Fuji is depicted in yellow and Edo in red.

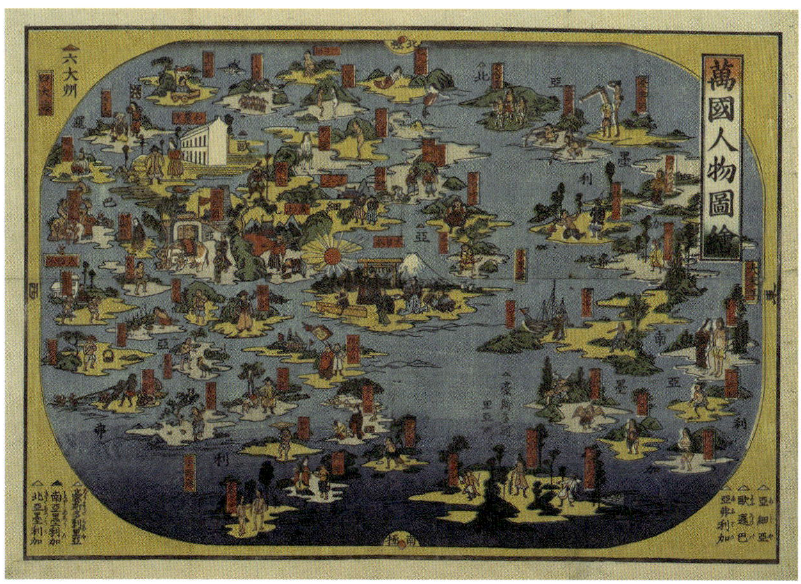

PLATE 13. *Map of the World and its Peoples.* Late Tokugawa period. Gifu Prefectural Library.

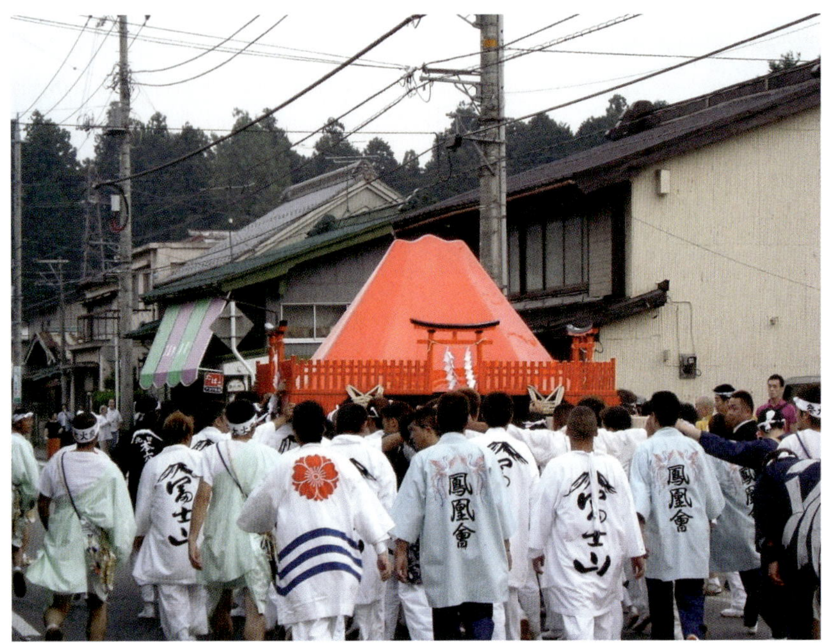

PLATE 14. Parading Fuji-shaped *mikoshi* during Fujiyoshida Fire Festival. Photo by author.

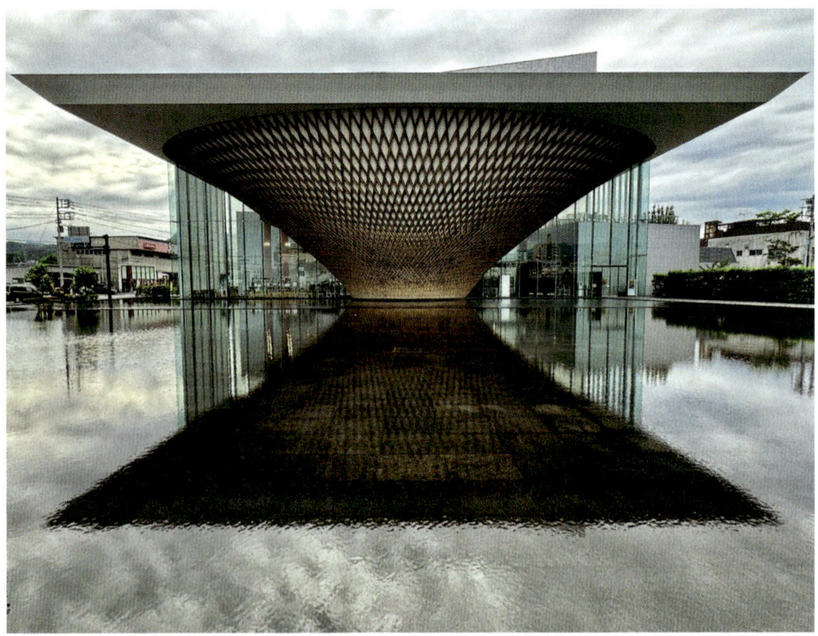

PLATE 15. Mount Fuji World Heritage Centre, Shizuoka (in Fujinomiya). Photo by Philip Ellway.

licenses) as his inferiors.[22] Shirakawa licenses in hand, the oshi asserted their newfound status by wearing ritual vestments and interfering with the shrine's festivals; and parallel to what happened in Kamiyoshida, the Sengen Shrine priest, along with his oshi ally, filed a lawsuit that forced the other oshi to return their licenses.[23]

This setback did not spell the end for Shirakawa influence. Thanks to lobbying by the lineage's doctrinal chief, the shogunate agreed to allow the Shirakawa to continue providing the oshi with instruction in kami worship even though the lineage was not permitted to issue official licenses that could be used to challenge the position of the shrine's *kannushi*.[24] So for the remainder of the Tokugawa period, most of the Kawaguchi oshi remained Shirakawa disciples even if at the same time they held Yoshida licenses that officially placed them beneath the *kannushi*.[25] In Kamiyoshida, too, increasing numbers of oshi attached themselves to the Shirakawa lineage, with far more becoming Shirakawa rather than Yoshida disciples from the late eighteenth to the mid-nineteenth centuries.[26]

As oshi and shrine priests used their Shinto affiliations to jockey for power on a local level, ideologically motivated Shintoists promoted their vision of a "purely" native form of kami worship on a national one. Many of them had ties to the Yoshida and Shirakawa lineages, and a significant number drew upon a form of kami worship melded with Confucian teachings called Suika Shinto, which, from its founding in the seventeenth century, rejected Buddhism and preached absolute devotion to the emperor.[27] From the late eighteenth century, Shinto advocates were especially influenced by the strain of thought known as *kokugaku*, variously translated into English as "National Learning," "National Studies," and "Japan Studies." It also goes by the term "nativism," although, as Mark Teeuwen succinctly puts it, "not all nativism was Kokugaku, nor was nativism all there was to Kokugaku."[28] Broadly speaking, participants in the National Learning movement investigated ancient Japanese literature and mythology to discern (and in the process, invent) what made Japan unique. And to them, what distinguished the land of the kami most of all was the imperial lineage descended from the sun goddess Amaterasu.

The promoters of Shinto, whether National Learning adherents or not, believed that to properly worship the kami required knowing the proper kami to worship. At Fuji this meant replacing both Great Bodhisattva Sengen with Great Kami Sengen and the goddess Kaguya-hime—thought in the medieval period to be a manifestation of Great Bodhisattva Sengen—with the goddess Konohanasakuya-hime (Princess of the Flowering Tree Blossoms).

Konohanasakuya-hime had impeccable credentials as far as those steeped in the ancient imperial myths were concerned. According to the *Record of Ancient Matters* and *Chronicles of Japan*, she married Ninigi, the grandson of Amaterasu, and discovered she was pregnant in just one night, drawing the suspicion of her husband. Offended by Ninigi's distrust, she retreated into a hut and set it on fire, claiming that any child of her husband would emerge unscathed. Three children were born safely, proving that Konohanasakuya-hime was a virtuous wife.[29] The goddess's ability to resist fire and deliver healthy children would seem to make her a good fit for Fuji, a volcano that could potentially erupt but was also a source of life-giving water. Yet Fuji is nowhere to be found in the *Record of Ancient Matters* and *Chronicles of Japan*. In fact, while stories linking Kaguya-hime to Fuji have ancient roots, the earliest works depicting Konohanasakuya-hime as a Fuji deity date to the seventeenth century. These include one written by the famous neo-Confucian scholar Hayashi Razan (1583–1657).[30] An advisor to four Tokugawa shoguns and the head of the government's Confucian academy in Edo, he, like other neo-Confucians, was a fierce critic of both Buddhist teachings and Daoist beliefs in immortals. He therefore not only identified the deity of Fuji as Konohanasakuya-hime in a record of his journey from Edo to Kyoto but also argued in a volume about the shrines of Japan that the legends invented by Buddhist priests about Kaguya-hime and Fuji were false.[31]

Origin tales of temples and shrines (*engi*) show that, as the Tokugawa period wore on, Kaguya-hime increasingly made way for Konohanasakuya-hime, who came to be worshipped at multiple shrines around Fuji. The popularization of Konohanasakuya-hime as the mountain goddess explains why she, not Kaguya-hime, graces the *One Hundred Views of Fuji* produced by Katsushika Hokusai in the 1830s.[32] The Shugendō complex at Murayama held onto Kaguya-hime until the middle of the nineteenth century, but Konohanasakuya-hime eventually triumphed there as well.[33]

By the end of the Tokugawa period, then, the Shintoization of Fuji was well underway. As more and more oshi around Fuji became disciples of the Yoshida and Shirakawa lineages, many of them embraced Konohanasakuya-hime and propagated Shinto prayers and purification rituals (*harai*).[34] Some Fuji devotees also adopted the antiforeign stance promoted by nativists. They included Rokugyō Sanshi (1765–1841), the Fujidō leader who helped Tatsu defy the ban on women climbing Fuji by accompanying her to the summit in 1832. Sanshi was on friendly terms with the National Learning scholar Takada

Tomokiyo (1783–1847); and although, as Janine Sawada notes, "Fujidō was not a channel of *kokugaku* in any formal sense," he shared the nativist view that Japan had been corrupted by foreign influences.[35]

The Shintoization of Fuji worship had its limits, however. Oshi may have promoted Shinto prayers and rituals and made Konohanasakuya-hime an object of worship on their kami altars, but Fujikō continued to embrace beliefs and practices with Buddhist roots.[36] And while pilgrims may have encountered Konohanasakuya-hime and Shinto rituals at Sengen shrines around Fuji, they continued to climb a mountain crowded with Buddhist icons, terminology, and institutions. Until the fall of the Tokugawa shogunate, ridding Fuji of Buddhism was easier said than done.

The Politics of Climbing Fuji

As nativists in the late Tokugawa period dreamed of eliminating the doctrines and practices of Buddhism, they faced the reality of Westerners muscling their way into the Land of the Gods. To make matters worse, the reviled foreigners wanted to climb sacred Fuji. A simple reason was the one reportedly given by George Mallory (1886–1924) when asked why he wanted to climb Mount Everest: "Because it's there." But Fuji was not just "there." It was everywhere—on consumer items and maps, in literature and art—and in the minds of millions of Japanese who saw Fuji as both a cosmic mountain and a national one. Foreign diplomats who resided in Japan in the 1850s and 1860s were well aware of Fuji's national importance; and when they submitted requests to the shogunate to climb the mountain, they knew it was a way to exercise their political will.

Bakufu officials initially resisted the requests—which was understandable not only for the immediate political implications but in the context of the decades-long threat that the Western powers had posed to Japan. In the late eighteenth and early nineteenth centuries, the Russians tussled with the shogunate over control of Sakhalin Island and the Kuril Islands to the north and east of Ezo (modern-day Hokkaido); and in 1808 the British sailed into Nagasaki Harbor, where they promptly took two of the resident Dutchmen hostage and used them as bargaining chips to get supplies from the militarily ill-prepared Japanese. Over the coming decades, the British and Americans intermittently hassled the shogunate to engage in trade and to provision whaling ships that hunted in the North Pacific, testing its policy of strictly limiting contact with foreigners and calling into question its ability to defend the realm.

Britain's humiliating defeat of China in the first Opium War (1839–1842) served as a stark warning.[37]

Concern about the foreign threat was shared by ordinary Japanese. What is now Hokusai's most famous print in the *Thirty-Six Views of Fuji* series, *Under the Wave off Kanagawa (The Great Wave)*, signifies as much, according to Christine Guth, who infers that the viewpoint is that of "an invisible foreign ship," and that Hokusai's waves, both here and in other works, reflect Japan's "vulnerability to foreign invasion and its power to resist it."[38]

The arrival of Perry in 1853 made the foreign challenge more immediate than ever. After cajoling officials into accepting a letter from President Millard Fillmore (1800–1874), Perry promised to return the following year. In the interim, the shogunate hastily built the coastal defenses depicted on the new edition of Akiyama's map and hard-liners called on the shogunate to meet the Americans with force. However, pragmatists in the government were fully aware that Japan's defenses were no match for America's military technology, so they ultimately decided to initiate diplomatic relations when Perry returned.[39]

Hokusai's protective waves notwithstanding, foreigners—and specifically Westerners—entered Japan in increasing numbers after Perry's arrival, and nearly all made a point of singing the praises of Fuji. These included Townsend Harris (1804–1878), the first American consul general to Japan, who was entrusted with negotiating a treaty with the bakufu after Perry's second mission in 1854. The Treaty of Amity and Commerce Between the United States and Japan (1858), often called the "Harris Treaty," served as a model for other "unequal treaties" signed between Japan and the Western powers. Among other things, it ensured that Americans could practice Christianity within Japan and that multiple cities (including Edo) would be opened to American residents according to a set schedule. It also stipulated that the Japanese government could not unilaterally raise tariffs on foreign goods beyond the amounts agreed upon in the treaty and that "Americans committing offenses against Japanese shall be tried in American Consular courts and, when guilty, shall be punished according to American law."[40] It was this last provision of "extraterritoriality" for US citizens that made the treaty particularly unequal in favor of the Americans.

In the journal he kept during his time in Japan, Harris describes his view of Fuji during a trip into the interior of the Izu Peninsula: "It is grand beyond description; viewed from this place [the village of Yugashima] the mountain is entirely isolated and appears to shoot up in a perfect and glorious cone, some ten thousand feet high; while its actual height is exaggerated by the absence of any neighboring hills by which to contrast its altitude." As a world traveler, he applied a firsthand comparative perspective unavailable to Tokugawa Japanese, adding, "In its majestic solitude it appeared even more striking to me than the celebrated Dwhalagiri [Dhaulagiri] of the Himalayas, which I saw in January, 1855."[41]

Harris's Dutch interpreter Henry Heusken (1831–1861) was even more effusive in his praise of Fuji. Like Harris, he situated Fuji's appeal within a global context. In his journal he notes there are taller mountains in the world, but then declares, "I don't think anything in the world will ever equal its beauty." He adds, "In spite of myself I pulled the reins of my horse and, carried away by an outburst of enthusiasm, I took off my hat and cried: 'Great, glorious Fujiyama!' Glory forever to the mountain of mountains of the Pacific Sea, which alone raises its venerable brow covered with eternal snow amidst the verdant countryside of Nippon [Japan]!"[42] In 1861, just a few years after he wrote this entry, Heusken was murdered by samurai known as *shishi* ("resolute samurai"). Rallying under the banner "Revere the emperor, expel the barbarians" (*sonnō jōi*), *shishi* took it upon themselves to protect "the verdant countryside of Nippon" from foreigners like Heusken.[43]

From the beginning, it was not enough for some Westerners to admire Fuji from a distance; they wanted to climb it as well. A multivolume report on Perry's mission published by order of the US Congress in 1856 included an exchange between Perry and Japanese officials in which the commodore flatly insisted that "as the nations know each other, the Japanese will permit Americans to go anywhere, to Mount Fuji, all over the country."[44] Here we see an early indication that going to Fuji was perceived by Westerners as an important litmus test of their ability to travel throughout Japan. The fact that Fuji was not just a famous place but a towering mountain gave it special appeal. This was an age when mountain climbing, in the words of Bernard Debarbieux and Gilles Rudaz, "functioned as a metaphor (an athletic conquest stood in for a legal one) and as a metonymy (the conquest of a summit stood in for that of an entire region) for territorial appropriation itself."[45] Whereas Jikigyō imagined Fuji to be the heart of a cosmic bodyscape, summiting Fuji with Western bodies was, symbolically at least, a way to incorporate Japan into a globalized bodyscape dominated by the West.

That was how it was viewed by Rutherford Alcock (1809–1897), the first British consul general in Japan and the first Westerner (and presumably the first non-Japanese of any kind) to climb Fuji. In his memoir of his time spent in Japan, Alcock describes his successful journey to the mountain in

September 1860. He emphasizes that the treaty recently signed by Britain and Japan ensured that "the head of a Diplomatic mission" had "*the free right to travel all over the empire*" (italics his).[46] However, the shogunate desired "to limit and restrict, as far as possible, all locomotion of foreigners, and all intercourse, commercial or social, with the natives," such that, before his trip to Fuji, "a journey in the interior, undertaken for the avowed purpose of recreation and observation,—and out of the beaten track, in the exercise of a treaty right, was as yet an unheard-of thing."[47] The plan to travel to Fuji was therefore a means to put the treaty right to the test—a way not only to escape his "*quasi* state prison in Yeddo [Edo]," but to see with his own eyes "whether the excitement and hostility towards foreigners, in consequence of the newly-contracted foreign relations, and departure from the ancient policy of seclusion and isolation, did or did not exist—away from the centre of government."[48]

Alcock's expedition was also a means to one-up the other foreign powers as they competed for influence in Japan. Fully aware of Alcock's intent, and not wanting to be outdone, the French ambassador appealed for permission to climb Fuji just a month and a half after Alcock submitted his request. Officials rejected his petition, however, because the ambassador planned on traveling with a Prussian even though the Prussians at that point had no treaty rights in Japan.[49] Thanks to the treaty with the United States, they did approve Townsend Harris's request to climb Fuji, although poor health apparently kept him from fulfilling his plan.[50]

Politics were also at the forefront of the minds of the officials who initially resisted Alcock's request to go to Fuji. Their reluctance made sense. In the first place, as Alcock notes in his memoir, they considered it unseemly for a man of his rank "to make the pilgrimage, limited by custom if not by law to the lower classes."[51] More important, allowing the British envoy to travel into the interior would open the bakufu to further criticism from those already angry that it had not done more to resist foreign pressure. (This is why, although they eventually acceded to the request to climb Fuji, officials adamantly refused to allow foreigners to visit highly sensitive sites like the shrines at Ise, where the imperial ancestor Amaterasu was worshipped.)[52] It would look particularly bad if any harm came to Alcock—a valid concern given the recent spate of attacks against both foreigners and the officials blamed for accommodating them. In fact, Alcock had earlier chastised the shogunate for not doing more to protect members of foreign legations from violence.[53] Hanging especially heavy over the heads of Japanese officials was the bold assassination of Senior Councilor Ii Naosuke (1815–1860) outside Edo castle mere months before Alcock submitted his request. The *shishi* who killed the councilor condemned him for authorizing the Harris Treaty, which allowed foreigners not only to live in Japan but to practice Christianity as well.[54]

When the government finally acceded to Alcock's request, it insisted on providing a substantial escort—an unwanted expense but one designed to keep Alcock and his party safe and under a watchful eye. Alcock bemoaned the fact that the cortege amounted to at least one hundred men and thirty horses.[55] His own party consisted of only eight men. In addition to other members of the British legation, these included a botanist and a lieutenant in the Indian navy "provided with a few instruments for the purpose of scientific observations."[56]

In the age of Western imperialism, dominating the world entailed measuring and categorizing it; and the inclusion of the botanist and the instrument-bearing lieutenant was a clear example of science following the flag. John Gould Veitch (1839–1870), the botanist who accompanied Alcock, belonged to a prominent family of British horticulturalists who studied, imported, and cultivated plants from around the world.[57] During the expedition he carefully recorded, and in some cases collected, the species they encountered along the way.[58] As a result, one of the more common trees on the slopes of Fuji, known as *shirabe* in Japanese, is referred to in English as Veitch's Silver Fir and goes by *Abies veitchii* in the Linnean system of classification used by scientists around the world.

Once they had reached Fuji's summit, the lieutenant put his instruments to work to determine its altitude, latitude, and longitude, as well as the size of the crater and the temperature of the air (54°F in the sun).[59] Alcock and his men also reportedly raised the British flag, fired twenty-one gunshots, sang "God Save the Queen," and toasted their achievement with champagne.[60] Here was a blatant instance of scientific efforts working in concert with imperial aims, which took on added significance by happening on a mountain as celebrated as Fuji.

High-ranking shogunal officials tended to regard Fuji worship with either indifference or disdain, so although they initially resisted Alcock's request to travel to Fuji because of concerns about trouble that might occur en route, they were not particularly bothered by the irreverent behavior of Alcock and his party once they got there.[61] And while there were Japanese who objected from the start—for example, an official in the Mito domain wrote in his diary that "the people have been enraged . . . and say there's no doubt the Mountain

Deity will not allow the barbarous curs to approach"[62]—many locals at the base of Fuji, including oshi, shrine priests, and *shugenja* (Shugendō practitioners), did not oppose the climb. In fact, the priests and *shugenja* in Ōmiya and Murayama went out of their way to welcome Alcock; and learning that he and his men planned to go to the villages of Subashiri, Yoshida, and Mishima after the climb, residents in Suyama inquired if the Englishmen could stop in their village as well. Explaining this receptive attitude was a mix of simple curiosity and a desire for the income that the foreigners and their Japanese retinue might provide. There may have been concerns about their barbaric ways, but the anticipation of entertainment and economic benefit initially prevailed.[63]

Views soured after it became clear that the foreigners were more a burden than a boon. Unlike free-spending pilgrims, the stingy British brought their own food, so they had no need or inclination to buy what the locals had to offer. The Japanese who accompanied them failed to spend much as well. To make matters worse, communities around the mountain and along the rest of Alcock's route had to provide horses and porters in addition to repairing the roads on which he and his retinue traveled. Some also objected that the British failed to follow the standard protocols of pilgrims, such as engaging in purification rituals at local shrines and avoiding the consumption of meat.[64]

Alcock's trip to Fuji may have been a financial letdown for those who lived around the mountain, but it made money for publishers in Edo. Among the works featuring Alcock and his party was *Humorous Pilgrimage to Fuji (Kokkei Fuji mairi)*, a comic travel journal in ten volumes by the popular author Kanagaki Robun (1829–1894). In the ninth volume appears a woodblock print showing Alcock on a horse accompanied by a party of samurai, other Westerners, and porters (see fig. 6.1). As Miyazaki Fumiko notes, the rest of the volume is unconnected to Alcock's journey, but Robun must have slipped in the print to help sales. She also points out that the image of Alcock portrays him in a dignified manner without any apparent hostility. It can therefore be seen as yet one more expression of Japanese taking pride in Fuji by portraying it as the object of foreign desire.[65]

Other images that appeared in widely circulated broadsheets (*kawaraban*) express a much more negative and sensationalist view. Perpetuating a false rumor that a divinely ordained typhoon had stopped the foreigners from reaching the summit, one of them shows storm clouds blowing members of Alcock's party down the mountain (see fig. 6.2). Projecting out of the storm is a *tengu*

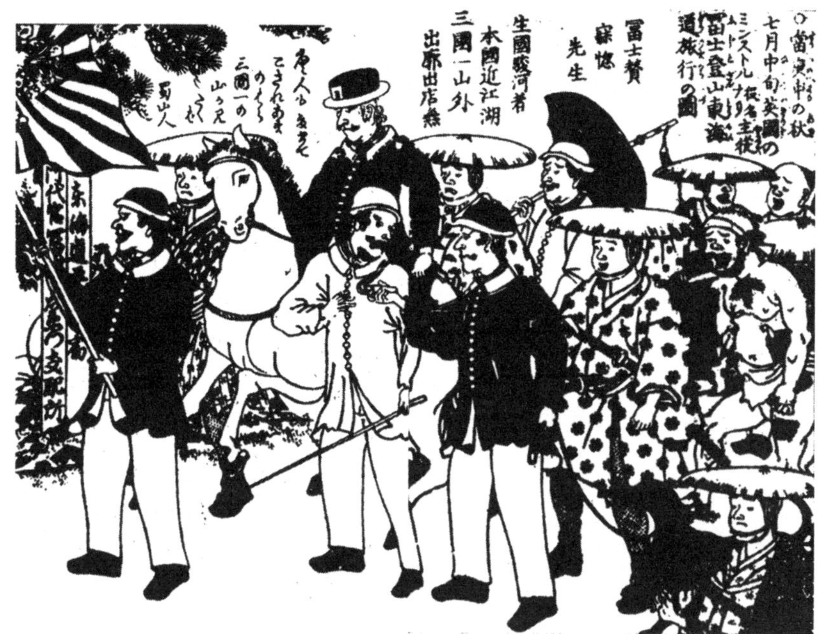

FIGURE 6.1. Alcock goes to Fuji. In Kanagaki Robun, *Humorous Pilgrimage to Fuji* (*Kokkei Fuji mairi*). 1860.

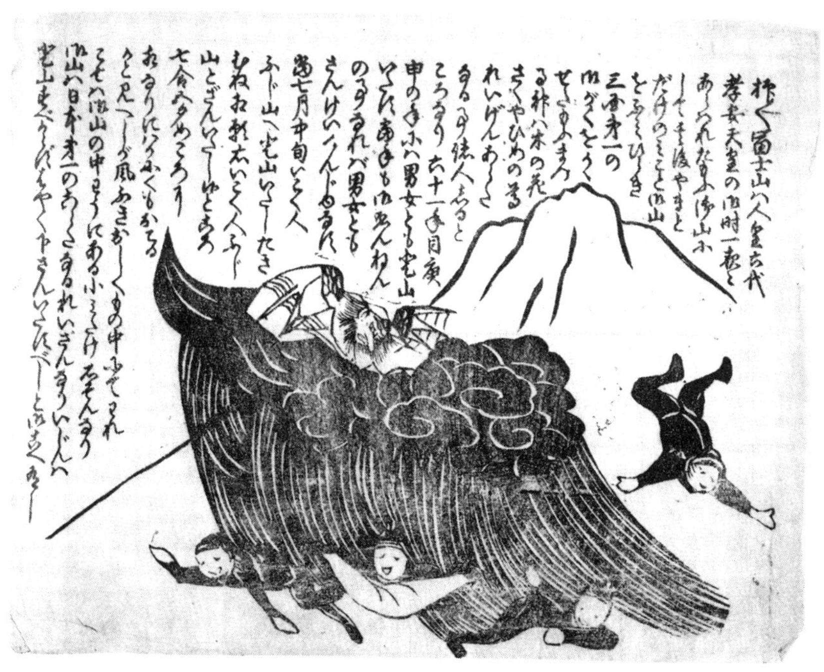

FIGURE 6.2. Storm blows Alcock and his men down Fuji. Anonymous broadsheet. 1860.

(mountain goblin) who proclaims, "This is the sacred mountain of Japan. Aliens! Go down the mountain at once."[66] It would not be so easy to rid Fuji—and the rest of Japan—of the British and their fellow Westerners, however.

The shogunate in fact reinforced foreign fascination with Fuji by including scrolls depicting the mountain among its diplomatic gifts.[67] These scrolls cemented the connection not only between Fuji and Japan but also between the stately mountain and the Tokugawa regime. A Kanō school scroll sent to US president James Buchanan in 1860, *Cranes Flying over Mount Fuji* (*Fuji hikaku zu*), shows Fuji with Miho-no-matsubara in the foreground (see fig. 6.3). The scroll additionally features cranes, which were auspicious in general but were also associated with the Tokugawa house in particular, symbolism that was probably lost on the Americans.[68]

Likely because of the unstable political situation and the physical threat posed by *shishi* such as those who killed Heusken, six years passed after Alcock's expedition before other foreigners followed his lead; but in 1866 and 1867, members of the Dutch, Swiss, American, and British delegations finally repeated his achievement by summiting Fuji.[69] The 1867 expedition of Harry Parkes, at that point Britain's ambassador to Japan, was especially notable, since he was accompanied by his wife Fanny, the first foreign woman to make it to the top of Fuji.

Lady Parkes's well-publicized climb was historic, not just on its own terms, but because it added pressure on authorities to bring an end to the prohibition on women climbing to the summit of Fuji—and by extension, other sacred mountains in Japan. In 1860 the shogunate had endorsed requests for women to climb to Fuji's summit because it was a momentous *kōshin* year (the year of the monkey in the sixty-year cycle of the Chinese calendar and the one in which Fuji was supposedly born). But now that the shogunate had permitted a foreign woman to make the ascent in an ordinary year, it was unjust, in the eyes of many, to deny Japanese women the same privilege. Soon after the Parkes's ascent, the Kamiyoshida oshi informed the headquarters of the Yoshida Shinto lineage with which they were affiliated that it had become untenable to refuse the requests of female worshippers who wanted to climb to the summit.[70] Just a few years later, in 1872, the new imperial government announced that women were free to climb to the top of any mountain, including Fuji.[71]

Indirectly, then, a climb by a foreign woman helped the advancement of Japanese women. To those who wanted to expel foreigners from the Land of the Gods, of course, it was just one more insult to Japanese pride.

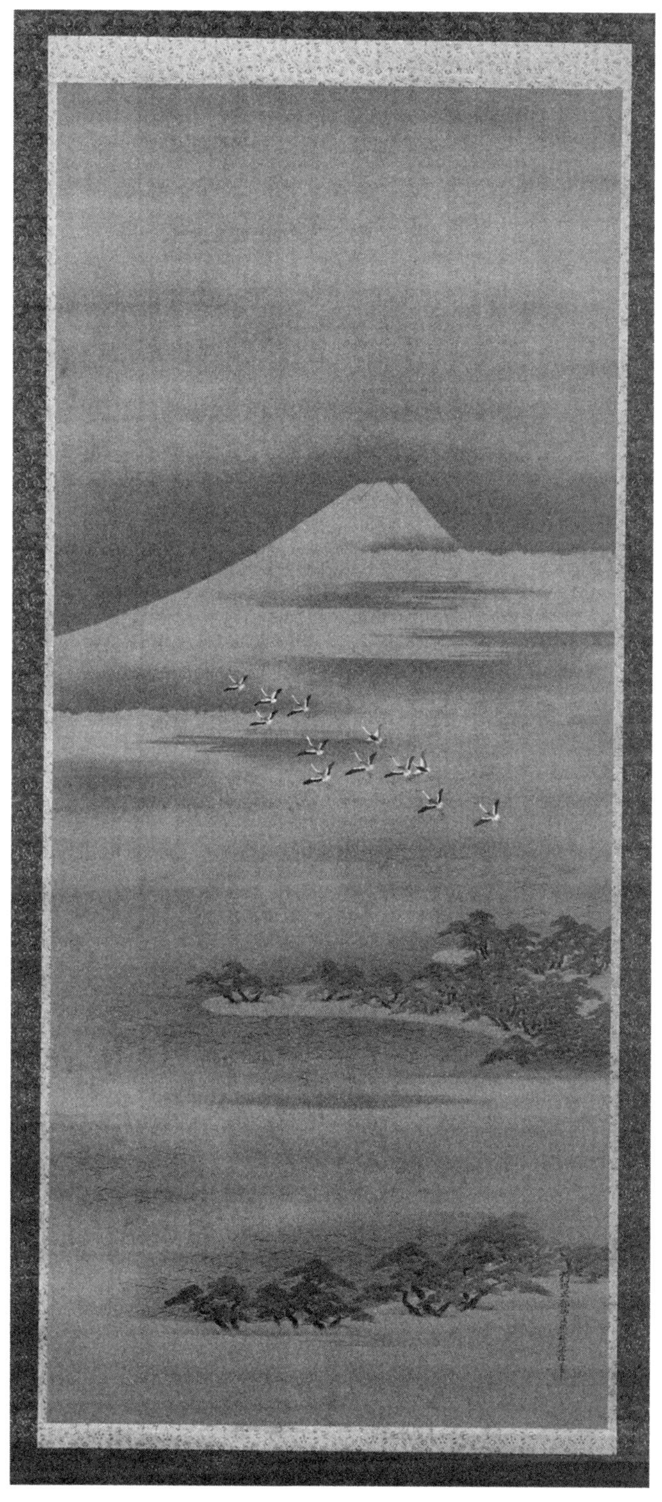

FIGURE 6.3. Kanō Tōsen. *Cranes Flying over Mount Fuji.* 1859.
Mount Fuji World Heritage Centre, Shizuoka.

The Meiji Restoration and the Separation of Kami and Buddhas

In the years leading up to its overthrow, the bakufu made efforts at reform, such as cutting the amount of time daimyo had to reside in Edo and conscripting commoners for military service,[72] but these did not change the underlying problem that it was unable to muster the resources it needed for a fully national response to the foreign peril. Increasing numbers of *shishi* saw the shogunate as inept and out of date; and in the 1860s, influential *shishi* determined that it had to be replaced if Japan were to meet the challenges of the day.

An alliance of *shishi* from Satsuma, Chōshū, and other rebellious domains brought the regime to its knees in 1867. That same year, fifteen-year-old Prince Mutsuhito ascended the throne as emperor. Acting at the direction of antishogunal leaders, in 1868 he declared a "restoration" (*ishin*) of imperial governance and the start of a new era called Meiji, "enlightened rule," that would last as long as he remained on the throne (until his death in 1912). The new emperor moved from Kyoto to Edo, which was promptly renamed Tokyo, meaning "eastern imperial capital." The emperor rarely exercised authority on his own initiative, however. He may have been sovereign and the descendent of the sun goddess Amaterasu, but it was a clique of samurai revolutionaries dominated by those from the Satsuma and Chōshū domains who devised and implemented the policies that transformed the realm.

In the years immediately following the fall of the shogunate, Meiji leaders pronounced a dizzying array of reforms, including the replacement of feudal domains with centrally administered prefectures, the abolition of samurai status, and the formation of a conscript army. Government officials and their allies in the private sector instituted measures designed to industrialize Japan as fast as possible. From cutting samurai topknots to disciplining soldiers and factory workers, the actions taken by political and economic elites—and, increasingly, the populace at large—shaped a nation that could not only defend itself against the imperial powers but also create an empire of its own.

Reforming Japan entailed learning from the very foreigners who threatened Japan. Although those critical of the shogunate's accommodation of the Western powers embraced the slogan "Revere the emperor, expel the barbarians," they realized they would not be able to guarantee Japan's independence without acquiring the tools that made the barbarians so powerful. In 1861 future Meiji oligarch Itō Hirobumi (1841–1909) participated in an act of arson against the British legation; but just a year later he decided to study in Britain and, upon his return, encouraged his compatriots to adopt the best that Britain and other foreign countries had to offer.[73] Then, remarkably for a government that had barely taken power, a significant number of Meiji officials decided in 1871 to travel through the United States and Europe under the leadership of courtier Iwakura Tomomi (1825–1883). The so-called Iwakura Mission failed in its goal of revising the unequal treaties, but its members learned a tremendous amount from the countries they visited.

Learning did not amount to naïve and obsequious mimicry. Far from it. Borrowing technologies, institutions, and ideas from the West was selective— and riddled with conflict, as disagreements erupted between different interest groups both in and out of government about what was worth importing from the West and what was better left alone. And while Meiji leaders and large numbers of ordinary Japanese wanted to make a clean break with the recent Tokugawa past, they drew inspiration from the much more distant—and therefore malleable—past of an ancient Japan in which the emperor presumably reigned supreme. The Meiji government initially adopted the governing organs of the imperial state as it had existed a thousand years earlier, resuscitating the long-moribund Council of State (Dajōkan) and Council of Divinities (Jingikan) in order to achieve the so-called "unity of rites and rule" (*saisei itchi*).

Meanwhile, radical nativists (many of them adherents of National Learning) saw an opportunity to make their vision of an unsullied and devout Japanese past into a contemporary reality. The imperial household was quickly purged of Buddhist rituals and paraphernalia, and the Council of State ordered the "separation of kami and buddhas" (*shinbutsu bunri*) nationwide. Although the order did not call for the abolition of Buddhism, nativists instigated a movement to "destroy the buddhas, eliminate Shākyamuni" (*haibutsu kishaku*). How thoroughly Buddhism was persecuted varied by region. In some places, like Kyoto, local authorities went out of their way to encourage the destruction of temple property and the forced laicization of Buddhist clerics. In others, the anti-Buddhist fervor was less dramatic or even absent. Across the entire country, however, the order to separate kami and buddhas forced ritualists to make an unprecedented choice between Shinto on the one hand and Buddhism on the other.[74] Practitioners of Shugendō, the form of mountain asceticism and worship that had long combined Buddhist-derived practices and beliefs with devotion to local kami, fell between the cracks. Banned outright by the new government, Shugendō would never fully recover. Practitioners who did not want to return to lay life had to affiliate with either Shinto or Buddhist institutions.[75]

The impact of the Shinto agenda on Fuji worship was swift and profound. Major Sengen shrines around Fuji and the mountain itself were purged of just about anything to do with Buddhism. In tangible terms, this meant the removal and/or destruction of Buddhist scriptures, icons, and even whole buildings. At Kamiyoshida's Sengen Shrine, a hall for Buddhist fire rituals (*goma*) managed by a nearby temple was destroyed, as were a Buddhist bell tower and the bell inside. Another casualty of nativist zeal was a gate containing a pair of Buddhist guardian kings (*niō*). In a dramatic and spiteful flourish, a group of unidentified townspeople (who likely included radicalized Shinto oshi) took a decapitated head from one of the kings and brought it to an oshi house where a pillar engraved with the title of the Lotus Sutra was located out front. They knocked over the pillar and placed the king's head on the pedestal where it had stood.[76]

Zealots destroyed Buddhist paraphernalia and structures on the mountain too. These included a commemorative hall and stone monument—this one, too, displaying the title of the Lotus Sutra—at the site just above the fifth station where the famous Buddhist monk Nichiren (1222–1282) is said to have deposited copies of the Lotus Sutra along with a prayer for the safety of Japan (then threatened by the Mongols).[77] The most prominent of the Shintoists intent on ridding Fuji of Buddhism was Shishino Nakaba (1844–1884). He was an adherent of a strain of National Learning started by Hirata Atsutane (1776–1843), a scholar who gained fame and followers for developing a full-fledged Shinto theology, including novel explanations of the afterlife, that was both national and cosmic in scope. After the Meiji government put Shishino in charge of the Sengen shrines in Ōmiya, Kamiyoshida, Kawaguchi, Murayama, and Subashiri in 1874, he led local officials up the mountain to remove and/or destroy the Buddhist icons that remained.[78] Some statues from the Yakushi Buddha (Medicine Buddha) hall on the summit were rescued and placed in an altar at the Fuji Takasago Sake Brewery in today's Fujinomiya City. There they continue to stand guard over vats of sake brewing beneath their feet (see fig. 6.4).

The appointment of Shishino as chief priest (*gūji*) of the Ōmiya Sengen Shrine, and as administrator for the four other Sengen shrines, is representative of the early Meiji state's efforts to bring ritual establishments (especially Shinto shrines) under its supervision. In 1870 the new regime expropriated shrine and temple lands that had been exempt from taxation during the Tokugawa period; and in 1874, as part of a nationwide land reform, it listed any ritual precincts (*keidai*) without clear records of private ownership as state land (*kan'yūchi*). To protect themselves from taxation—and, it seems, under pressure from local officials—even those with records of private ownership

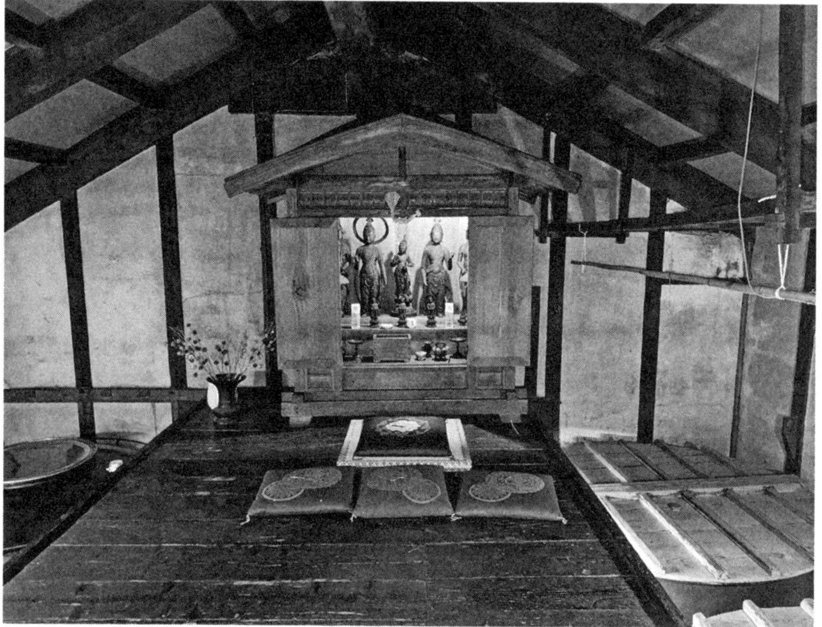

FIGURE 6.4. Altar with Buddhist statues rescued from Fuji's summit at Fuji Takasago Sake Brewery. Vats of sake are located underneath. Photo by Philip Ellway.

decided that their precincts should be put in the hands of the state. Most shrine and temple precincts, including those around Fuji, were registered as state land.[79] Meanwhile, the government disempowered Buddhist institutions by putting an end to the Tokugawa bakufu's policy of requiring all households to formally affiliate with a particular temple. The government also declared in 1872—the same year it announced that it would no longer enforce bans on women climbing sacred mountains like Fuji—that, as far as it was concerned, Buddhist monks and nuns were free to marry and eat meat.[80] In this and other ways, the government withdrew support from the Buddhist establishment while at the same time subordinating it to the institutions and rules of the new state.

Even though Buddhist temples were registered as state land, the government stayed out of their internal affairs. Not so with Shinto shrines. In the years following the Meiji Restoration, nativist officials worked to put these now "purely" Shinto institutions in active service to the state, organizing them into different ranks according to ancient models. The Ministry of Divinities (Jingishō), established in 1871 to replace the Council of Divinities, also worked to standardize Shinto rites throughout Japan. Officials in Tokyo took particular interest in major Shinto shrines, putting men like Shishino in positions of authority at the expense of existing priestly lineages.[81]

Shishino used his position not only to rid Fuji of anything to do with Buddhism but also to organize and reform Fujikō along Shinto lines. At the urging of a group of oshi from Kamiyoshida, he established an umbrella group for Fujikō that would eventually become the religious organization Fusōkyō (*Fusō* stands for Fuji and *kyō* means "teaching").[82] According to the group's doctrine, the Fuji deity Konohanasakuya-hime was to be worshipped together with three older, "creator" kami who occupied a central position in the pantheon of deities embraced by Hirata school Shintoists. Teachings of Kakugyō and Jikigyō that accorded with this vision were incorporated into Shishino's doctrine while others were discarded.[83]

Instrumental to Shishino's efforts was a charismatic Fuji devotee, Itō Rokurōbei (1829–1894). In 1870 Itō was possessed by the Sengen deity and later became head of a Fujikō known as Maruyamakō (meaning "Circular Mountain" *kō*). Renowned for his austerities and healing powers, he was arrested in 1873 and again in 1874 by the police at a time when the Meiji state was cracking down on what it considered superstitious and disorderly practices.[84] Shishino, a creature of the Meiji establishment, and Itō, a man hounded by, and highly critical of, that same establishment, would seem to be odd bedfellows. Yet each had something to offer the other: Shishino could protect Itō from the scrutiny of the state while Itō could draw thousands of believers into the fold. They also shared the nativist aim of purging Buddhist beliefs and practices from Fujikō, although, despite their efforts, there were groups belonging to Fusōkyō that continued to hold onto them.[85]

After Shishino's death in 1885, Itō and his followers broke with Fusōkyō and registered with the Home Ministry's Bureau of Shinto Affairs as an independent religious organization known as Maruyamakyō. Despite winning official recognition by the state, Itō continued to be critical of the Meiji regime and particularly its campaign to remake Japan along Western lines. He disapproved not only of the emperor's government but the emperor himself, writing that he "is to worship the sun and moon above, and to preserve the country below. He has forgotten this."[86] Yet after Itō died in 1894, Maruyamakyō lost both its vitality and critical edge, turning into a largely docile shadow of its former self. The organization survived, but it is now a minor religion with a nominal membership of about ten thousand.[87]

The Fujidō network of worshippers followed a similar trajectory. As discussed in chapter 4, a young Fujidō member boldly submitted a petition to the shogun's government in 1847 asking for official recognition of the movement's teachings, including those concerning the Age of Miroku that had supposedly begun in 1688 with a cosmic upheaval (*furikawari*). Wary of any doctrine and/or behavior that posed an alternative to the status quo, officials responded not with recognition but repression. The attempt to stifle Fujidō had little practical effect, however, as it continued to grow in popularity over the next several decades. Then came the Meiji Restoration and with it the hope that direct imperial rule would usher in the conditions necessary for the utopian fulfillment of the Age of Miroku.[88] Although a number of Fujidō members were quickly disillusioned, others sought to align the movement's teachings, which already had a nativist tinge, with the goals of the imperial state.[89] Prominent among them was Shibata Hanamori (1809–1890). Like Shishino, he was an adherent of Hirata school National Learning, and he accordingly worked to Shintoize the Fujidō movement. Whereas "Jikigyō Miroku and his successors assumed a religious supreme being that transcends the emperor's and Amaterasu-ōmikami's authority, namely the original father and mother," for Shibata, as Miyazaki Fumiko notes, "the authority of the emperor was viewed as absolute."[90] Shibata acquired growing influence in the Fujidō organization, which in the 1870s was renamed Jikkōsha (True Practice Association) and then in 1882, under Shibata's leadership, Jikkōkyō (True Practice Teaching). Today, like Maruyamakyō, it is a minor religion in Japan.

The nativist officials who tried to win ordinary Japanese to their Shinto cause during the first years of the Meiji regime were soon sidelined by pragmatists looking not to Shintoize Japan but to institute economic and political changes that would enable it to compete with the Western powers. In the early 1870s the government instituted a Great Promulgation Campaign with a Great Doctrinal Academy (Daikyōin) that was dominated by radical Shintoists but also employed Buddhist priests and even entertainers as doctrinal instructors (*kyōdōshoku*). The doctrine they were supposed to preach comprised three main tenets: "(1) respect for the gods, love of country; (2) making clear the principles of Heaven and the Way of Man; and (3) reverence for the emperor and obedience to the will of the court."[91] Yet instructors—who included Itō and Shibata—were also given manuals telling them how to address much more specific and seemingly nonreligious topics like "taxation, conscription, compulsory education, and the solar calendar."[92]

Broad injunctions to love Japan and revere the emperor could be embraced by a wide swath of Japanese, yet the campaign ran into trouble early on as Western-derived innovations like the solar calendar drew suspicion and bureaucrats withheld funding.[93] A major blow occurred in 1875, when Buddhist sects withdrew their support for the campaign. That was also the year that, in the face of strong popular resistance, the government was forced to repeal its ill-conceived ban on the Buddhist practice of cremation, which had been condemned as both unfilial and unhygienic but proved impossible to extinguish.[94] Meanwhile, Shintoists engaged in theological arguments and battles over ritual ground that undermined the state's goals. Deciding that involvement in matters to do with "religion" was more trouble than it was worth, the government ordered in 1882 that priests of the higher-ranked shrines extricate themselves from the Great Promulgation Campaign and refrain from "religious" practices like officiating at funerals.[95] Finally, in 1884, the state pulled the plug on the campaign altogether.[96]

To the disappointment of nativists, the effort to turn Japan into an exclusively Shinto nation failed, but it changed Fuji worship in two significant and long-lasting ways. First, it instigated the destruction of Buddhist structures and other items on and around Fuji. Second, it accelerated the Shintoization of Fujikō through umbrella organizations like Fusōkyō and Jikkōkyō. Today the main Sengen shrines are dedicated to the worship of Konohanasakuyahime, not Great Bodhisattva Sengen; and the summit, under the control of Fujisan Hongū Sengen Taisha (in Fujinomiya), remains a thoroughly Shintoized space. Along the main road leading to Kitaguchi Hongū Fuji Sengen Jinja (the Sengen Shrine in Fujiyoshida), visitors pass by the foundation stones where the Guardian King gate once stood (see fig. 6.5). In its place is a sign explaining why it is no longer there. Few stop to read it.

JAPAN EXPERIENCED tremendous upheaval in the final years of the Tokugawa period and those that immediately followed the Meiji Restoration. Facing the greatest foreign threat to Japan since the Mongol attacks of the thirteenth century, the shogunate tried to preserve Japan's independence while avoiding war with the Western powers. The policy of accommodation was viewed as a sign of weakness by hard-liners determined to keep the insolent barbarians at bay. Acting in the name of the emperor, revolutionary samurai overthrew the shogunate, installing a regime that appealed to an ancient order as it reformed Japan to meet the challenges of the future. This gave nativists an opening to fulfill their long-held dream of reviving a seamless union of emperor, kami worship, and the Japanese people that had supposedly prevailed in the distant past. To some extent their campaign to Shintoize Japan succeeded. Immediately after the Restoration, the new government ordered the separation of

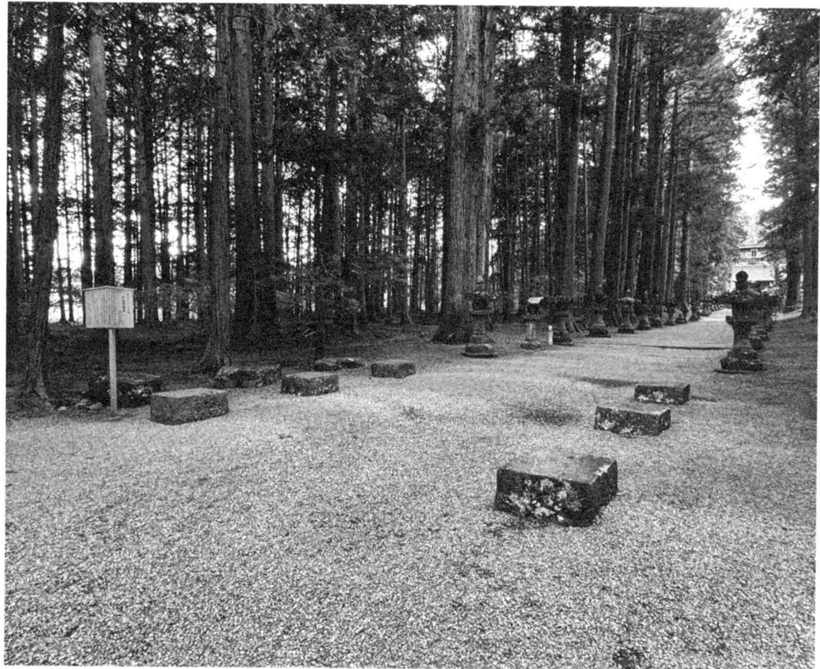

FIGURE 6.5. Foundation stones where Buddhist Guardian King gate once stood on the approach to Kitaguchi Hongū Fuji Sengen Jinja in Fujiyoshida. Photo by author.

kami and buddhas, which at times translated into outright persecution of Buddhism. A major pilgrimage site that also symbolized Japan, Fuji made an especially attractive target for Shintoization.

Yet the nativist vision of expelling foreigners from Japan never came to pass. Bent on modernizing Japan as quickly as possible, the early Meiji government instead welcomed large numbers of Westerners into the country, many of whom, like their predecessors in the years leading up to the Restoration, made a point of climbing world-famous Fuji. Going forward, Japanese would treat the beloved mountain less as the home of a specific Shinto deity than as a unifying, yet at times divisive, national symbol—one that, like the imperial lineage and the nation itself, transcended the boundaries of any one creed.

7

A Nation and Mountain in Flux

FOR THE REMAINDER of the nineteenth century and up to the end of World War II, Shinto shrines at Fuji served the government's goal of inculcating Japanese in "state teachings" (*kokkyō*) that centered on loyalty to the emperor. In this respect, Fuji and its shrines were representative rather than unique, since the propagation of state teachings occurred at ritual sites throughout Japan, including those on and around other sacred mountains.

But Fuji was not just any place. Embodying the essence of Japan in the eyes of both Japanese and foreigners, it was a potent force for national cohesion that far exceeded the limits of state-managed Shinto. A central feature of the nation-state is its claim to a defined territory with a shared past. Fuji loomed large in both. The nation-state develops institutions and technologies to meddle in many aspects of people's daily lives, such as the education of children. Through literature, art, clothing, and numerous consumer items, Fuji had already entered into the intimate corners of those lives (see fig. 7.1). People could also come into physical contact with the mountain by traveling to it. The same could not be said for the emperor, whose body could scarcely be seen, much less touched. And while one might entertain doubts about the sanctity and continuity of the imperial lineage, there was no denying the material existence of Fuji from the misty past to the present.

Fuji was not only a unifying symbol of the nation but also a high-profile stage where changing relationships among nature, religion, science, gender, empire, and the state were put on display. These relationships were often fraught with contradictions and strife. Even as Fuji represented continuity and consensus, it exhibited the tensions, upheavals, and conflicts of a rapidly changing world. Consider the attempt by the husband-and-wife team Nonaka Itaru (1867–1955) and Chiyoko (1871–1923) to spend the winter of 1895–1896 on Fuji's summit to study its weather. Members of the educated elite,

162

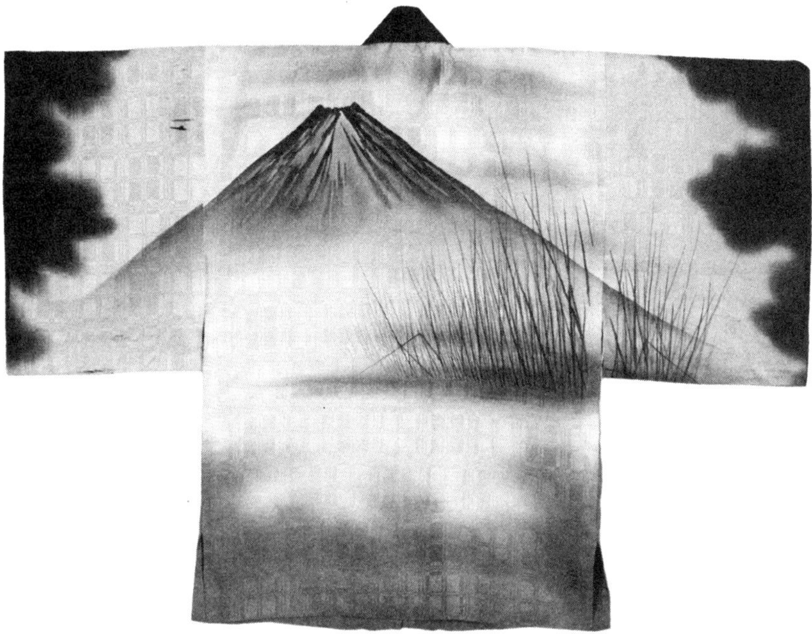

FIGURE 7.1. Hand-painted kimono. Meiji period. Andrea Aranow Textile Design Collection. Textile Hive.

they viewed their work as a contribution to global meteorology and to the nation, aligning the goals of their project with those of a state that wanted to make Japan both a scientific and imperial power. Yet Chiyoko's involvement defied the expectations of womanhood promoted by that same state—even if she cast it as a way to support her husband—making her akin to Tatsu, the female Fujikō member who had defied the ban on women climbing to the summit sixty-three years earlier.

Because it was as tangible as it was impressive, Fuji also became a focus for those in and out of the imperial government who wanted to foster patriotism among the subjects of the newly imperial nation. The mountain appeared prominently in schoolbooks intended to educate children to be industrious citizens and loyal subjects and was especially useful to late-Meiji ideologues who claimed that the Japanese people were distinguished by a love of nature, both deep seated and unique.

Meanwhile, the mountain attracted a growing number of climbers unattached to Fujikō teachings and practices. Teachers and military commanders organized climbs for the children and soldiers in their charge. Climbs were also arranged by newspapers and hiking clubs committed to the modern idea that healthy citizens were key to a healthy nation. By the end of the Meiji period (1868–1912), worldly hikers far outnumbered Fujikō pilgrims, and upon reaching the summit they encountered not only shrines and torii (gates signifying sacred space) but also a post office, a new feature of the modern state. Both foreigners and Japanese commented on the mix of old and new, with some bemoaning the new amenities and infrastructure that were transforming the landscape on and around the mountain.

Before and during World War II, Fuji symbolically anchored an empire whose boundaries shifted over time. Thongchai Winichakul coined the term "geo-body" to describe the territoriality of a nation expressed through maps, which teach people to accept constructed national boundaries as perfectly natural.[1] Fuji might also be described as a "geo-body"—in this case "geo-" referring to "geological" rather than "geographical"—but one of an axial kind. Even as the extent of Japanese territory fluctuated wildly during the twentieth century, Fuji remained a fixed point around which the empire grew—and then collapsed.

After Japan's defeat in 1945, Fuji's role as a symbol of Japan—at this point just a nation, no longer an empire—remained as strong as ever, but this did not mean everyone in postwar Japan viewed Fuji the same way. In 1951 Kim Tal-su (1919–1997), a Korean who had emigrated to Japan at the age of ten, published a short story set in a village at the base of Fuji to protest discrimination against people of Korean descent.[2] Around the same time, a decades-long political and legal battle broke out over who should control Fuji's summit, the Shinto shrine Fujisan Hongū Sengen Taisha (the Ōmiya Sengen Shrine) or the state. The fight had roots in local competition dating back to the Tokugawa period, but it became a national flashpoint for debates over the role of religion, particularly Shinto, in postwar Japan. It was precisely its status as totem, or "geo-body," of the nation that made Fuji a site of intense conflict over what that nation had been and should be.

Weathering Fuji

On October 1, 1895, Nonaka Itaru began recording meteorological phenomena on Fuji's summit with the ambitious goal—thanks to an alarm clock—of taking measurements every two hours, day and night, for an entire year. In an age when the Japanese press eagerly celebrated tales of scientific discovery in dangerous, exotic places, his project offered something unique: a bracingly raw view of the most familiar of mountains. Fuji, after all, was anything but distant.

Easily seen from the streets of the capital and climbed each summer by thousands, it was a national icon. Yet no one had endured the long, harsh winter on Japan's highest summit. Beginning in autumn, heavy snows covered the traces of human presence on the upper reaches of the volcano and effectively turned this well-trafficked peak into a terra incognita. Nonaka's mission captured the imagination of a public eager to see a Fuji not seen before.[3]

The attempt to take weather readings on snowbound Fuji exemplified the growing Japanese participation in scientific pursuits with global reach. In the Tokugawa period there developed a thriving tradition of empirical inquiry along with a professional class of naturalists who took part in a range of social practices that, as Federico Marcon describes it, "secularized nature by transforming what was once the enchanted realm of unfathomable divine forces and metaphysical principles into a multiplicity of 'objects' that could be grasped and manipulated through protocols of observational, descriptive, representational, and reproductive techniques."[4] It is thanks to the spread of such objectifying techniques that we have the wealth of data about the Hōei eruption in 1707. In the years immediately leading up to and following the Meiji Restoration, Japanese naturalists increasingly engaged with Western ideas, institutions, people, and practices, situating (and transforming) their work in a globalizing context.[5] For example, seismology developed in Japan as a collaboration between foreigners and Japanese counterparts, quickly making the earthquake-prone nation into a leader of the field. Japan's prominence in seismology was encouraged by the Ministry of Education, which, in 1892, created an Imperial Earthquake Investigation Committee (Shinsai Yobō Chōsakai) that published summaries of its members' findings in English (and initially in French too). Among the committee's projects was a volcanological survey of Japan, which included a volume on Fuji, at this point understood as part of a larger complex of volcanoes extending north to Fuji from the Ogasawara Islands about one thousand kilometers south of Tokyo.[6]

In addition to geologists, scientists in fields ranging from biology to meteorology turned their attention to Fuji. As they went about their specific forms of investigation, they reinforced a broader division between humans (or culture) on the one hand and nonhumans (or nature) on the other. This split sundered the Fujikō vision of a cosmic bodyscape in which Fuji's body was integral to those of all creatures, especially humans. At the same time, the bodies of these scientists came into direct contact with the mountain's, whether through the tephra and changing weather patterns on its summit or through the vegetation on its slopes. This bodily contact is nowhere clearer than in the experience of Nonaka Itaru and his wife Chiyoko, who, without Itaru's approval, joined him a couple of weeks into what was originally a solo mission.

The descendant of a samurai lineage and son of a judge in Tokyo, Itaru had left the security of medical school in 1889 (to the chagrin of his father) to study high-altitude meteorology, and in 1894 he approached the Central Meteorological Station technocrat Wada Yūji with his plan to spend the winter on Fuji's peak. By proposing this dangerous (many would say foolhardy) plan, Itaru clearly aimed both to contribute to the field of meteorology and to participate in a global mountaineering tradition populated by gentleman-adventurers. Wada acceded to Itaru's request, providing him with the necessary instruments and training for his mission.[7]

Over the previous decades, first in Europe and increasingly elsewhere, meteorology had evolved from an amateur endeavor into a systematic science, such that "the meaning of a locality changed from its status as an exclusive end of investigation to a specimen in a larger entity, a point on the grid," as Vladimir Janković puts it.[8] By the time modern meteorology took hold in Japan, it had developed standard techniques—such as the creation of weather charts and maps—for comparing local phenomena on a grid that was national and increasingly global in scope. The government consolidated the application of these techniques in the Central Meteorological Station, established in Tokyo in 1875 under the auspices of the Home Ministry and linked to a network of local observatories by telegraph in the 1880s, making the practice of meteorology an instrumental part of the Meiji regime's growing technocracy.[9]

Unlike the laboratory-based sciences, meteorology was a body of knowledge and practice produced largely "*inside* its own subject matter."[10] It shared with other field sciences a lack of control far greater and a boundary between subject and object far more permeable than that experienced in the confines of the laboratory. This made it susceptible to forces that complicated the formation of a tidy body of knowledge all its own. In fact, alpine meteorology's unconfined nature is what made it attractive to adventure-seeking Itaru—and, as we will see, unexpectedly accessible to his wife Chiyoko—in ways that a bureaucratized lab science never could.

The physical demands of his mission made Itaru dependent on local resources and expertise. In preparing for his mission, Itaru befriended Satō Yōheiji, owner of the Tarōbō inn on Fuji's lower skirt. Satō shared his considerable experience with all things Fuji, including quickly changing weather patterns, the construction of stone shelters, and the use of snowshoes. This is a reminder that the exploits of cosmopolitan adventurers are only made possible by the often-unsung locals who assist them.[11]

What Itaru brought to the table was a scientific outlook that made Fuji into a specimen for high-altitude meteorology. We see this outlook in action during the two exploratory climbs he made in January and February of 1895.[12] Itaru's accounts of both ascents begin by noting the exact times of his departure from Tokyo and his arrival in Gotenba (see map 8.1 for the train route), thus locating his climbs in the methodical framework of railroad timetables. This modern division of time, along with a topographical map, accompanied Itaru up the mountain. Tellingly, Itaru chose not to apply the metric scale to gauge his upward progress but instead used the well-established convention of referring to Fuji's stations. This enabled him to colonize the traditional system for signifying elevations on Fuji by indexing it to the centigrade scale devised by Anders Celsius (1701–1744) a century and a half before.

Upon his arrival at each station, Itaru used a watch and thermometer to record the time and temperature, creating a quantified world in which his body not only moved but *felt*. Here is a description of the third station on his February climb:

> I arrived at the third station at 8:55 a.m., and when I looked back, it appeared that rain was trailing from a band of clouds in the direction of the Tarōbō. Gusts of wind from the summit drove fine particles of snow down the snowpack from the station above, making it seem just as if it were really snowing. Not only was it difficult to breathe, it was impossible to look forward, and I wondered if the snowy wind hitting my face would pierce right through me. Here the temperature was 1.7 degrees [Celsius].[13]

Itaru provides no cultural markers to orient the reader in this passage. Gone are the summertime slopes of Fuji, dotted with shelters and the climbers who used them. In their place, we encounter a bodily experience available to anyone in the right meteorological conditions, bracketed by clockwork on one end and the Celsius scale on the other. Itaru's text links the most immediate and intimate of physical sensations to the most abstract and universal of measurements, advancing an extensive—and highly *intensive*—project of bodily enlightenment through the practices of meteorological science.

Yet to cast Itaru as an emissary of science who appropriated local knowledge for cosmopolitan ends captures only one part of the story, which included his wife Chiyoko. To Itaru's surprise, Chiyoko climbed to the summit on October 12, joining him just eleven days into his machinelike regime of taking weather readings every two hours, day and night.[14] She had already assisted Itaru by coordinating the purchase and transport of supplies for the observatory and living quarters (primitive wooden structures connected by a passageway and covered with stones) that laborers were hired to build on the peak, but unbeknownst to him, she had surreptitiously stockpiled provisions for two people instead of one. In her serialized diary *Fuyō nikki* (*Fuyō*, meaning "lotus blossom," is an epithet for Fuji, while *nikki* means "diary"), published in installments from January 7 to February 1, 1896, in the newspaper *Hōchi shinbun*, Chiyoko explains that she worried Itaru could not possibly keep up such a demanding effort on his own, adding, "Even with my useless body, I thought I should accompany him so I could at least prepare his meals."[15] Her diary also includes a poem in which she laments her "useless" female body—only to reveal that it houses a powerful spirit:

My body may not be a rugged man's,
but my striving spirit
is second to none.[16]

In making this claim, she resembled the recently deceased Matsuo Taseko (1811–1894), a headstrong nativist poet who lamented her "weak woman's body" even as she actively participated in efforts to bring down the shogunate in the name of the emperor.[17]

Chiyoko refused Itaru's entreaties for her to go back down the mountain, strategically assuming the role of dutiful wife to assert her own will. She also put this strategy to work in a letter that she wrote to Wada Yūji before she made her unapproved climb, virtuously announcing, "To protect one's household is the way of a woman." Toward the end of her letter she informs Wada that she is sending the membership fee to join the all-male Meteorological Association (Kishō Gakkai) in the hope that a woman might be admitted.[18]

Turning a tale of heroic masculinity into one of heroic domesticity, Chiyoko took charge of household chores in extremely challenging circumstances. By the end of October the observatory was encased in so much ice and snow that, when two members of Gunji Shigetada's expedition to the Kuril Islands in 1893 and 1894 paid the couple a surprise visit, they had to enter the snowbound shelter rear end first through a small window.[19] According to *Fuyō nikki*, it was through this same window that Chiyoko and Itaru exited on the emperor's birthday on November 3, braving the bitter cold so they could prostrate themselves in the ice and snow in the direction of the imperial palace in Tokyo.[20]

Over the coming weeks, the thin air, poor hygiene, inadequate nutrition, lack of exercise, and perilously low temperatures (any liquid not within a foot of the stove immediately froze) took a terrible toll on Chiyoko and Itaru. As Chiyoko put it, "Working in a dark cave like small bugs and lacking adequate food, our bodies and minds grew feeble."[21] Both suffered severe headaches, loss of appetite, and edema; and by early December, Itaru—feverish and hardly able to move—was in especially bad shape.[22]

Chiyoko held up better than Itaru, so it fell to her to take over the task of recording weather data, according to the newspaper *Yomiuri shinbun*.[23] In fact, interviews conducted decades later with the Nonakas' son and others who knew them well revealed that as early as October, Chiyoko had begun calibrating instruments and taking readings.[24] The Nonakas' own accounts skirt this sensitive subject. Both merely note that the meteorological work continued uninterrupted during Itaru's decline. Yet Itaru's words indicate that he regarded his wife more as teammate than helpmate. An example is his description of the edema that afflicted Chiyoko. Using terminology usually associated with the battlefield, he notes in his account that "this illness was the root cause of my comrade's defeat (*mikata haiboku*)."[25] Apparently Itaru viewed his wife as a fellow soldier in a war against a common foe. In fact, the memoirs of both Nonakas make clear that bodily suffering, combined with a determination to endure their battle with the elements, was an inescapably shared experience that diminished, if not entirely eliminated, gender distinctions.

By the time another intrepid pair (two local men delivering greetings from family and friends) visited the couple on December 12, Itaru and Chiyoko looked so wretched that the visitors broke into tears at the sight of them. Itaru stoically urged them not to reveal their plight, yet upon their return to Gotenba, the men leaked the bad news, and a rescue effort quickly ensued. Led by Wada, a team of porters climbed to the summit. At first Itaru and Chiyoko refused to budge, but Wada said he had orders to retrieve them, so they finally acquiesced to being carried down the mountain.[26] When they arrived at Satō's inn on the evening of December 23, the weary party encountered a throng of family, friends, journalists, and curiosity seekers eager to greet the newly famous couple and their rescuers.[27]

While Jikigyō had sacrificed his body to save humanity, Itaru and Chiyoko were ready to sacrifice theirs for the advancement of science and the nation. Although the Nonakas did not fulfill their mission of taking weather readings on Fuji throughout the winter, they contributed to scientific progress through their suffering bodies. No sooner had they arrived at the inn than a Tokyo University pathologist raced to Gotenba to give the couple a thorough medical exam. The Nonakas proved to be willing patients—one, for the sake of their immediate health, and two, to transform their bodily failure into potentially useful data for the field of high-altitude medicine. In his report to the Meteorological Association (published in the association's journal), Itaru offered the results of the medical exam in full, down to such details as the quantity and quality of his urine.[28]

He also catalogued numerous mistakes concerning nutrition, hygiene, heating, and so on, and suggested remedies ranging from the provision of lemon water to the building of a corridor for regular exercise. Simple measures like these would ensure the success of future research on Fuji, an optimistic outlook reinforced by the observation that "even the weak body of a woman can handle it."[29]

Their ambitious effort and dramatic rescue made the Nonakas instant celebrities. They were even the subject of a play performed in Tokyo in February 1896.[30] But how their story was told differed significantly depending on who did the telling, both right after their rescue and decades later. Wada, for instance, wrote an article appearing in the January 5, 1896, issue of *Taiyō* magazine that barely mentions Chiyoko, noting only that she was a virtuous woman "standing at her husband's side."[31] Chiyoko was not content to remain quiet. Just two days after the *Taiyō* article appeared, she began publishing her serialized diary in the *Hōchi shinbun*. In it she challenges Wada by including her bold letter to him; implies, if never outright declares, that she engaged in scientific work; and puts Wada on the spot by trumpeting his promise to Itaru to build an observatory on Fuji's peak (a promise he had made to convince Itaru to abandon his mission).[32]

While Wada's account is matter of fact and scientific in tone, Chiyoko's is rich in metaphors, emotions, and classical allusions, situating her in a time-honored literary tradition shaped to a great degree by women, including, for example, the famous Murasaki Shikibu, author of *The Tale of Genji*.[33] Like other members of this tradition, Chiyoko included numerous poems in her work. These were composed by her and her husband as they suffered on Fuji, and although some express heroic determination, others reveal despondency and self-doubt. Here is one written by Itaru:

> A half shell split from the world,
> I wonder, after a hundred years,
> will someone pick me up?[34]

Through poems like this one we see Chiyoko and Itaru not simply as data-collecting machines but as flesh-and-blood humans, making their heroic attempt to spend the winter on Fuji all the more impressive.

Chiyoko's celebration of poetry in such daunting circumstances inspired Ochiai Naobumi, an accomplished poet and "national literature" (*kokubungaku*) activist. Drawing mainly on Chiyoko's diary but including excerpts from Itaru's publications, he issued a book-length account of the Nonakas' experience in the fall of 1896 called *Snows of the Lofty Peak* (*Takane no yuki*; see fig. 7.2). The

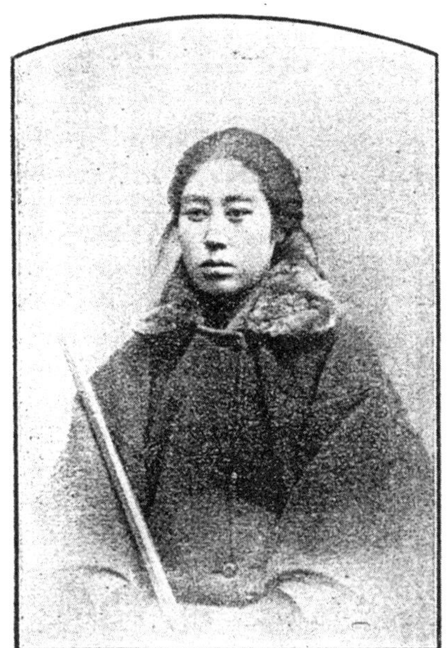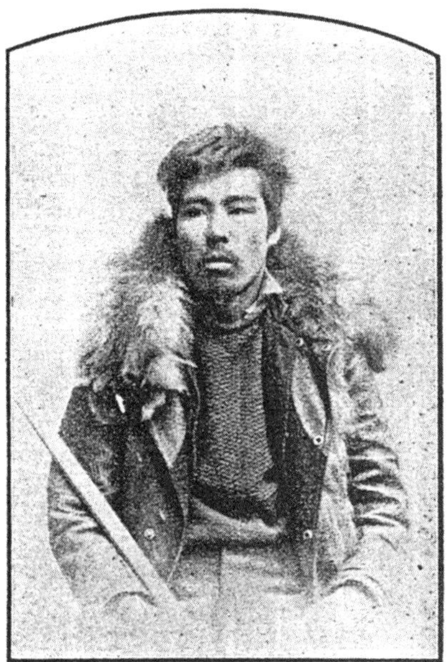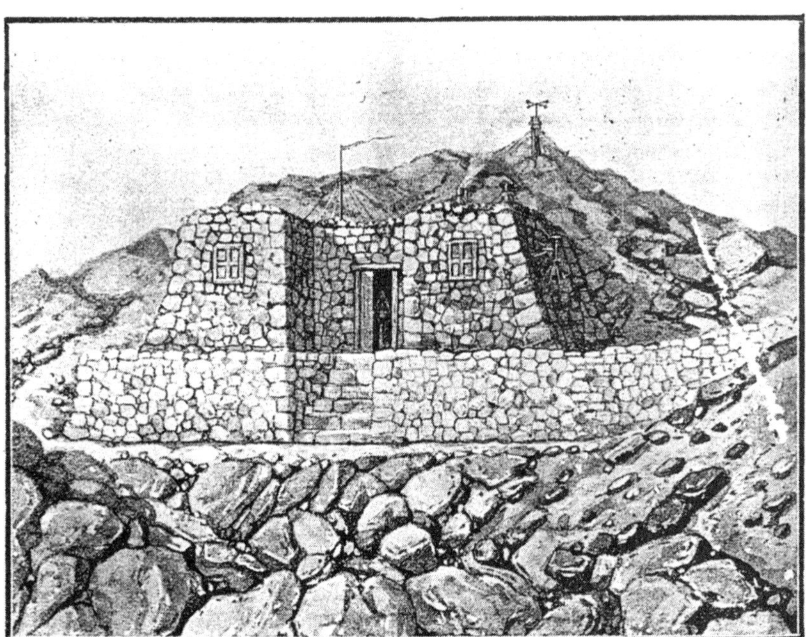

FIGURE 7.2. Nonaka Chiyoko and Itaru, and the Fuji Weather Observatory. In Ochiai Naobumi, *Snows of the Lofty Peak* (*Takane no yuki*). 1896.

Nonakas' story was an ideal choice for Ochiai, who used it to promote interest in Japanese literature while providing his nascent publishing house with a book that would sell.[35]

The Nonakas' story continued to be told orally and in writing as the decades passed. Frederick Starr (an American anthropologist with a special interest in Fuji worship) included an account of the Nonakas in his 1924 book *Fujiyama: The Sacred Mountain of Japan*, the first comprehensive book on Fuji in English.[36] And in 1948 Hashimoto Eikichi published *The Summit of Mount Fuji* (*Fuji sanchō*), a novelistic version of the Nonakas' tale clearly meant to inspire a defeated people by celebrating a Japanese man and woman who had endured the unendurable.[37] Tales about the couple also continued to circulate among meteorologists who worked at the permanent weather station that was established on Fuji's summit in 1932 (and that Itaru visited along with his youngest daughter when it opened).[38] Prominent among these was Fujiwara Hiroto, who wrote historical novels—including one about Jikigyō's dramatic death on Fuji—under the pen name Nitta Jirō (a reference to Nitta Shirō, the retainer of shogun Minamoto no Yoritomo who, according to legend, encountered the fierce Fuji deity in the Hitoana Cave). Nitta thought Hashimoto failed to give Chiyoko enough credit, so he published his Chiyoko-centered *Person of the Lotus Blossom* (*Fuyō no hito*) in 1971, making sure to emphasize her role in taking weather readings early into their stay on the summit. Many years later, in 2006, Chiyoko and Itaru's accounts of their experience were republished in a single volume aimed at general readers.[39] The Meteorological Agency closed its Fuji observatory in 2004 because most of the data it had been collecting could be gleaned by satellites (although researchers in a number of different fields have continued to use the facilities during the summer months). In the closure's aftermath, the Nonakas' story was timely once again.[40]

What has made their tale so compelling to different generations of readers is its multifaceted nature. One can paint Itaru and Chiyoko as members of Japan's cosmopolitan elite who treated Fuji as a site for the advancement of high-altitude meteorology and medicine. As shown by their choice to leave the confines of their shelter and prostrate themselves in the direction of the imperial palace on the emperor's birthday, they also ardently embraced their role as loyal subjects. Yet they cannot be reduced to emperor-revering drones acting in the service of science and the state. Chiyoko's decision to join Itaru on the summit posed a stark challenge to gender norms, and even as they worked with the most up-to-date scientific equipment, both made sense of their experience by writing poetry in a literary tradition stretching back over a thousand years. Their poetry made them all the more human, and thus their suffering and fortitude all the more impressive, in the eyes of the public.

Taking into account the many-sided quality of the Nonakas' story illustrates that the conceptual boundaries drawn between subject and object, humans and nonhumans, nature and culture, exist more in the mind than in the way the world actually operates and is experienced. Like all humans, scientists are embodied, making them full and often unwitting participants in the world they study. Too often people lose sight of the fact that the dichotomy between humans and nature is conceptual rather than absolute, making it easier to treat nonhumans, including Fuji, as mere tools.

Splendid Creation of Nature

A year before the Nonakas tried to spend the winter on Fuji, a book called *On the Japanese Landscape* (*Nihon fūkeiron*) hit the shelves amid the nationalistic fervor of the First Sino-Japanese War (1894–1895). Its author, Shiga Shigetaka (1863–1927), argued that Japan's natural beauty was a defining feature of the nation and its people. Comparing the Japanese landscape with those of other countries, Shiga aimed not only to establish Japan's uniqueness but also, in Richard Okada's words, "to demonstrate unimpeachably, through scientific terminology and poetic passages, the superiority of the Japanese landscape and, in the process, to solidify the image of Japan as an autonomous entity with a clear internal identity and image that [could] serve to unite its citizens in the face of possible armed foreign intervention."[41] Shiga's work thereby contributed to an ideological process by which, as Julia Adeney Thomas puts it, "nature had begun to acquire the reflective surface that would make it a narcissistic mirror for the nation."[42]

What made Japan particularly admirable (and enviable), according to Shiga, were its volcanoes, Fuji foremost among them.[43] Continuing the venerable tradition of emphasizing Fuji's celebrity overseas, he cited evidence of foreign admiration that dated back centuries. He also shared this quote from his contemporary John Milne (1850–1913), an English geologist in the employ of the Meiji government who played a key role in the invention of the modern seismograph and the development of the field of seismology: "Not only do we find a vast number of native books describing this mountain, but every book treating of Japan which has been published in foreign countries, always finds occasion to mention the 'peerless Fuji.' In consequence of its height, the symmetrical curvature of its slopes, and its solitary grandeur, Fuji has become one of the most famous mountains of the world."[44]

Shiga attributed the beauty of Fuji's shape to the "logarithmic curve" that extends from the base to the summit.[45] In doing so he followed the example set by Milne just after a great earthquake had rocked the Nōbi Plain (where the city of Nagoya is located) in 1891. Speaking before the Seismological Society of Japan, Milne credited the relative stability of the castles in Nagoya and Ōgaki to their "Fuji-san curve." In a second lecture he returned to this theme, waxing lyrical that "every engineer who passed it [Fuji] ought to lift up his hands and thank nature for that beautiful monument of mathematics crystallized before him. It was perfect in its curves; it was one of those figures that could be discussed by x's and y's."[46] While Jikigyō Miroku and his successors taught that Fuji was the Original Father and Mother of the cosmos, Milne saw it as a mathematical source of inspiration for engineers looking to save buildings from collapse. To Shiga, the mountain's mathematical, and therefore absolute, perfection made it an unparalleled instance of the unique beauty that was intrinsic to the Japanese nation—setting aside the fact that Fuji, like the nation, was far less stable than it might appear.

Although Shiga displayed no personal inclination to climb mountains, he encouraged his compatriots to experience Japan's rugged terrain firsthand. Joining him in this effort were the members of the Japan Alpine Club (Nihon Sangakukai), founded in 1905 thanks largely to the efforts of Kojima Usui (1873–1948), a banker who energetically promoted mountain climbing in numerous articles and books. The group consisted for the most part of "doctors, lawyers, bankers, and students from the most selective imperial universities, along with a sprinkling of movie stars and imperial princes," but it played an outsized role in the development of a "modern alpine complex" for imperial Japan.[47]

Foreigners were closely involved in the creation of this complex, which enabled adventurers to discover Japan in a whole new way. Prominent among them was Englishman Walter Weston (1861–1940), a Christian missionary and avid mountaineer who helped Kojima start the Japan Alpine Club. Weston was the celebrated author of *Mountaineering and Exploration in the Japanese Alps*, published in 1896.[48] Referring to the mountain ranges that run down the spine of central and northern Honshu, the term "Japanese Alps" (*Nihon Arupusu* in Japanese) had been coined by William Gowland (1842–1922), an English mining engineer hired by the Meiji government, but Weston played a key role in popularizing the term both in Japan and abroad.

People living among these mountains had long made use of their resources, and pilgrims had regularly visited their sacred peaks. But at the start of the twentieth century, much of this territory remained uncharted, making it illegible and inaccessible to outsiders. Kojima and other members of the Japan Alpine Club therefore viewed the newly named Japanese Alps and other mountainous regions of Japan as terrain to be explored and subsequently shared with the public via literature, science, and art.

Inspired to a large extent by the work of English polymath and mountain enthusiast John Ruskin (1819–1900), Kojima drew on all three of these modes of apprehending the world to develop a novel type of travel writing—and specifically, "mountain writing" (*sangaku bungaku*)—that expressed what he saw as a more authentic appreciation of nature. Kojima's work found a wide audience, contributing to a broader "discovery of landscape" in late Meiji Japan that, in the words of Karatani Kōjin, involved a "relentless defamiliarization of the familiar" to bring "into existence landscapes which, although they had always been there, had never been seen."[49] Drawing on geology, biology, and other sciences to defamiliarize the familiar was often a tricky enterprise given that the boundaries between science, literature, and art were contested and in flux.[50] It also opened Kojima to criticism from members of the literary establishment who found his use of scientific language jarring. Sometimes he, too, felt he had missed the mark. A few years after the publication of a book on his travels to Fuji (1905), he voiced his regret that "it was only 'commentary on nature,' and was 'just like going over the forty-eight traditional winning techniques in *sumo* without actually wrestling.'"[51] Although Kojima later regretted what Nobuko Fujioka calls the "detached, scientific descriptions" that he used to write about Fuji, it is understandable that he felt the need to employ them in order to create a fresh perspective on one of the best-known mountains on earth.[52]

Mountain enthusiasts like Kojima encouraged all Japanese to take up mountaineering, but like their counterparts in the West, they especially targeted the nation's youth. As Kären Wigen notes, "Whether through the Boy Scouts in Britain, the *heimat* movement in Germany, or the American hiking clubs that multiplied after the Civil War, rugged country was increasingly cast as a place to fortify both physical strength and native-place pride—and, by implication, to enhance young people's fitness for imperial rule."[53] Shiga accordingly included an appendix to his book (replete with passages lifted from an English guide to traveling in "wild countries") in which he exhorted his compatriots to climb mountains and specifically encouraged teachers to "work hard to instill among their students a spirit for mountain climbing."[54]

Fuji was a popular destination for teachers and their students. The *Yamanashi nichi nichi shinbun*, the main newspaper for Yamanashi Prefecture, paid special attention to the growing number of female students who made the ascent up Fuji, publishing articles on their climbs in the summers during and soon after the Russo-Japanese War (1904–1905).[55] One appearing in an issue from August 1905 contrasts the current fad of schoolgirls summiting Fuji with the former prohibition on women climbing the mountain.[56] That same month the illustrated magazine *Fūzoku gahō* published a special issue on Fuji with an account of a voyage up and around the mountain by a group of magazine staff members. The account notes that they met groups of students, both male and female, along the way, and one of the photographs appearing in the front matter of the magazine shows schoolgirls resting on the summit.[57]

Media outlets actively encouraged students to climb Fuji in addition to reporting on them. The *Yomiuri shinbun* newspaper created the Fuji Mountain Climbing Club for Students (Gakusei Fuji Tozankai) in 1902 (three years before the founding of the Japan Alpine Club), whose stated purpose was to "cultivate the vigor of the nation's people."[58] Another aim was to promote scientific knowledge. A notice about the club published in 1903 emphasized that students going to Fuji would be accompanied by experts in fields that included meteorology, physics, biology, geology, and hygiene.[59] The celebrated Nonaka Itaru was among them.

Fuji was a popular destination for students and other Japanese not only because it was so famous but because it was relatively easy to reach. Unlike the remote summits of the Japanese Alps, it had hosted climbers for centuries, and modern railroads made it more accessible than ever. The easiest climbing routes to reach by train were initially those located to the east and south of Fuji thanks to the opening of the Tōkaidō line's Gotenba station and Suzukawa station (now Yoshiwara station) in 1889. Travel to Yoshida at the northern foot of the mountain became much easier with the opening of a train station in nearby Ōtsuki in 1902, which was linked to the town by horse-drawn railway in 1903.[60] This comparatively easy access, combined with the fact that, unlike the more rugged peaks of the Japanese Alps, it could be scaled without specialized mountaineering equipment or training, made Fuji all the more attractive to climbers.

In response to the early twentieth-century boom in mountain climbing, the governors of Shizuoka and Yamanashi prefectures worked with local officials and businesses to update and expand the infrastructure created in the Tokugawa period. Generations of pilgrims had followed well-worn paths up the mountain and even enjoyed the convenience of climbing stairs on the final approach to the summit. They had also stopped at shelters along the way to rest, eat, and drink, while those with enough money could hire porters to carry their

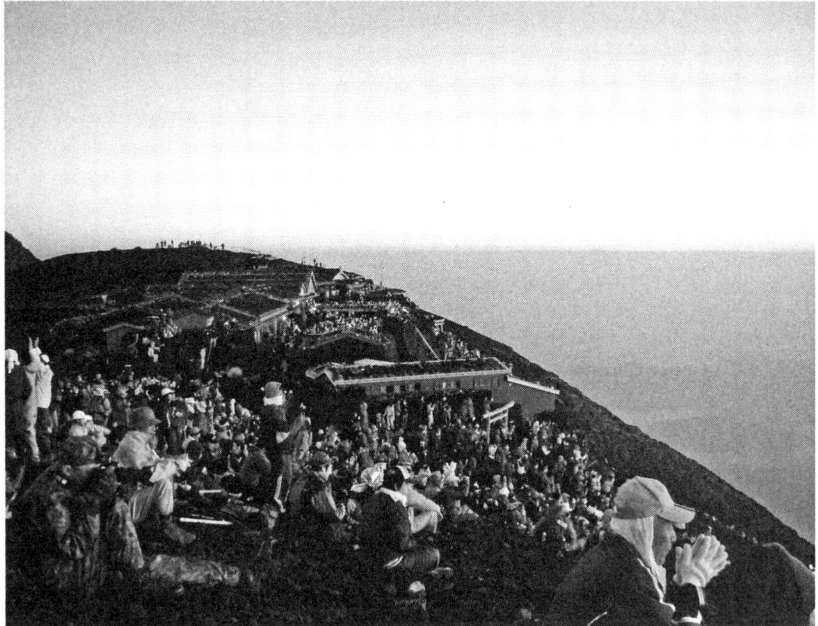

FIGURE 7.3. Crowds greet the dawn on Fuji's summit. Photo by author.

belongings. Thanks to lobbying by the Yamanashi governor, from 1906 climbers could now visit a post office opened at the eighth station where the trails from Yoshida and Subashiri met.[61] In 1907 a first aid station and a new Fujisan Hotel opened there too. Starting in that same year, climbers willing to pay a fee could also place telephone calls from the eighth station.[62] Not to be outdone, the governor of Shizuoka lobbied to open a post office on the summit where the trails from his prefecture converged. The summit post office began operating in 1907, and a public telephone was installed in 1908.[63] The phone was especially useful to those running the summertime weather station that had been established by the government's Central Meteorological Station in 1889.[64] Weather instruments, a post office, a telephone line: all these state-sponsored installations demonstrated that Japan was a nation rapidly progressing into the future even as it celebrated the national heritage embodied by Fuji (see fig. 7.3, 7.4, 7.5).

Although members of the Japan Alpine Club supported the creation of a "modern alpine complex," too much development risked destroying the natural beauty they wanted fellow Japanese to appreciate. Kojima was so concerned about rampant building projects on and around Fuji that he submitted a

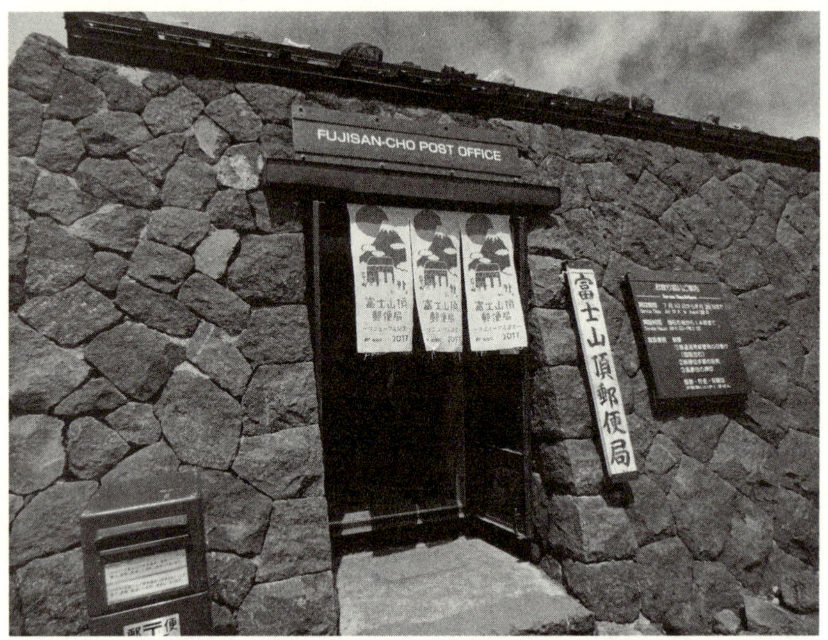

FIGURE 7.4. Post office on Fuji's summit. The banners hanging in the doorway commemorate the renovation of the office in 2017. Photo by author.

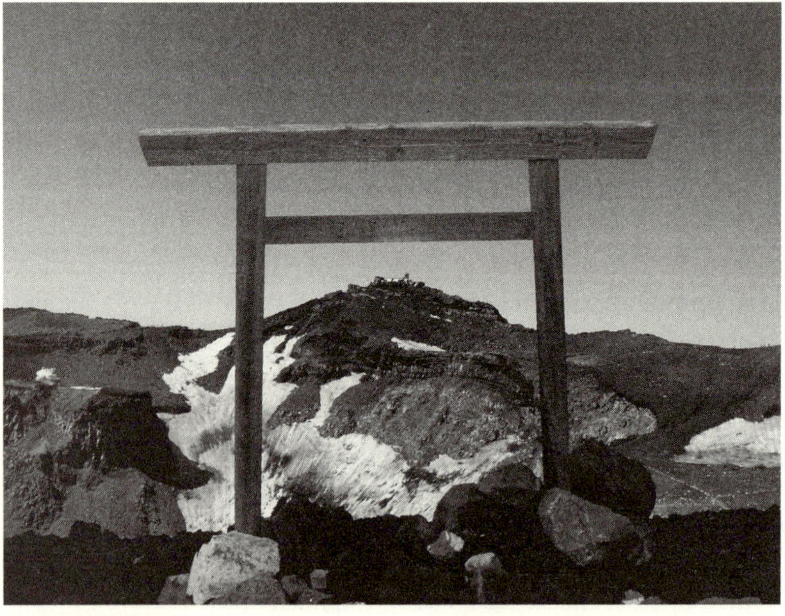

FIGURE 7.5. Torii at the edge of Fuji's crater with weather station in the distance. Photo by author.

serialized article to the *Yomiuri shinbun* in July 1909 called "On the Protection of Mount Fuji."[65] In it he lamented the modern conveniences that had already marred this "splendid creation of nature" and warned against future development plans that included an alpine railway to the summit.[66] The railway never materialized, but local communities continued to make improvements along the routes traveled by climbers, including road repairs at the base of the mountain and the installation of safety features like iron railings and stone walls on the way up to the summit.[67] Kojima may have wanted to preserve Fuji as a "splendid creation of nature," but profit-minded locals wanted to make the mountain as accessible as possible.

As Japan modernized at a breakneck pace, Fuji became a focal point for both Japanese and foreigners who wondered at the mix of old and new. Walter Weston's *The Playground of the Far East* (1918)—for which Shiga Shigetaka wrote a glowing forward—includes his account of a climb with his wife to Fuji's summit in 1914. Toward the end he notes the "many startling and suggestive contrasts of ancient and modern ways," emphasizing that "nowhere else does one meet the old and the new jostling one another so violently, without apparent objection or incongruity in native eyes." He elaborates:

> There, the unromantic materialism of the twentieth century stretches out its hand across a thousand years and draws the tenth century to its side with all its old-world dreams and communings. Almost at the very door of the most sacred shrine on this holy peak, the post-office banner flutters in the breeze to beckon the tired, but triumphant, pilgrim to dispatch to the four corners of the Empire the picture post-card that shall announce his successful toil. And as at early dawn you turn from a surprised contemplation of the most up-to-date installation of modern meteorology on the crater's edge, your astonished eyes are arrested and held with reverent interest by the shivering limbs and the adoring gaze of some aged pilgrim, whose white-clothed form enshrines the glowing devotion of a primeval worship paid in all sincerity to the splendours of the Rising Sun.[68]

The assemblage of "ancient and modern ways" described by Weston also took the form of two giant wooden Fujis in Tokyo. In 1887 entrepreneurs built one in Asakusa Park with a lightning rod on the summit and replicas of the fifty-three stations of the Tōkaidō highway arranged at its base. Electric lights were added a year later.[69] Unlike the Fujizuka built by worshippers, this Fuji was designed for entertainment, yet on the opening day, Fujikō members climbed the wooden structure while chanting the name of Amida Buddha (clearly showing that the Shintoization of Fuji worship described in chapter 6 had its limits).[70] A fierce typhoon destroyed this wooden Fuji, but the *Yomiuri shinbun* created another one in 1916 as the highlight for an exhibition in Tokyo's Ueno Park aimed at women and children. Modeled on the mountain-opening ceremonies (*yamabiraki*) that took place at the base of Fuji to initiate the climbing season, the opening ceremony for the newspaper's Fuji (one-hundredth the size of the original) featured a series of prayers and a climb by white-clad Fujikō members who carried tinkling bells and banners that fluttered in the breeze.[71] While these devotees may have been perfectly sincere, their actions served commercial ends, and to some extent this was nothing new. Worship of the mountain had generated profits for hundreds of years, especially for the communities at its foot. Yet here a different dynamic was at play: Fuji worship turned into the sort of "tradition-as-spectacle" that enriched purveyors of nostalgia in a world that was rapidly leaving behind the Fujikō and numerous other holdovers from the Tokugawa period.

For centuries, many had climbed Fuji more in the spirit of tourism than worship, but they had done so largely within a framework determined by the customs, beliefs, and dictates of pilgrimage. In the twentieth century this framework collapsed. Although white-garbed Fujikō continued to go on pilgrimage to Fuji, they were increasingly joined and then outnumbered by those who climbed the mountain for the purpose of recreation. Articles in the *Yamanashi nichi nichi shinbun* from the 1910s describe people racing up—and skiing down—Fuji's slopes. Some even played baseball on the summit.[72] Mountaineer Yokoi Haruno, who referred to Fuji as a "preschool" (*yōchien*) for mountain climbing in a magazine article from 1918, noted in a geography journal published two years later that those who summited Fuji before the Russo-Japanese War mainly belonged to Fujikō, but then Japan was gripped by "mountain-climbing fever" (*tozan netsu*), and pilgrims in traditional white garb soon found themselves in the minority, with Fuji "completely transformed into a park."[73] Fuji would, it turned out, become the centerpiece of the Fuji-Hakone National Park established by the government in 1936, five years after Japan's National Parks Law was passed.[74]

Fuji's transformation from the heart of a cosmic bodyscape into the core of a national park did not mean it ceased being sacred. Climbers who eschewed the traditional practices of Fujikō pilgrims might still pray at the Sengen Shrine on the summit. And like mountain climbers in other parts of the world, they may well have experienced a feeling of transcendence and connection to a higher power that could be described as "spiritual" or "religious." Meanwhile, as a multitude of folk beliefs and practices were relegated to the category of "superstition" (*meishin*), ideologues in imperial Japan crafted "a modern and unified Japanese Spirit of certain, albeit mystified, form," as Gerald Figal puts it;[75] and Fuji, whether through state-managed Shinto or, more importantly, through the work of popular writers like Shiga Shigetaka, was caught up in a process that required demystification on the local level to make room for mystification on the national one. Faith in Fuji as the Original Father and Mother could be cast aside, but if the nation was sacred, then Fuji, the embodiment of the nation, was sacred too.

Fuji Goes to School

Fuji's conspicuous and long-standing place in the land of Japan as well as the hearts and minds of its people made it irresistible to those who wanted to inspire a love of country. One of the most effective ways to do this was through the systematic education of Japan's youth, an impressionable and captive audience in state-run schools. Unlike Shinto shrines that were visited only now and then, these institutions provided a constant, pervasive, and intensive means to transform children into productive and loyal subjects. (Around Fuji, they also happened to provide regular employment for the large number of oshi who became schoolteachers.)[76]

When the Meiji government established compulsory schooling in the early 1870s, its goal was largely utilitarian, but as the mandated education announced on paper became a lived reality for a growing number of school-age Japanese (the attendance rate went from about 28 percent in 1873 to more than 90 percent at the beginning of the twentieth century),[77] a growing chorus of ideologues both in and out of government argued that too much emphasis was being placed on pragmatic skills at the expense of "moral education" (*tokuiku*). As a result, in 1880 the Education Ordinance was revised to put a greater focus on ethics instruction (*shūshin*); and an "emphasis on promoting imperial loyalty and indigenous morality" also appeared in teaching guidelines issued by the Ministry of Education in 1881.[78] Ideological efforts intensified in the 1890s amid concerns that schoolchildren (and Japanese in general) still lacked a fully developed "sense of nation" (*kokuminteki no kannen*) even as Japan took an increasingly important role on the world stage.[79] The impulse to develop national sentiment along with moral education resulted in the 1890 Imperial Rescript on Education bequeathed by the emperor. Read out loud on special occasions and memorized by generations of schoolchildren, the rescript emphasizes classic Confucian values like filial piety along with devotion to the emperor, respect for the Meiji Constitution (promulgated in 1889), and military service. It also refers to the glory of the *kokutai*, often translated as "national polity" or "national essence" but whose characters (国体) literally mean "national body."[80] An organic body consists of multiple parts, but ideally, all work together to ensure the survival of the body as a whole. By teaching children the term "national body," the government drove home the message that, although the Japanese people might be divided into different parts, their singular focus should be the health of the nation. Heading this body, of course, was the imperial lineage, the divine and presumably unbroken dynasty of Japan.

As an integral and prominent part of the national body, Fuji emerged in teaching materials aimed at schoolchildren of all ages, including songbooks, geography texts, and reading primers. It also appeared in art manuals used by instructors to teach students how to draw. A ceremonial recitation of the Imperial Rescript conveyed state ideology in the most blatant, rigid—and to a child, probably the most boring—of ways. It was arguably more effective to instill in Japan's youth a tacit, and therefore more deeply ingrained, "sense of nation" as they went about the practical tasks of learning to read, write, and draw. This was where Fuji proved its worth.

Educators often taught children to take pride in Fuji (and thus the nation) by emphasizing its worldwide fame. This dynamic is clear in a songbook issued by the Ministry of Education in 1881 that includes a song called "Fujisan." Set to a melody by Haydn, the lyrics bluntly state, "Foreigners gaze up admiringly. So do Japanese. [Fuji] is our pride."[81] A guide for teachers that accompanied the songbook explains that "Fujisan" was designed to "stimulate patriotic feelings in its listeners."[82] To teach that Fuji was admired by foreigners was no empty claim. As John Milne noted, accounts written by travelers to Japan almost always included paeans to the country's supreme peak. Appreciation of Fuji had also spread through photographs and woodblock prints, which played a major role in the development of impressionism and postimpressionism in the West.[83] Hokusai's *Under the Wave off Kanagawa* (*The Great Wave*) became especially popular—and influential—around the world.[84] Through the Japanese-inspired decorative arts known as Japonaiserie—which included textiles, ceramics, glassware, and wallpaper—Europeans and Americans also reproduced images of Fuji on a mass scale.[85]

Japanese schoolchildren learned that what made Fuji so widely admired in Japan and around the world was its impressive height, its beautiful form, or both. A reading primer published in 1900 to teach small children the *hiragana*

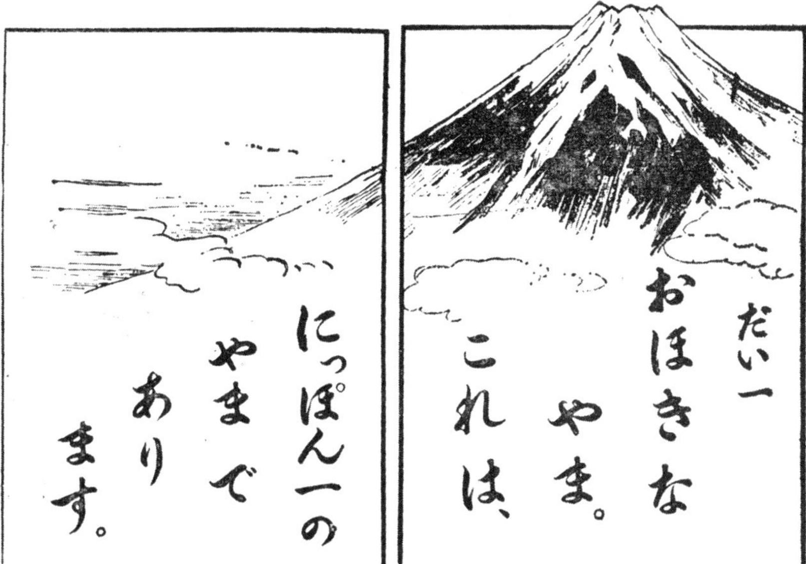

FIGURE 7.6. "Japan's greatest mountain." In *National Language Reader for Elementary Schools* (*Kokugo dokuhon, jinjō shōgakkō yō*), vol. 2. 1900.

syllabary includes a picture of Fuji rising above the clouds. Underneath are these simple words: "A big mountain. This is Japan's greatest mountain." To accentuate Fuji's majestic size, the picture extends across two frames and even projects above one of them (see fig. 7.6).[86]

Another textbook issued in the same year for older children points out not only its height, reinforced by the fact that snow can be seen on the summit all year long, but also that it is shaped like a beautiful white fan, adding that "there is no other mountain like it in the world." In addition, it uses Fuji to link the distant imperial past to the modern imperial present by noting that "it takes about four hours by steam train from Tokyo to reach the base of Fuji in Suruga Province," thus situating the mountain—and the train ride—in the imperial geography of ancient Japan.[87]

In the Tokugawa period, Fuji's image suffused everyday life through books, woodblock prints, and a variety of consumer objects ranging from combs to cups. The schoolbooks of imperial Japan went a step further by encouraging children to create their own versions of the mountain. Reading primers published in 1900 and 1903 show small boys using piles of dirt to make mini-Fujis.[88] The one from 1903—the year the Ministry of Education centralized the production of

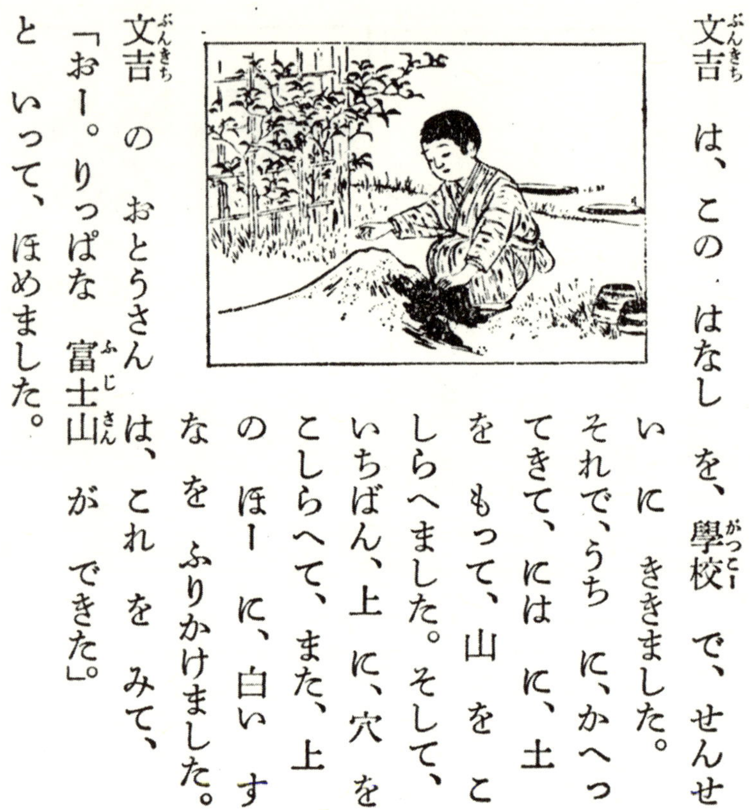

文吉（ぶんきち）は、この・はなしを、學校（がっこう）で、せんせいに ききました。それで、うち に、かへってきて、にはに、土をもって、山をこしらへました。そして、いちばん、上に、穴をこしらへて、また、上のほーに、白いすなをふりかけました。文吉（ぶんきち）のおとうさんは、これをみて、「おー。りっぱな富士山（ふじさん）ができた」。といって、ほめました。

FIGURE 7.7. Bunkichi creates a mini-Fuji in his family garden. In *Elementary Studies Reader* (*Jinjō shōgaku dokuhon*), vol. 4. 1903.

schoolbooks—informs us that young Bunkichi was inspired to create his own Fuji after learning about the mountain from his schoolteacher. Taught that the real Fuji has a "great hole on the very top," the boy duly creates a hole—that is, the crater—on top of his own (see fig. 7.7). Bunkichi earns the praise of his father, who exclaims, "Ah, you've made a splendid Mount Fuji!"[89]

Children also drew pictures of Fuji in the classroom. Manuals produced between 1895 and 1941 for teachers to use in drawing lessons include templates of Fuji that progress from simple versions in the lower grades to more elaborate and naturalistic ones for advanced students.[90] The rudimentary versions take the form of the stylized, three-peaked Fuji that originated in the Kamakura period (1185–1333) and appeared over the centuries in everything from Fuji mandalas (sacred maps) to business insignia.[91]

Schoolbooks from the 1890s and early 1900s also did their part to encourage Japan's youth to climb Fuji. One published in 1893 includes a chapter on Fuji that provides a naturalistic description of the mountain, explaining that its upper slopes consist of volcanic rocks with no vegetation and that there is snow at the summit even in summer. It also lists the main climbing entrances and notes that climbing becomes especially difficult after the eighth station. It further states that the crater is about one *ri* (almost four kilometers) in circumference, and that because it is twelve thousand *shaku* (feet) high, the summit is often shrouded in clouds and fog. When it is clear, however, climbers can enjoy a view "incomparable in the world." Immediately following this chapter is a series of verses that praise Fuji, with two of them calling the mountain not just a symbol of Japan but the very *sugata* ("figure," "form," or "shape") of the nation.[92]

Absent from the verses and the preceding chapter is any mention of the ritualized terrain that long defined pilgrimage to the mountain. The climax of the short narrative is the stupendous view, not communion with the Fuji deity either in the caves at the base or during the climb to the top. The chapter casually notes that two of the peaks on the summit are named after the buddhas Shākyamuni and Dainichi, but otherwise the mountain appears devoid of religious belief and practice. That includes Shinto. As far the authors of the textbook were concerned, there was no need for a deity to mediate the relationship between Japanese and the glorious "figure of the nation."

A chapter appearing in a textbook published in 1900 likewise provides information about the climbing trails, porters, rest stations, and other aspects of climbing Fuji. It also includes visceral descriptions of the bodily hardships involved in the climb, especially above the seventh station, where "the air gets progressively thinner, breathing is more labored, and sweat drips like rain." The chapter adds that the cold is so severe that it "penetrates the body." Emphasizing the physical challenge makes the payoff—the majestic view—all the greater. The chapter mentions the shrine on the summit, but only to stress that climbers standing in the shrine precincts can gaze "to the south where the land meets the sea along the Tōkaidō, and upon mountain peaks that look like clods of dirt, lakes like tiny ponds, and long rivers like silver threads."[93] Nowhere do we see any indication of Shinto belief and practice. Having pushed one's body to the limits, why pray to a god when you could enjoy the perspective of one?

An essay on climbing Fuji in a textbook from 1903 similarly emphasizes both the hardships of the climb and the subsequent rewards of a spectacular sunrise and a sweeping view of the mountains, rivers, and ocean below.[94] The essay is accompanied by verses praising the mountain's splendor, one of which explicitly links the modern technology of Meiji Japan to the appreciation of natural beauty:

> Viewed from a train window,
> Fuji's sublime figure,
> projecting above the clouds,
> its peak always snowy white.[95]

While the essay depicts the grueling physical effort of climbing Fuji, this poem celebrates the mechanized ease of viewing it from a train window. By pairing an essay that depicts the bodily effort of climbing Fuji with a verse about gazing at it from a train, the 1903 textbook reflects the global movement to promote the character-building endeavor of mountain climbing along with a modern infrastructure to support it.

Students continued to climb Fuji in large numbers in the following decades, but after 1903, essays about climbing disappeared from schoolbooks. Perhaps by this point the practice was so ingrained that bureaucrats in the Ministry of Education decided there was no need to continue encouraging it in writing. It was sufficient to hammer home the message that grand and beautiful Fuji was not just the pride but the embodiment of the nation. From 1910 to the 1940s, children learned a song composed by the popular children's author and folklorist Iwaya Sazanami (1870–1933) called "Fujisan" or "Fuji no yama" (both mean Mount Fuji).[96] The lyrics read:

> Its head projecting above the clouds,
> looking down on mountains all around,
> hearing the thunder rumbling below,
> Fuji, the greatest mountain of Japan.
>
> Towering into the blue sky above,
> its body wearing a kimono of snow,
> with hems of mist unfurling wide,
> Fuji, the greatest mountain of Japan.[97]

Fujikō taught that the parental body of Fuji was at the heart of a cosmic bodyscape. In these verses, Fuji's kimono-clad body serves as a personification of the national one. A manual distributed to teachers explicitly says so. Commenting on the song's significance, it states that Fuji's greatness is akin to "the national body" (*kokutai*) and "national strength" (*kokuryoku*), and its beauty to "the spirit of the nation's people" (*kokuminsei*). As if the point were not made clearly enough, it adds, "Through Mount Fuji, this song truly describes the land of Japan, the national body of Japan, and the national spirit of Japan."[98] Children continued learning the song after Japan's defeat in World War II (although it was divorced from the discredited ideology of the "national body"), and today a "melody road" at the foot of the mountain plays the tune when cars pass over it.

Iwaya's song celebrated the nation and its people, but the expansion of the Japanese empire complicated the picture of what counted as "Japan" and what did not. Nationalism narrowly focused on the home islands (*naichi*) was in tension with a more cosmopolitan imperialism that treated conquered territories as part of Japan proper and its people as fellow subjects of the emperor (although, to paraphrase George Orwell, some were more equal than others.)[99]

We see this tension at work in textbooks published after 1895, when the treaty that ended the First Sino-Japanese War (1894–1895) awarded Taiwan to Japan. The fact that Taiwan had mountains taller than Fuji challenged Fuji's status as Japan's greatest mountain, and rather than dodge the issue, textbooks published in the following decades deal with it head on. One from 1910 states that "the highest Japanese mountain is Mount Niitaka [in Chinese, Yushan] in Taiwan," which, it continues, is about 13,070 *shaku* (feet) high, making it about one thousand feet taller than Fuji. Acknowledging that "Fuji fell to second place after Taiwan became our territory," it goes on to praise the snow that graces Fuji year round. Like essays appearing in earlier textbooks, it also observes that Fuji is as beautiful as a white fan. Therefore, "it must be said that this is the greatest mountain in the nation, or more so, the most famous mountain in the world."[100]

This message appears again in a language book from 1920. In a chapter called "Japan's High Mountains" that takes the form of a conversation between two brothers, the older brother says that Fuji is the highest mountain in the home islands. When asked by the younger brother what the tallest mountain in *Nihon* (Japan) might be, he responds that it is Niitaka in Taiwan. He goes on to clarify that the second- and third-tallest mountains are also in Taiwan, making Fuji the fourth-highest Japanese mountain. We then learn that the Himalayas are the loftiest mountains in the world. And yet, according to the older brother, "it's not always the case that the highest mountains are the most famous." When his brother asks which one is the most famous, he duly replies, "That's Mount Fuji."[101] By distinguishing fame from height, the dialogue attempts to resolve any tension between a Japanese identity determined by the home islands, as embodied by Fuji, and one encompassing territories—and as a consequence, even taller mountains—overseas. In short, Japanese could have their empire and Fuji too.

That empire grew rapidly during the early twentieth century. In the wake of its war with Russia in 1904–1905, Japan tightened its hold on Korea and fully annexed the peninsula in 1910. Allied with Britain, the United States, and France during World War I, Japan then won the rights to Germany's concession on the Shandong Peninsula in China and its island territories in the Pacific. In the early 1930s, it seized all of Manchuria; and during the Second Sino-Japanese War (1937–1945), it conquered portions of northern and coastal China and Southeast Asia.

Japan's rapid military expansion is reflected in a reading primer for small children from 1933 that juxtaposes a picture of Fuji with one of a battleship. The text accompanying the pictures reads, "Tarō drew a picture of a battleship. Hanako drew a picture of Mount Fuji" (see fig. 7.8). It is hard to imagine a more barefaced link between the soaring incarnation of Japan and the military expansion of the Japanese empire. Note, too, the gender difference: boys look to military action abroad while girls attend to the home front.[102]

As war intensified later in the 1930s and the early 1940s, the state promoted devotion to the national body more than ever. A scaled-up campaign to cultivate the "spirit of the nation's people" emphasized the divine status of the emperor and, by extension, all of Japan, breathing new life into state-promoted Shinto.

Verses in a reading primer and songbook used from the early 1940s describe Fuji as "a deified mountain" (*kami no yama*), breaking with the secular depictions of the mountain appearing in educational materials until that point.[103] Yet even here, there is no mention of the kami Konohanasakuya-hime or any other individual deity or religious practice. That level of specificity would disrupt the seamless union of Fuji with the national body and imperial state. The verse immediately following Fuji's description as a "deified mountain" in the primer reverts to form with these simple words: "The people of the world look up to Japan's greatest mountain."[104]

Students in imperial Japan learned that "the beauty of the national spirit takes the form of Fuji, and reverence for Fuji's beauty is the beauty of the national spirit," creating, as Abe Hajime puts it, "a circular logic" that was difficult to escape.[105] This is not to say that children, or Japanese in general, were sponges always ready to absorb a constant flow of nationalist ideology, whether focused on Fuji or not. "As long as one attends to those who produce it, ideology will appear in every corner, filling the villages, classrooms, and households

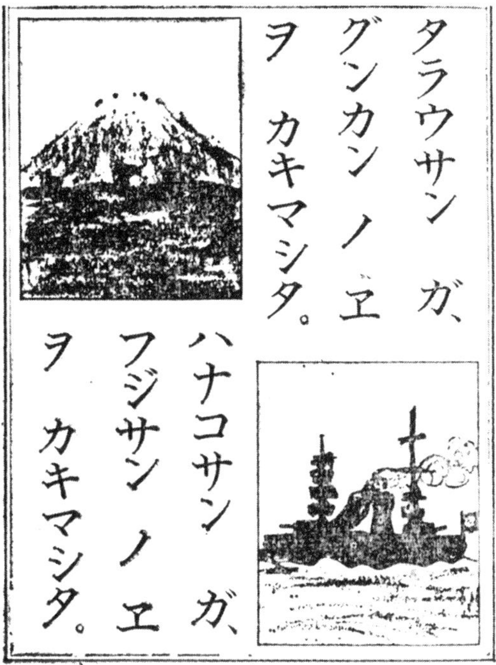

FIGURE 7.8. "Tarō drew a picture of a battleship. Hanako drew a picture of Mount Fuji." In *Elementary National Language Reader* (*Shōgaku kokugo dokuhon*), vol. 1. 1933.

with billowing vapors of influence," observes Carol Gluck, who adds that the ideologues of imperial Japan "were constantly taking the public pulse, usually to find that it was weak in national spirit."[106] The work of anxious ideologues was never done, which was why they turned again and again to Fuji, a mountain as equally familiar as it was beautiful, to achieve their aims. Judging the impact of that work is difficult, but a mountain as universally admired as Fuji was the perfect tool to foster both a national and imperial spirit.

Whose Fuji?

While Japan was at war in the 1930s and 1940s, Fuji's role as the sacred icon of a sacred nation became all the more important. It loomed especially large in the paintings of Yokoyama Taikan (1868–1958), a staunch nationalist and famous representative of *Nihonga* (Japanese Painting), a style that drew on

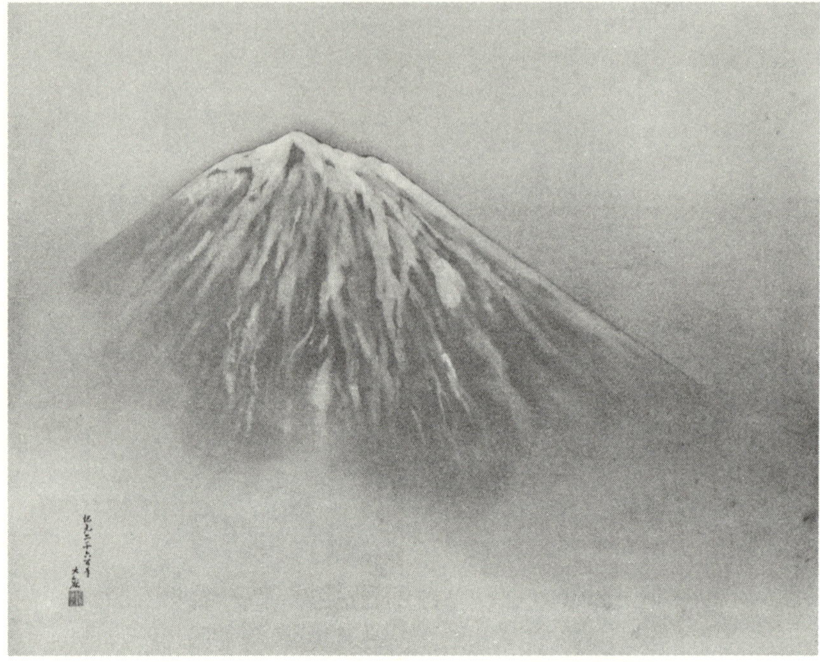

FIGURE 7.9. Yokoyama Taikan. *Mount Fuji.* 1940. Tokyo Fuji Art Museum.

both traditional and modern techniques to produce a distinctively Japanese form of art. Taikan was, it is fair to say, obsessed with Fuji; and as John Rosenfield observes, "Taikan's published works contain nothing more overtly propagandist than his countless views of Mount Fuji—as though painting it over and over again he was reciting a patriotic mantra."[107] Taikan put his money where his brush was by donating the proceeds from the sale of twenty paintings (ten of them of Fuji) for the purchase of fighter planes in 1940 (see fig. 7.9). Three years later he founded the Japan Patriotic Art Society (Nihon Bijutsu Hōkokukai) in support of the war.[108]

Fuji also served as the perfect stage for gestures of imperial devotion. In 1934 the seventy-odd members of the Mount Fuji National Flag Hoisting Association (Fujisan Kokki Keiyōkai), among them officials from several ministries, flew what was reportedly the largest rising sun (*hinomaru*) flag ever made (at 367 square meters) on the summit.[109] In 1940, the year Japan celebrated the founding of the empire twenty-six hundred years earlier by the mythical Emperor Jimmu (see fig. 7.10), nearly three thousand people climbed to the summit from Yoshida one day in July to pray for the longevity of the emperor.[110]

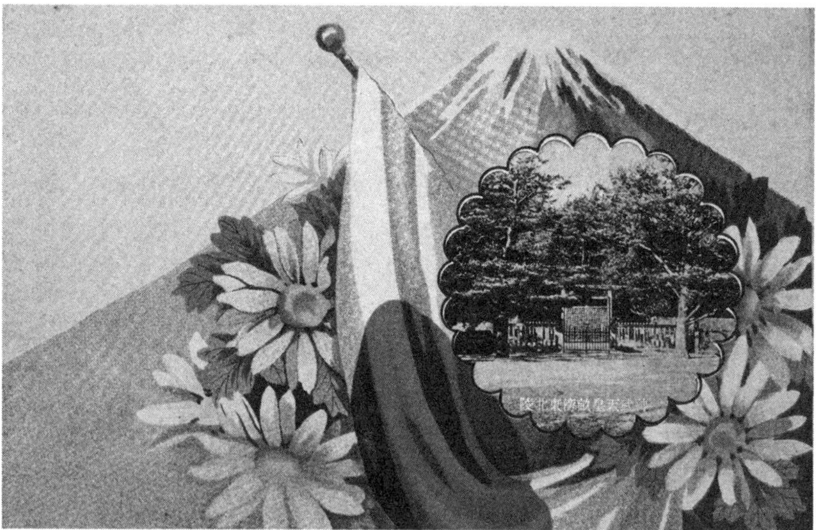

FIGURE 7.10. Postcard commemorating the twenty-six-hundredth anniversary of the founding of the empire by the mythical Emperor Jimmu. 1940. Private collection of Matsushima Jin. In front of Fuji appear the Japanese flag, a photo of the supposed tomb of Jimmu, and chrysanthemum flowers, which symbolize the imperial household.

Two years later the renowned architect Tange Kenzō (1913–2005) published a prize-winning plan to build a commemorative complex for the "Greater East Asia Co-Prosperity Sphere"—in other words, Japan's empire—at the foot of Fuji, "the loftiest example of nature" in Japan. Featuring a Shinto shrine and memorial to the war dead, it was to be connected to Tokyo via a "Greater East Asia Highway," enabling travelers from the capital to reach this "hallowed ground" in about an hour.[111] The complex was never built, but Tange redeployed major elements of his design in his master plan for Hiroshima's Peace Memorial Park, completed in 1954.[112]

Fuji saw a surge in climbers in 1940.[113] However, from 1941, the year Japan went to war with the United States in the Pacific, the number of climbers declined, falling precipitously from 1943, when a bus that took climbers from the center of Yoshida to the start of the Fuji climbing trail ceased operating and the post offices at the eighth station and the summit closed.[114] By that summer, a string of losses overseas had made Japan increasingly desperate, and from 1944, the United States brought the war home by launching a massive bombing campaign of Japan's major cities. In a cruel twist, American pilots

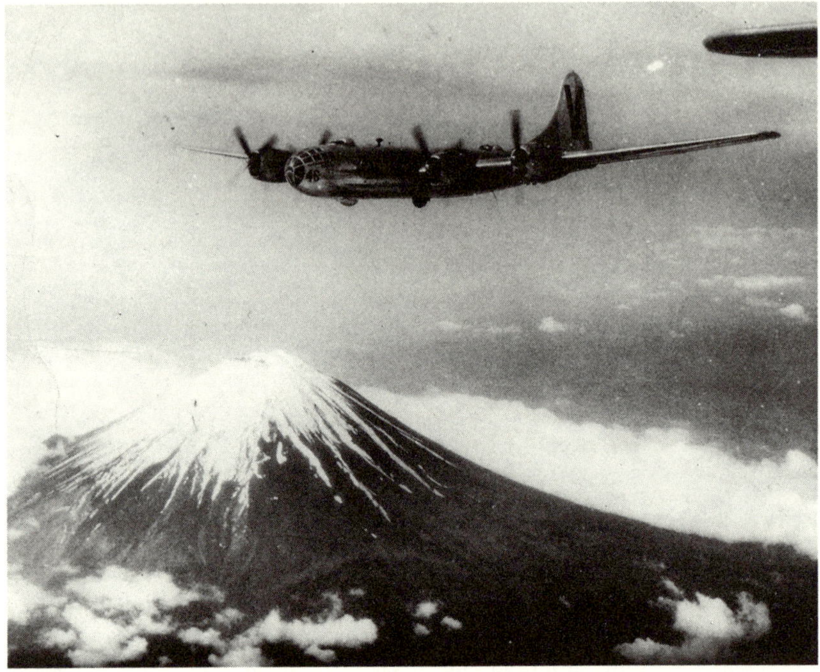

FIGURE 7.11. B-29 bomber over Mount Fuji. 1945. United States Army Air Forces. Wikimedia Commons.

relied on Fuji, the symbol of Japanese unity and strength, to orient themselves on the approach to Tokyo (see fig. 7.11).[115]

Soon after the atomic bombings of Hiroshima and Nagasaki in August 1945, the war came to an end and the Allied (mainly American) Occupation of Japan (1945–1952) began. The United States promptly incorporated Fuji into insignia for its occupying forces and took over military training grounds located at its foot.[116] To this day US marines stationed at the base of the mountain call themselves the "Fuji Samurai." While the US military appropriated Fuji both physically and symbolically, some Occupation officials thought images of the mountain should be suppressed in Japanese media because of its association with Japanese nationalism and militarism. One American censor went so far as to order director Makino Masahiro (1908–1993) to eliminate any shots of Fuji from his 1946 film *Sophisticated Wanderer* (*Iki na fūraibō*) even though it was set at the mountain's base.[117]

Yet Fuji could not be censored out of existence and efforts to do so were limited and short lived. While it was easy enough to ban editorials that celebrated the glories of the Japanese empire, it was impossible to order millions of Japanese to avoid looking at Fuji in the distance or to give up the innumerable everyday items on which images of the mountain appeared. Adding to the absurdity of an attempt to censor images of Fuji was the fact that the United States had decided to keep Hirohito (r. 1926–1989), the reigning Shōwa emperor and rallying point for imperial Japan, on the throne.[118]

It did not take long for Fuji—once again Japan's tallest mountain thanks to the loss of the empire—to resume its familiar role as the nation's "geo-body" standing firm amid sudden change. Now it was common to underscore that Fuji stood for a Japanese people at peace rather than a national body at war. An article appearing in the national daily *Asahi shinbun* in 1953, for example, explicitly refers to Fuji as a "symbol of peace" while pointing to the troubles caused by the US military on and around the mountain.[119]

At the same time, Fuji served as an object of resentment for those excluded from the dominant vision of a peaceful yet ethnically homogenous Japan. "In the Shadow of Mount Fuji" (*Fuji no mieru mura de*), a short story published in 1951 in the leftist magazine *Sekai*, was written by Kim Tal-su, a Korean who had come to Japan as a child. He was one of several million Koreans who had moved to Japan, either voluntarily or through coercion, before and during the war. Most returned to Korea soon after war's end, but about six hundred thousand remained. Once exhorted, and often forced, to serve the emperor as soldiers in the imperial army and laborers in Japanese factories and mines, the so-called *zainichi* (Korean residents of Japan) were quickly identified as aliens who had no place in an ethnically defined postwar Japan. Koreans had always been treated as second-class citizens, but at least in the imperial era they had *been* citizens with certain attendant rights, including the ability to vote and run for public office. In the years immediately following the war, these rights were taken from them; and in 1952, just a little over a week before the Occupation officially ended—and while the Korean War (1950–1953) raged on—the Japanese government confirmed in no uncertain terms that Koreans, along with Taiwanese, had lost their Japanese nationality altogether.[120]

Kim's story pairs the discrimination suffered by *zainichi* with that endured for centuries by Japan's hereditary outcasts known as *burakumin,* who traditionally engaged in activities, like tanning and performing cremations, that were seen as polluting. The Meiji government legally abolished their outcast status, making them the supposed equals of other Japanese, but hostility and discrimination remained.

Along with three other Koreans, the unnamed protagonist of the story is invited by a friend, Iwamura Ichitarō, to stay with his *burakumin* family at the base of Fuji. At one point the protagonist thinks to himself that the Koreans and their hosts are "kindred spirits" because they all "faced disdain and persecution."[121] It turns out that the desire for solidarity is tragically one-sided. Everyone gets along well until it comes to light that the protagonist and other visitors are in fact Korean. Iwamura's sister and mother grow cold and abruptly announce that it is time for bed. The next day the protagonist goes hunting, and the story ends with these dramatic words: "I lifted the gun, and took aim at Fuji. I pulled the trigger with every last bit of my strength. The gunshot shattered the mountain's silence. I fired again and again in rapid succession. Then I went on and on, shooting at Mount Fuji like a man gone mad."[122] Kim has his protagonist literally—and in the end, futilely—target Japan's greatest mountain to criticize a Japan that may have been at peace but marginalized anyone who disrupted the fantasy of a unified people working to build a better future.

Decades later, the second-generation *zainichi* author Yi Yang-ji (1955–1992), who grew up at the base of Fuji, followed Kim's example by using the mountain to reflect on the place of those of Korean descent in Japan. Whereas Kim's story is bleakly pessimistic, in her 1989 essay "Mount Fuji" (*Fujisan*), Yi admits to a more ambivalent "love-hate" relationship with Fuji as someone neither exclusively Korean nor Japanese. At one point she describes sitting in a train in Korea taking in the landscape of her ancestors and declares, "I love Korea. I love Japan. I love both countries." The essay ends with her imagining Fuji, eyes closed, as Korean mountains stretch outside the train window.[123]

Fuji's role in perpetuating the fantasy of a harmonious and homogenous Japan after World War II explains why *zainichi* authors used the mountain to challenge that fantasy. Its status as the symbol of a unified Japan also fueled a decades-long conflict over who controlled the summit, which was triggered by the Occupation policy of separating religion (particularly Shinto) from the state.[124] The stakes in this conflict were material and local as well as ideological and national. Communities on the northern side of the mountain, joined by organizations ranging from the All Japan Tourism Association (Zen Nihon Kankō Renmei) to the National Parks Association (Kokuritsu Kōen Kyōkai), argued that the summit properly belonged to the Japanese people and should be public, that is, state, property. Fujisan Hongū Sengen Taisha (located in Fujinomiya and from here on referred to as Sengen Shrine; see fig. 7.12), along with the Association of Shinto Shrines (Jinja Honchō) and other allies, countered that Fuji's summit properly belonged to the shrine, not the state, since it was the shrine's main object of worship.

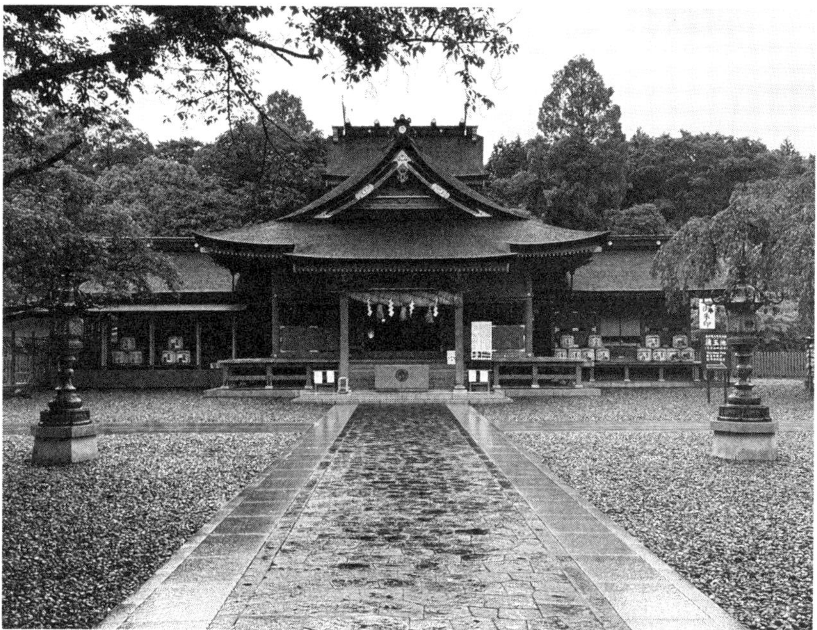

FIGURE 7.12. Fujisan Hongū Sengen Taisha in Fujinomiya. Photo by Philip Ellway.

Before and during the war the summit was, in fact, considered the "inner sanctuary precincts" (*okumiya keidai*) of Sengen Shrine, but it was simultaneously considered the property of the state, like the precincts of other Shinto shrines and Buddhist temples throughout Japan. This posed a problem once the Occupation issued its "Shinto Directive" in December 1945 to dismantle what the Japanese government had considered a "national teaching" that transcended the private realm of religion, but which the Occupation viewed as an ultranationalistic state religion that had helped drive Japan to war.[125] The directive therefore forbade state support for Shinto in any way, shape, or form, with the separation of religion and the state applying to other "religions, faiths, sects, creeds, or philosophies" as well.[126] Article 89 of the new constitution, promulgated in November 1946, clarified that "no public money or other property shall be expended or appropriated for the use, benefit or maintenance of any religious institution or association, or for any charitable, educational or benevolent enterprises not under the control of public authority."[127]

In 1939 the government had already decided to cede ownership of state-owned ritual precincts to Buddhist temples, but now the state also had to transfer precincts to Shinto shrines.[128] It was additionally forced to end a custodial forest system that had enabled temples and shrines to profit from sales of timber in the forests put under their charge. Returning the ritual precincts immediately surrounding temples and shrines was a relatively straightforward affair, and it was usually clear when a forest should be put in the hands of the state rather than the religious institutions that had overseen them. How to deal with entire sacred mountains was more ambiguous, but with the blessing of William Bunce (1907–2008), the chief of the Occupation's Religions Division, the committees set up by the Finance Ministry to handle the disposition of state-owned religious lands took a generous approach, for the most part handing them over to the temples and shrines that claimed them.[129]

Fuji was another story. It was not just a sacred mountain but the nation's mountain. It had also long been the source of conflict. In the Tokugawa period, Sengen Shrine and communities around Fuji clashed over issues ranging from who could collect money from pilgrims on the summit to whether women could climb beyond the lower slopes. As recently as 1917, the shrine had asserted its claim to Fuji from the eighth station to the summit by ordering Yamanashi Prefecture to abandon its police and first aid facilities at the eighth station. The prefecture refused.[130]

Although Sengen Shrine's control was never absolute—and as noted above, sometimes directly challenged—officials in both the Tokugawa and Meiji governments had generally acknowledged the shrine's authority over the upper reaches of Fuji. Tokugawa Tadanaga (1606–1634), daimyo of both Kai and Suruga, recognized Sengen Shrine's managerial jurisdiction over Fuji from the eighth station up, which was validated by the Tokugawa government's Office of Temples and Shrines in 1779.[131] In the Meiji period, the shrine lobbied the imperial state to confirm its claim, and finally succeeded in 1899 when the governor of Shizuoka, representing the central government, recognized everything above the eighth station as the shrine's inner sanctuary.[132] This did not change the fact, however, that Fuji's summit, along with the shrine precincts at the mountain's foot, continued to be registered as public rather than private property.

The shrine relied on the legal precedents established in Tokugawa and Meiji Japan when it appealed to the postwar government to affirm that it owned Fuji's upper reaches. To support its claim, it emphasized the religious character of the mountain, which was considered the "god body" (*goshintai*) of Konohanasakuya-hime. In a booklet submitted to the Finance Ministry, it made the case that although the entire mountain was a god body, the territory above the eighth station was particularly sacred, noting that from here onwards climbing became all the more difficult and that the dramatic rock formations and fields of snow contributed to a feeling of "mystical splendor." Thus the religious character of the shrine's "inner shrine precincts" not only derived from history but was intrinsic to its very topography.[133]

Opposing the shrine were Yamanashi Prefecture, the All Japan Tourism Association, and the Nature Conservation Society of Japan (Nihon Shizen Hogo Kyōkai). They argued that it was wrong to "privatize" (*shiyūka suru*) the world-famous symbol of Japan, adding that Fuji was at this point more a site for recreation than for religious practice.[134] Although the Tokugawa and imperial governments recognized Sengen Shrine's authority over the peak, they chose to leave the summit in a political no-man's-land between the prefectures of Yamanashi (formerly Kai Province) and Shizuoka (formerly Suruga Province). Politicians and officials in Yamanashi feared nonetheless that designating the summit the private property of Sengen Shrine would put it firmly under the control of Shizuoka. This was of particular concern because plans were in the works to build a cable car from the Yamanashi side of Fuji to the summit, plans that would likely be thwarted if the Finance Ministry recognized the shrine's claim.[135]

The Finance Ministry's Central Committee to Dispose of Temple and Shrine Precincts (Jiin Keidaichi Shobun Chūō Shinsakai) was caught between the shrine and its opponents. It first decided that the shrine should own just a small portion of the summit deemed to be religiously significant, with the rest remaining in the hands of the state. After intense lobbying by the shrine's chief priest (*gūji*), Satō Azuma, the committee changed its mind, and toward the end of December 1952 recommended the reverse: that virtually all of Fuji above the eighth station should belong to the shrine and that only certain parcels of land—including the site of the weather station—should be designated the property of the state.[136]

The shrine's opponents found this new ruling unacceptable and intensified the conflict. On February 5, 1953, a raucous demonstration orchestrated by the governor of Yamanashi and funded by Yamanashi's prefectural assembly took place in the plaza outside Tokyo's Shinbashi train station. Protestors from Yamanashi and elsewhere chanted slogans such as "Protect Fuji by the will of all the nation's people" and "Absolutely oppose the disposal of Fuji." The protest prominently included Fujikō members and the heads of oshi lineages from Yoshida, who had transported their bright red, Fuji-shaped *mikoshi* (portable shrine) to Tokyo, hauling it from Shinbashi station to the Diet building (the seat of Japan's government), and on to the Finance Ministry. Demonstrators tried to storm the ministry but were held off by police. Meanwhile, tens of thousands of leaflets opposing the shrine's claim to Fuji dropped from an airplane circling over Tokyo.[137]

The Yamanashi governor and his business allies were motivated by economic gain, but that alone could not account for the passion exhibited by the shrine's opponents both in Yamanashi and nationwide. The decision to bring Yoshida's time-honored, Fuji-shaped *mikoshi* to the protest signified that this latest conflict over Fuji's summit had deep roots as far as those living at the base of the mountain were concerned. Meanwhile, across Japan, those in favor of designating the summit as state property were driven by the widespread belief that Fuji belonged to "the Japanese people in their entirety," as the prominent journalist Tokutomi Sohō (1863–1957) expressed in a message of support for the protestors.[138]

Under intense pressure, in December 1953 the Diet established a special inquiry into the decision to make Fuji's summit the property of Sengen Shrine. The investigating committee consisted of twelve Diet members, most of whom proved hostile to the shrine's position during two days of hearings in which they heard testimony from Chief Priest Satō Azuma as well as officials from the Finance Ministry, the Education Ministry, and the Welfare Ministry's National Parks Section. Also providing testimony were the head of the Association of Shinto Shrines, a professor of folk religion, two legal scholars, and the high-profile journalist and critic Abe Shinnosuke (1884–1964). Satō stressed the antiquity of the shrine's relationship to Fuji. He also assured the committee that, while the summit properly belonged to the shrine, "We will cooperate to the fullest extent with state law and look after the welfare of the masses," adding, "I want to make Fuji available to the citizenry and to the peoples of the world."[139] The chief of the National Parks Section noted that, although those who owned property in national parks had to abide by government regulations, negotiating with landowners could be problematic, so it was best to have "as little private land as possible" in national parks.[140] In contrast, officials from the Finance and Education ministries took the shrine's position. So did the secretary-general of the Association of Shinto Shrines, Yoshida Shigeru (not to be confused with the prime minister of the same name), who used the opportunity to blame the Occupation policy of separating Shinto from the state for creating the current conflict.[141]

While Satō and his allies claimed that making the summit shrine property would present no obstacle to public access, others contended that ownership by the state would in no way impede religious practice. A socialist politician on the Diet committee, Kitayama Airō, brought up the example of sun worship multiple times, arguing that revering the sun did not require owning it.[142] Other committee members who opposed the shrine argued that handing the summit to the shrine amounted to transferring a deity, which translated into official—and illegal—support of religion.[143] Committee member Furuya Sadao, a socialist from Yamanashi, went a step further by pressing a secularist agenda, noting with approval that "primitive religion," including worship of sacred mountains, was "gradually disappearing among the people."[144] Abe Shinnosuke took a secularist approach as well, stating it was wrong for the state to support "primitive religion."[145] Taking aim at Yoshida's critique of separating religion and the state, he argued that transferring a national asset as important as Fuji to Sengen Shrine, even if said to be done in the name of protecting religious freedom, was tantamount to reviving the "unity of rites and rule" (*saisei itchi*) that had plagued imperial Japan. "The issue at the heart of this problem is whether or not we truly foster the spirit of democracy in Japan," he concluded, thus framing a territorial dispute as a battle for the soul of the nation.[146]

Satō showed no interest in ideological grandstanding. Unlike Yoshida, he made clear that he was happy to play by the rules of the postwar order, which included the twin principles of religious freedom and separation of religion and the state. Keeping his eyes on the prize, he stressed that the shrine's goal was simple: "to maintain ownership as of old in accordance with the new law."[147] But while Satō and his supporters argued that the law was on their side, the Diet committee ultimately disagreed, concluding that Fuji, "loved by the people as the symbol of Japan," properly belonged to the state.[148] As a result, the Finance Ministry decided against handing the summit to the shrine, and in 1957 announced it would transfer management of Fuji above the eighth station to the Cultural Treasures Preservation Commission (Bunkazai Hogo Iinkai).[149]

This did not mark the end of the conflict. Sengen Shrine took the government to court, arguing that other sacred mountains had been transferred to religious institutions without any problem, so there was no good reason not to recognize the shrine's claim to Fuji's summit. The case slowly made its way through the court system, finally reaching Japan's Supreme Court.[150] Citing the constitutional obligation to protect both property rights and religious belief, the court ruled in 1974 that, aside from small portions reserved for the weather station and a couple of other public facilities, the summit properly belonged to the shrine. It took another thirty years before the Finance Ministry provided a property deed, since the paperwork required a legal address and, despite the shrine's victory, the location of the summit remained in limbo between Shizuoka and Yamanashi prefectures. In 2004, however, the shrine at last convinced the Finance Ministry to provide a deed without a fixed address, two years before the shrine was to celebrate its founding twelve hundred years earlier.[151] Negotiating the tension between religious freedom and the separation of religion and the state—a tension evident anywhere the two values are both espoused—the shrine had succeeded in its long campaign to parlay Tokugawa privileges into postwar property rights and secure control over the summit of Fuji—national symbol, natural wonder, and incarnation of a god.

IN A NOVEL by Natsume Sōseki (1867–1916) published in 1908, *Sanshirō*, the novel's namesake converses with one of the other main characters, Professor Hirota, on a train ride to Tokyo. At one point Hirota cynically observes, "Oh yes, this is your first trip to Tokyo, isn't it? You've never seen Mt. Fuji. We go by it a little farther on. It's the finest thing Japan has to offer, the only thing we've got to boast about. The trouble is, of course, it's just a natural object. It's been sitting there for all time. We certainly didn't make it."[152] Shiga Shigetaka and like-minded nationalists argued that Japan was superior because of a physical terrain in which Fuji reigned supreme. Yet here Sōseki points to the absurdity of taking pride in something that has "been sitting there for all time." By emphasizing that the Japanese did not make Fuji, "the only thing we've got to boast about," he delivers a critique of the inflated national pride that had grown all the stronger after Japan's victory in the Russo-Japanese War.

It is only in physical terms, however, that the Japanese did not make Fuji. Culturally speaking, they had refashioned the mountain over many centuries, from a kami of epidemics and realm of immortals into the embodiment of a nation and a school for mountain climbing. As we have seen, this transformation was anything but smooth. Fuji's powerful charisma made it the focus of conflicting agendas, both material and ideological. Locals continually competed for control over Fuji, and the fact that the mountain symbolized a supposedly unified nation made it a flashpoint in battles over the form that nation should take.

It is also important to remember that Japanese were not alone in transforming Fuji. For well over a thousand years, ideas, institutions, and practices from East Asia molded Fuji's image in Japan and abroad. As Japan rapidly developed into an imperial power in the nineteenth and twentieth centuries, Fuji acquired worldwide fame, making it a truly global mountain. Chapter 8 will show how Fuji attained global status not only in the ways people represented the mountain but in how they physically exploited it, entangling Fuji in the complex processes responsible for the Anthropocene, the "Epoch of Humans," with which all living things must now contend.

8

A Global Mountain on a Human Planet

OVER THE PAST two centuries Fuji has become a global mountain. Standing at the intersection of the Pacific, Eurasian, North American/Okhotsk, and Philippine plates, it has always been connected to the tectonic forces that shape the planet. And for over a thousand years it has been imagined according to continental literary conventions and religious beliefs. After they arrived in Japan in the 1500s, globetrotting Europeans drew Fuji into wider networks of appreciation and interpretation. Yet it was really only in the nineteenth century that the mountain's fame exploded beyond East Asia. From this point on, Fuji became a global icon. Meanwhile, imported objects, practices, and ideas—ranging from the Berlin blue used by Hokusai and Hiroshige to create their woodblock prints to the spread of mountain climbing as a sport rather than a form of pilgrimage—helped refashion how Japanese people viewed and experienced their cherished peak. As we saw in chapter 7, Fuji also served as an important means to foster a nationalistic spirit and sense of common purpose that were central to Japan's development as a world power.

Fuji also became a global force through the ways it was materially exploited. As both Japanese and non-Japanese spread modern visions of Fuji, they took advantage of its physical resources—particularly its water—to produce tea, paper, textiles, and other goods that drew the mountain into a global system of commodities exchange. And while ideologues like Shiga Shigetaka deployed Fuji as a means to bolster pride in a "nature-loving" and, more specifically, "mountain-loving" nation, the army used the grasslands at its foot for military exercises that were integral to the growth of an expanding empire. The more Fuji was simplified into a national totem that linked Japan's past and present, the more it was caught up in fast-changing and increasingly world-wide networks of international trade and armed conflict.

The globalization of Fuji drew its role as a geological actor into processes that made humans into geological actors as well. A growing number of voices in and out of academia have suggested that human impacts on Earth systems are now so extensive that they have given rise to the Anthropocene, the "Epoch of Humans."[1] In 2024 the International Union of Geological Sciences (IUGS) rejected a proposal from the interdisciplinary Anthropocene Working Group (AWG) to add this epoch, with a start date in the mid-twentieth century, to the Geological Time Scale.[2] This decision does not change the fact that human activities—especially since the "Great Acceleration" of the mid-twentieth century—have had profound effects on the planet. These include climate change, mass production of synthetic compounds, homogenization and extinction of flora and fauna, and remaking of landscapes and waterways on a colossal scale. I see value in using the term "Anthropocene" to emphasize the degree to which these impacts now shape the planet, while keeping in mind that certain portions of humanity bear much more responsibility than others. The rich consume far more resources and energy than the poor, and a select number of nations have contributed so much to anthropogenic change that they can be considered geological actors in their own right. The exploitation of Fuji's resources helped make Japan into one of them.

Commodified Fuji

Humans have lived and manipulated habitats around Fuji for millennia, but the use of Fuji's resources has greatly intensified during the last few hundred years and expanded dramatically from the late nineteenth century. Stone Age hunter-gatherers around Fuji, like their counterparts in other parts of the world, appear to have regularly lit fires to create grassland or open woodland environments hospitable to hunting and the growth of edible plants. Starting from the Yayoi period (600 BCE–300 CE), agriculturalists cultivated rice paddies at lower elevations and grew millet, barley, soybeans, and other upland crops at higher ones. During the Tokugawa period (1600–1868), population growth and the rapid expansion of a money economy drove those living around Fuji, as elsewhere, to maximize agricultural output and produce specialized goods for sale in distant markets. Farmers in the village of Arakura to the north of Fuji went so far as to use the mud and silt from diverted slush flows (*yukishiro*) to create arable land on top of lava.[3] They also dug a

3,681-meter-long tunnel to channel water from Lake Kawaguchi.[4] Villages on the eastern side of the mountain depended on a 1,280-meter-long aqueduct built in the seventeenth century to convey water from Lake Ashi, and farmers near Suruga Bay converted much of the wetland that surrounded the Ukishimanuma (Ukishima Lagoon) into rice paddies.[5] Because of the local geology, topography, and climate, there were limits to how much upland communities around Fuji—and particularly those to the north of the mountain—could depend on agriculture to make ends meet. They therefore turned to other money-making enterprises, such as forestry and the packhorse trade. Northern communities especially relied on the production of silk. In fact, the tunnel connecting Lake Kawaguchi to Arakura was largely financed by silk producers, and much of the water drawn from the lake was used to dye silk.[6] Meanwhile, thanks to a milder climate, the southern villages cultivated tea. They also manufactured paper known as *Suruga hanshi* that was prized for its high quality. Silk, tea, and paper—these three commodities drew Fuji into trade networks that were first national and, starting toward the end of the Tokugawa period, global in scope, transforming local ecologies in the process.

Silk

Communities at the northern foot of the mountain have been making "Kai silk" for at least a thousand years. But, reflecting a nationwide trend, it was in the Tokugawa period that all aspects of silk production—from harvesting mulberry leaves to weaving finished cloth—rapidly expanded.[7] The silk industry that grew in the shadow of Fuji was household based and depended largely on the labor of women who raised silkworms, spun and dyed thread from their cocoons, and wove cloth. Using clear, cold water from Fuji allowed them to create silk with an exceptionally fine texture.[8] Some villages, like Shimoyoshida, focused mainly on weaving finished textiles. Others, including Narusawa and Yamanaka (now called Yamanakako), supplied raw fiber from the cocoons they raised.

Silkworms feed ravenously on mulberry leaves (see fig. 8.1), so local farmers planted mulberry trees around homes, along riverbanks, on ridges between cultivated fields, and in woodland clearings. From the early nineteenth century, an increasing number of fields were turned into mulberry orchards. Demand was strong enough that some villagers invested the time and effort to plant trees on top of lava flows, which required breaking up basalt and covering it with dirt.[9] Even so, the production of mulberry leaves and the cocoon-spinning worms that fed on them could not keep up with the voracious demand of silk

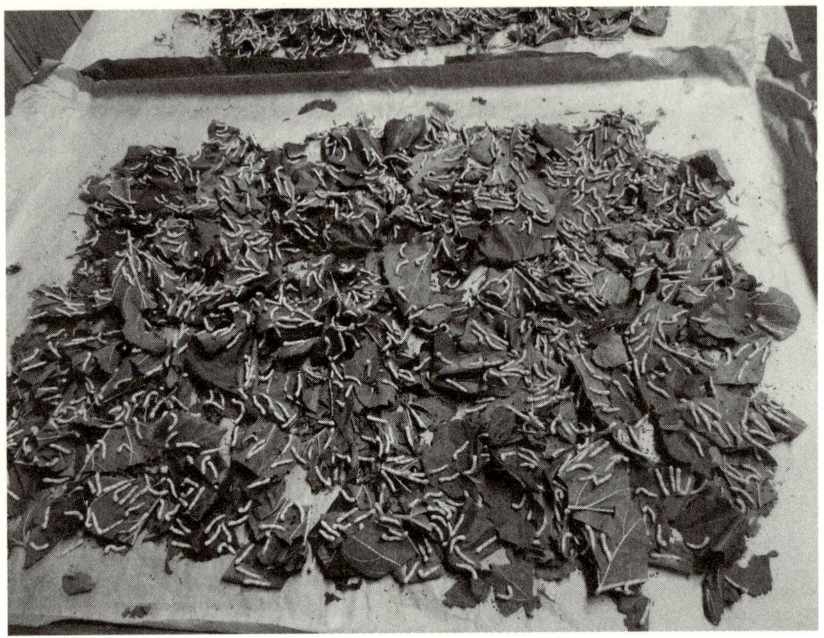

FIGURE 8.1. Silkworms feeding on mulberry leaves. Photo by author.

spinners and weavers, who relied on large amounts of raw fiber imported from other parts of Kai Province and even farther afield.[10] Dyers, meanwhile, took advantage of plants that grew on and around Fuji, using the tannin-rich berries and bark of Japanese alders to make a black dye and gromwell roots to make a purple one. Yet they too depended heavily on material imported from beyond the region—in their case, indigo grown in western Japan.[11] Thus a local ecosystem composed of mulberry leaves, Fuji water, and the laboring bodies of both silkworms and people (especially women) produced goods for national markets and was at the same time vulnerable to fluctuations in the cost of raw materials determined by those same markets.

When Japan was forced to open trade relations with the West in the 1850s, the local silk industry reached new heights as a result of foreign demand, particularly for raw fiber that could be turned into finished textiles abroad. A disease called pebrine had decimated silkworms in Europe, and the combination of the Taiping Rebellion (1850–1864) and the Second Opium War (1856–1860) had disrupted exports from China. This made Japanese silk a sought-after commodity, and raw silk fiber quickly became Japan's number one export and remained so right up to 1940.[12] With both foreign and

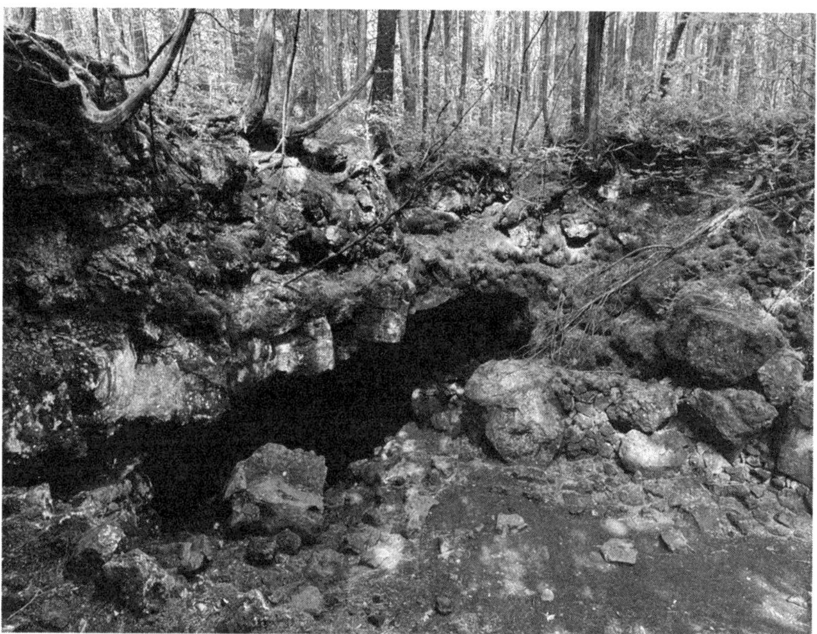

FIGURE 8.2. Entrance to cave in Aokigahara forest where locals stored silkworm eggs. Photo by Philip Ellway.

domestic demand increasing the value of Japanese silk, farmers at the base of Fuji expanded the cultivation of silkworms and devoted more land to mulberry trees in the years following the Meiji Restoration.[13] Toward the end of the nineteenth century they began to store silkworm eggs in caves in the Aokigahara lava field, taking advantage of their cool and steady temperatures (see fig. 8.2).[14] Local weavers, meanwhile, responded to the changing times by manufacturing silk for Western-style umbrellas.[15]

Mechanically powered looms were slow to catch on in communities at the base of Fuji, but their numbers boomed in the 1910s thanks to electricity (generated by water originating from Fuji) and the rapid growth of Japan's economy during World War I.[16] A newspaper article from 1920 observed that many locals had "abandoned the agriculture, commerce, and charcoal business of their ancestors to become weavers, installing mechanically powered looms and enthusiastically producing Kai silk." Thanks to their newfound prosperity, according to the article, even those living in remote villages sported gold watches, gold rings, and gold teeth. They were also able to purchase fine silk garments and patronize brothels.[17]

Even as silk production at the northern base of Fuji was mechanized, it remained, for the most part, a cottage industry. Starting in the early Meiji period, textile mills were built in different regions of Japan, and this was true to some extent in the communities north of Fuji.[18] As late as the mid-1930s, however, more than 90 percent of textile manufacturers in the region had fewer than five looms and the average number of looms per manufacturer was two.[19] Even in the decades following World War II, the local textile industry continued to be dominated by small, mainly family-based workshops.[20] Despite efforts by local officials, politicians, and businessmen to lure larger companies to the region, as of 1970, up to 93 percent of local textile producers still relied on family labor, with only about 7 percent employing workers who were not family members.[21]

The fate of the textile business north of Fuji was inextricable from geopolitics, military conflicts, and global trade. The good times of the late 1910s came to an abrupt end in the post–World War I recession, but by the end of the 1920s, the silk industry had fully recovered.[22] Meanwhile, an increasing number of textile producers north of Fuji, like their counterparts throughout Japan, turned to rayon because the raw material used to make it (wood pulp) was far less expensive than silk thread.[23] Both the rayon and silk industries suffered during the worldwide depression in the early 1930s, but were then buoyed by exports to Japan's colonial possessions. Textile makers especially targeted Manchuria, which Japan had seized from China and turned into a puppet state in 1932.[24] When the tide turned against Japan's military during World War II, industry conditions deteriorated as laws targeting consumer items deemed peripheral to the war effort took their toll. The final blow for textile makers came when the national government confiscated thousands of looms—along with temple bells and the great metal torii that marked the entrance to Kamiyoshida's oshi district—to provide metal that could be used to manufacture weapons.[25]

The production of silk and rayon continued to be a mainstay of the local economy after World War II. Families bought new looms to replace the destroyed ones; and through the 1960s, locals in Narusawa and surrounding villages continued to store silkworm eggs in the Aokigahara lava caves.[26] Yet the industry could not withstand the forces of globalization. Raw silk and cheap textiles produced overseas flooded the market, leading to a precipitous drop in the number of local manufacturers from the 1970s. Now almost all the mulberry trees at the foot of Fuji have vanished; and instead of thousands of textile producers in the region, there are only several hundred remaining. These have managed to hold on by producing high-quality goods for niche markets. Although the textile industry is no longer a pillar of the economy, it remains a point of pride. Visitors arriving at the Fujisan train station in Fujiyoshida will encounter a weaving information center, and the city hosts a textile festival each October.

Tea

Like the manufacture of raw silk and textiles to the north of Fuji, the production of tea to the south of the mountain grew during the Tokugawa period and accelerated after the Meiji Restoration. Tea plants thrive in volcanic soils. Along with the adoption of new steaming and rolling techniques, this allowed the region south of Fuji to become a significant tea producer during the eighteenth century, when much of the tea was shipped to Edo.[27] The opening of Japan to trade with the United States, Britain, and other Western countries at the end of the 1850s provided a sudden boost to tea production throughout Japan, including in the shadow of Fuji, where entrepreneurs expanded tea fields and improved production methods.[28] Trade continued to expand after the Meiji Restoration, with now-unemployed samurai turning to tea cultivation to make a living.[29]

As foreign (and particularly American) demand for Japanese tea rose over the decades that followed,[30] tea growers opened new fields and applied fertilizer to increase output in existing ones. During the Meiji period, fertilizer consisted mainly of cut grass, but in the Taishō period (1912–1926), herring meal, soybean meal, and vegetable oil cakes were added to the mix. Farmers also adopted chemical fertilizers derived from fossil fuels, increasingly relying on them in the years following World War II.[31] It was also in the postwar period that the harvesting and processing of tea leaves were mechanized on a mass scale.[32] Automation, combined with the generous use of fertilizer, allowed the growing and processing of tea to continue to expand south of Fuji, where it remains an important agricultural product today (see fig. 8.3).

Fuji's mystique is often used as a marketing tool for the tea grown in the foothills below. When a new café opened at the Portland Japanese Garden in 2017, the Jugetsudō tea company sponsored an article in a local magazine titled "Jugetsudo's Exquisite Teas from Mount Fuji Now Pouring at the Portland Japanese Garden." The article/advertisement stresses that "the teas themselves are grown in the highlands of Shizuoka, close to Mount Fuji, where the pure water and dramatic swings in temperature from day to night impart the tea leaves with their excellent and unique flavors and aromas."[33] What goes

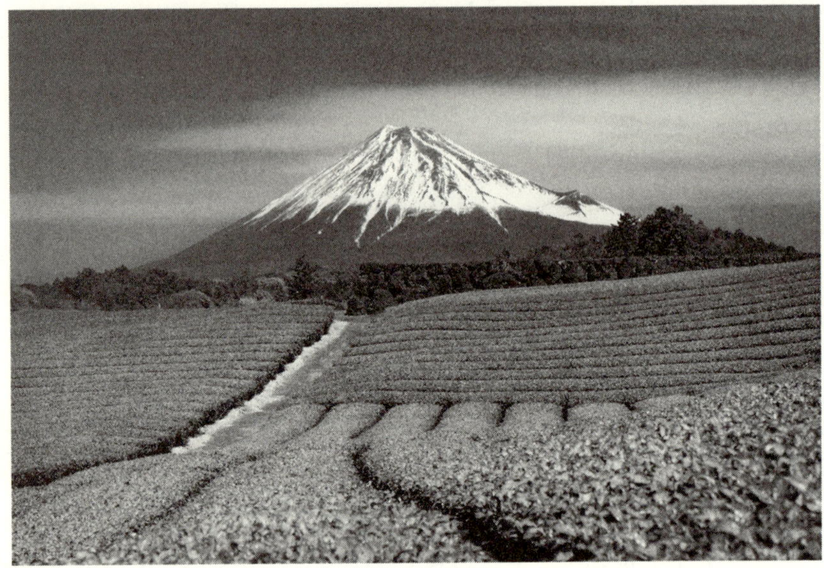

FIGURE 8.3. Fuji and tea fields. Alamy stock photo.

unmentioned is the contamination of local groundwater caused by the ample use of inorganic fertilizers.[34]

Paper and Pollution

As the cultivation of tea expanded south of Fuji, so did the manufacture of paper, which became the most economically important—and ecologically destructive—commodity produced in the region. Like the textile and tea industries, the paper industry depended on water from Fuji. Large-scale paper manufacturing in the region began in the late eighteenth century, when local farmers in what is now Fujinomiya City began using the bark of *mitsumata* (*Edgeworthia chrysantha*) to make sheets of paper called *Suruga hanshi*, which was prized in Edo, Osaka, and elsewhere for its high quality.[35] *Mitsumata* remained a key ingredient after the Meiji Restoration, so village leaders established large plantations of the shrub in the area. The cultivation of *mitsumata* and the manufacture of paper provided welcome employment for locals thrown out of work after the new regime, in 1871, closed the Yoshiwara post station along with fifty-two other official post stations along the Tōkaidō.[36]

Over the decades that followed, a series of entrepreneurs adopted new production techniques and machinery to turn what had been a home-based craft into a factory-driven industry.[37] By the early 1900s, machine-made Japanese-style paper (*washi*) was exported from the region to markets around the globe, including America and Britain. The letter paper produced by one factory won gold medals at the Lewis and Clark Centennial Exposition in Portland, Oregon, in 1905 and the Japan-British Exhibition in London in 1910.[38] Another company that produced paper napkins (which displayed, among other things, images of Fuji) won a silver medal at the same exhibition in London.[39]

As time passed, *washi* makers both exported their products to foreign countries and imported raw materials from abroad. From the 1920s, large amounts of abaca (manila hemp) in particular were imported from Japanese-owned plantations in the Philippines. The relatively cheap and easily processed fiber either replaced existing materials (such as *mitsumata*) or was combined with them.[40]

The Meiji period also witnessed the rapid growth of a Western-style paper industry that used wood pulp from entire trees instead of the bark of *mitsumata* or other plants traditionally used to make *washi*. With the help of foreign experts, Japan's first Western-style paper manufacturers were established in Osaka, Tokyo, Kyoto, and Kobe.[41] From the 1890s, local entrepreneurs and Tokyo-based investors built up a particularly large concentration of pulp and paper mills south of Fuji, thanks to the mountain's water and a generous supply of timber on its lower slopes. The largest paper company in the area was Fuji Seishi (Fuji Paper), which set up its first factory on the banks of the Urui River in Iriyamase Village (now part of Fujinomiya City) to take advantage of its hydropower.[42] It also received permission to harvest trees—mainly fir, hemlock, and spruce—from a nearby state-owned forest.[43] In his 1901 guidebook to Fuji, the celebrity meteorologist Nonaka Itaru warned climbers taking the route from Ōmiya Shrine (Fujisan Hongū Sengen Taisha) to watch out for tumbling logs.[44] Fuji Seishi and other paper companies established in the Meiji period benefitted from the opening of the Tōkaidō Railroad in 1889, which provided easy access to major markets (most importantly, Tokyo);[45] and they used nearby Suruga Bay to dispose of wastewater produced by their mills. This practice began as early as the 1890s, when local residents, worried about pollution discharged into the Urui River, pressured Fuji Seishi to build a wastewater channel from its paper mill to the mouth of the Numa River, which empties into the Tagonoura Inlet of Suruga Bay (see map 8.1).[46]

By 1932 Fuji Seishi had grown into the largest paper company in Japan, making 36 percent of the paper produced nationwide.[47] After the opening of its factory in Irimayase, it built several more in the region.[48] Between 1897 and

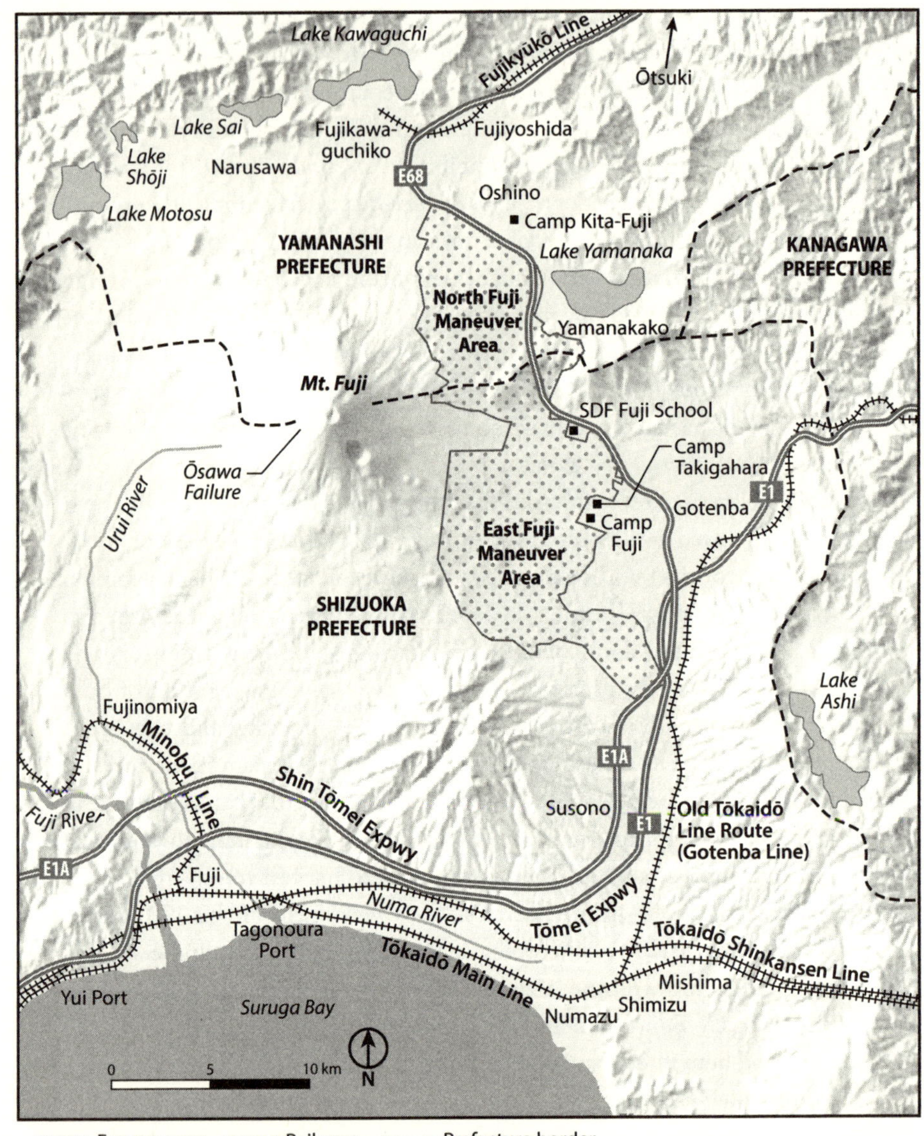

MAP 8.1. Fuji region in twentieth and twenty-first centuries.

1908, the company also established mills in Hokkaido to take advantage of its coniferous forests. After Russia relinquished the southern half of Sakhalin Island (called Karafuto by the Japanese) to Japan after its defeat in the Russo-Japanese War (1904–1905), Fuji Seishi, along with competitors Ōji Seishi and Karafuto Kōgyō, established operations there too.[49] Forests close to the Yalu

River (located between Korea and Manchuria) became another important resource for Japan's growing paper industry.[50] In 1933 Fuji Seishi, Karafuto Kōgyō, and Ōji Seishi merged into a giant conglomerate under the name Ōji Seishi, controlling over 80 percent of Western paper production in Japan and monopolizing the market for newsprint. As the Japanese paper industry expanded its reach in Hokkaido, Sakhalin Island, and the continent, it continued to flourish in the industrial zone south of Fuji—especially in the villages and towns that now constitute Fuji City, where pulp and paper mills continued to multiply and where Japan's second-largest paper company after Ōji Seishi, Daishōwa Seishi, was established through a series of mergers.[51]

Like the textile industry, Japan's pulp and paper industry suffered as war in Asia and then the Pacific intensified. Between 1932 and 1937, exports of Japanese paper doubled, thanks in large part to Japan's domination of the Chinese and Manchurian market; and the growth of rayon manufacturing, encouraged by a government looking to reduce dependence on imported cotton, increased the demand for pulp.[52] But in 1938 the government instituted price controls and in the following year began rationing the raw materials on which the industry depended. As the production of newsprint paper and other paper for nonmilitary uses plummeted, the government ordered factories south of Fuji to make goods for the armed forces such as electrical cables, materials for gunpowder, and propellers and engine parts for airplanes.[53] Meanwhile, forests on Fuji's skirt, like forests throughout Japan, were sacrificed to the war effort, with large swaths of pine trees cut down so that oil extracted from their roots could be turned into airplane fuel.[54] In this way, Fuji was entangled with Japanese militarism materially as well as ideologically.

The Japanese economy struggled during the first few years after Japan's defeat, and the devastated paper industry struggled along with it, but procurement spending by US forces during the Korean War (1950–1953) stimulated the economy, which then grew at a blistering pace during Japan's "high-growth" period from the mid-1950s to the early 1970s. The three companies derived from Ōji Seishi (the conglomerate was broken up during the US Occupation) prospered during this period along with hundreds of other paper and pulp companies created before and after the war. Production ramped up to the point that Japan became the world's third-largest producer of paper in 1959 and the second-largest in 1970 (today it is again third, after China and the United States).[55]

The feverish construction of new factories south of Fuji—mainly to produce pulp and paper but also to manufacture chemicals, auto parts, and other industrial goods—had significant environmental impacts, including the industrialization of the Ukishimagahara wetlands on the eastern side of the Tagonoura Inlet. From the seventeenth to nineteenth centuries, farmers had created water control systems to grow rice in the marshlands surrounding the Ukishima Lagoon, which shrank significantly over time.[56] In the 1930s a canal was built to divert water from the Numa River—which drained what was left of the lagoon into the Tagonoura Inlet—directly into Suruga Bay, sharply reducing the risk of flooding and opening even more land to cultivation.[57] It also allowed the postwar development of Tagonoura into a port suitable for vessels that could, among other things, supply the imported wood chips on which the Japanese pulp and paper industry increasingly depended.[58] This involved extensive dredging and the transformation of the western end of the Numa River into a shipping channel with a boat basin.[59] A second diversion canal for the Numa River built in the 1960s drained what remained of the lagoon.[60] Today the former wetland around the port of Tagonoura is an industrial zone consisting largely of warehouses and a large fuel depot. With the Ukishimagahara converted to agriculture and industry, only a small portion of the original wetland remains, confined for the most part to a minuscule nature preserve flanked by factories and a major highway.

In addition to reshaping the physical environment, the rapid growth of the pulp and paper industry severely polluted it. Despite recent efforts to limit toxic emissions and waste, pulp and paper making have long been among the most ecologically destructive activities on the planet. Their production consumes tremendous amounts of water and energy and relies on wood from either old-growth forests or monoculture plantation forests, sometimes referred to as "green deserts" due to their lack of biodiversity. It also releases significant amounts of pollution into water and the atmosphere, including sulfur-based compounds used in making pulp that contribute to acid rain, and chlorinated dioxins involved in the bleaching process that are highly toxic and bioaccumulate in the food chain (see fig. 8.4). Even paper recycling, widely viewed as ecologically responsible, creates problems because of harmful chemicals involved in the conventional deinking process, although recently progress has been made in replacing those chemicals with "bio-deinking" enzymes generated by bacteria and fungi.[61]

The contamination of air and water south of Fuji was particularly bad from the 1950s to the mid-1970s. Sulfuric smoke choked the air, and untreated effluent from pulp and paper mills poured into the Tagonoura Inlet, forming sludge that released foul-smelling gas and contaminated the bay with PCBs (polychlorinated biphenyls) that harmed local sea life, including

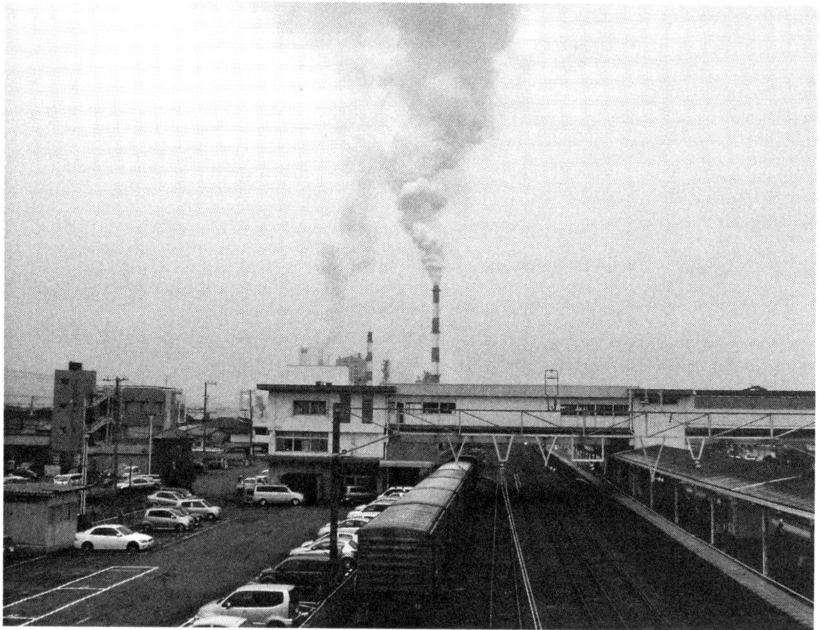

FIGURE 8.4. View of paper mill smokestacks from Fuji City train station. Photo by author.

the bay's famous *Sakura ebi* shrimp on which the livelihoods of local fisher-men depended.[62]

Pollution at Fuji drew national and even international condemnation. Because of the horrific pain and deformities they caused, as well as their out-sized impact on the legal and regulatory treatment of environmental pollution and its consequences, Minamata disease and "it hurts, it hurts disease" (*itai-itai byō*)—one caused by mercury poisoning in Minamata Bay, the other by cadmium poisoning in the Jinzū River Basin—were seared into public mem-ory as examples of the biological cost borne by humans and nonhumans for the rampant industrialization that turned Japan into a "toxic archipelago," to use Brett Walker's wording.[63] The pollution at Fuji's base is not as infamous today, but during the 1960s and '70s, it was a major focus for the media, due in part to the striking contrast between the splendid beauty of Japan's national icon and the unsightly vision of contaminated air and water below. The first paragraph of an article on Japanese pollution published in 1972 in the *New York Times* paints the following picture: "Today, whizzing by the sea at 150 miles an hour in a 'bullet' train, one marvels at the snow-swept cone of *Fuji-san* rising majestically from a plain covered with twisting power lines, jutting smokestacks, sprawling factories and feeder roads." The article then introduces Koda Toshihiko, a resident of Fuji City suffering from asthma caused by the smoke spewed by paper mills. Koda ruefully observes, "Iron and bronze are damaged by gas from the factories" and "students in the elementary schools grow up suffering from asthma. Plants wither and die. The birds around Mount Fuji are decreasing in number. They no longer visit the town." The article also describes the contamination of Suruga Bay, noting that locals could no longer swim in its "reddish-brown waters, once renowned for their crystal clarity."[64]

The polluted air and water south of Fuji also feature in the 1971 monster film *Gojira tai Hedora* (*Godzilla vs. Hedorah*). Originating in outer space, Hedora grows from the toxic sludge in Suruga Bay (the name "Hedora," also transliterated as "Hedorah," derives from *hedoro*, the Japanese word for sludge) and then on the smoke emitted by factories on shore. Godzilla ends up battling Hedora several times, and with the help of electrodes built by the Japanese military, finally vanquishes the monster by frying its body (composed of sludge) until only dust remains. Fuji appears multiple times in the film. At the start, the mountain looms in the distance with smokestacks in the foreground. Later there is a lingering shot of snow-capped Fuji after Godzilla has destroyed Hedora. At the end of the film, we see Hokusai's famous *Great Wave* woodblock print followed by this ominous question: "Will there be another one?" (*Soshite mō ippiki?*).[65] The implication is that there will be unless humanity cleans up its act.

In the years immediately before and after *Gojira tai Hedora* was released, citizen-activists south of Fuji, like their counterparts throughout Japan, put tremendous pressure on politicians and businessmen to combat pollution.[66] They organized study meetings and high-profile demonstrations, including an especially prominent protest attended by thousands of protestors in the summer of 1970 in Fuji City. In a vivid display of their anger, fishermen from nearby Yui Harbor (see map 8.1), alarmed by the drop in catches of shrimp, formed a flotilla of boats with large banners declaring, "Eliminate sludge pollution" (*hedoro kogai tsuihō*) and "Restore Suruga Bay" (*Suruga-wan o kaese*) to attend the protest. They and other protestors garnered supportive coverage in the national press and on television.[67]

Protests like this one, combined with successful litigation and the election of local politicians sympathetic to the antipollution cause, forced the paper and pulp industry not only to pay compensation for damage already done but also to take measures to mitigate pollution in the future.[68] Mills invested in new equipment to treat wastewater, leading, by the end of the 1970s, to a sharp reduction in particulates entering local rivers and Suruga Bay; and public and private funding was used to remove the built-up sludge from the Tagonoura Inlet and construct a facility to process it for final disposal (although it took a citizens' lawsuit to force Shizuoka Prefecture to collect the money owed by the major pulp and paper companies for this project). Meanwhile, pulp and paper mills, along with other factories in the region, installed equipment to reduce harmful airborne emissions, particularly sulfur dioxide.[69]

These local antipollution measures tracked with national ones. The Diet that convened in 1970 was nicknamed the "Pollution Diet" for its passage of numerous pieces of environmental legislation, and in the following year the government created an Environment Agency (upgraded to a ministry in 2001). Thanks to government directives and private initiatives, the quality of Japanese air and water improved dramatically in the decades that followed. So did energy efficiency and rates of recycling.

Despite progress on these fronts, numerous environmental problems continue to plague Japan. These range from the loss of plant and animal habitats to the prevalence of endocrine-disrupting chemicals in human and nonhuman bodies alike.[70] In the industrial corridor that extends along the northern shore of Suruga Bay, a particular concern is the depletion and salinization of local groundwater supplied in large part by Fuji. Business and government leaders were well aware of the problem by the 1960s, when research showed an alarming drop of the water table and the salinization of wells, although the severity of salt intrusion varied according to seasonal rainfall and the specific geology of each locale. Various industries were responsible, but thirsty pulp and paper mills were particularly to blame. In response, businesses and government officials developed pumping regulations and also cooperated in building an aqueduct to draw additional water from the Fuji River. Further efforts were made to increase water efficiency, including the use of recycled water.[71] Yet the overconsumption of groundwater remains a significant worry in the region. The problem is on vivid display in Mishima, an industrialized city located at the end of one of Fuji's lava flows and dependent on the water that runs through it (see map 8.1). Along with neighboring Shimizu and Numazu, Mishima gained notoriety for a citizen's campaign that stopped the construction of a sprawling petrochemical complex in the 1960s—in fact, the success of this campaign provided a model for the antipollution movement in Fuji City[72]—and in recent decades, enormous strides have been made in revitalizing the city's waterways and adjoining habitats thanks in large part to a

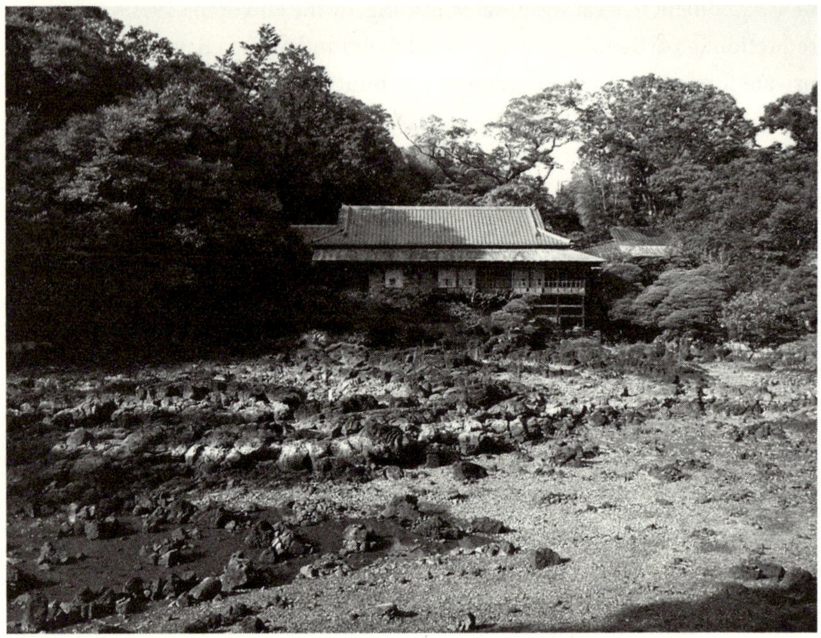

FIGURE 8.5. Typical state of Kohama Pond in Rakujuen Park, Mishima City. Photo by author.

nonprofit called Groundwork Mishima.[73] Nevertheless, visitors to the city's most popular tourist site, Rakujuen Park, can see that industrial consumption—especially by Toray Industries' Mishima plant—continues to have a major impact on the local water table.[74] The park was originally the estate of Imperial Prince Komatsu-no-miya Akihito (1846–1903), who built a villa next to a pond in 1890. Today the park is listed as a Japanese Natural Monument (*tennen kinenbutsu*), but the pond is usually empty, filling only during periods of extended rainfall (see fig. 8.5). Given the central importance of the pond to the grounds of the villa, the loss of its water has had a profound impact on both the ecological and cultural integrity of the site.

Pollution in Suruga Bay remains a problem too. The sludge may have been cleaned up, but mercury, PCBs, and other toxins saturate the tissues of fish and other organisms that inhabit its waters. Tenacious and lipophilic POPs (persistent organic pollutants, often referred to as "forever chemicals") accumulate in animals higher up the food chain—including humans. In the late 1990s, several decades after the sludge was removed and pollution controls put into place, substantial levels of POPs were detected in a range of sea life living in

Suruga Bay.[75] An article published in 2017 on the enormous amounts of POPs found in creatures living deep in the Mariana Trench awarded the bay the distinction of being "the only Northwest Pacific location with values comparable" to those in the trench.[76] This unenviable status was reinforced by a study published in 2022 on deep-sea sharks in Suruga Bay showing that these predators carry high loads of flame retardants (PBDEs, polybrominated diphenyl ethers) used in an assortment of manufactured goods.[77] Japan is often described as "postindustrial," but the contaminated sharks and other creatures of Suruga Bay would not know it.

Meanwhile, most of the forest planted around Fuji decades ago now goes unused and unmanaged, reflecting a trend seen throughout Japan. Japanese forests were decimated during World War II, and with the loss of timber from Sakhalin Island and other colonies in Asia, clear-cutting in the "home islands" (*naichi*) continued at a rapid pace in the years following the war. To replenish the wood supply, the postwar government subsidized the planting of timber trees such as Japanese cedar (*sugi*) and cypress (*hinoki*) on a massive scale.[78] It takes time, however, for these trees to reach maturity, and with the demand for wood soaring during the high-growth years, Japanese companies came to depend on imports. Starting in the 1950s, the paper and pulp industry began importing whole logs, but it then turned primarily to wood chips during the 1960s.[79] The industry's reliance on imported wood chips and pulp grew markedly over time. In recent years, only 15–17 percent of the pulp and wood chips consumed in Japan have been produced domestically.[80] Because it was cheaper to source wood abroad—particularly in Southeast Asia, where state patronage and a lack of regulations dampened prices[81]—Japan's construction industry turned to wood imports as well. Not only was it relatively difficult to harvest trees on Japan's steep and often hard-to-access slopes, but much of the country's forest consisted of small woodlots owned by families who lacked the means to invest in up-to-date equipment or pay rapidly rising wages that, compared to those in countries like the Philippines and Indonesia, made (and continue to make) Japanese labor prohibitively expensive.[82] As a result, from 1955 to 2000, industrial wood imports of all kinds (including chips, logs, pulp, plywood, and sawn lumber) went from about 5.5 percent of industrial wood used in Japan to over 80 percent. By 2021 the percentage had dropped to around 64 percent, but that was due less to an increase in domestic production of wood (which rose just slightly) than to an overall decline in demand for wood and wood products.[83]

By relying so heavily on imports, Japan has cast a long "ecological shadow" overseas and at the same time produced—through neglect—unhealthy,

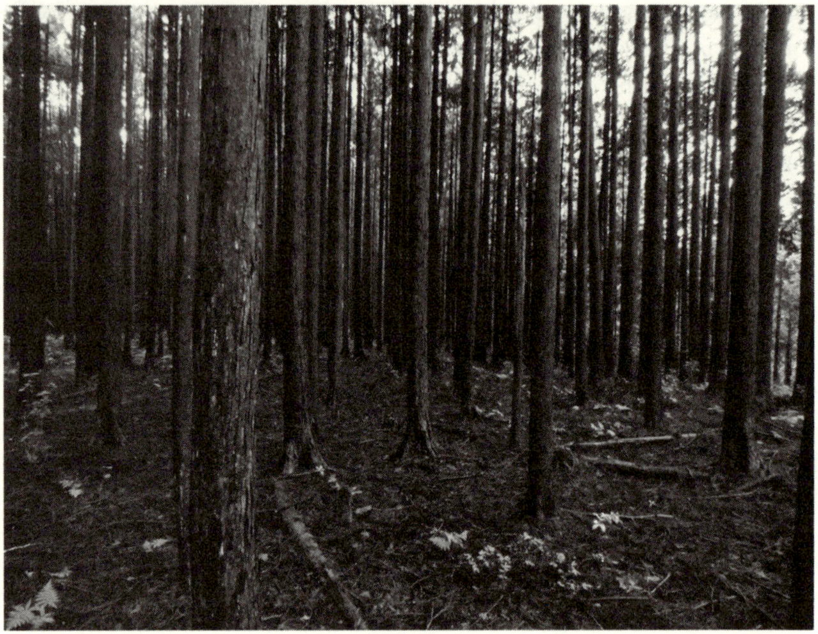

FIGURE 8.6. Plantation forest (a "green desert") at Fuji. Photo by author.

sun-deprived forests at home.[84] Since the 1980s, Japanese trading companies and the Japanese government have invited harsh criticism for encouraging the destruction of old-growth forests in Oceania and Southeast Asia.[85] Less noticed is the process of "dual decay" in which the destruction of forests abroad is tied to the underutilization of plantation forests in Japan, where the low cost of wood deters owners from maintaining, harvesting, and replanting them. Exacerbating the problem is the rapid aging of rural communities and the consequent shortage of working-age laborers. It might not seem to be a problem to leave forests alone, but the trees that comprise them were originally planted in crowded rows in anticipation that they would be regularly thinned. When thinning is neglected, not enough sunlight penetrates the canopy, leading to poorly developed root systems and a lack of ground cover. The result is a "green desert" that, in addition to lacking biodiversity, is vulnerable to erosion and landslides in Japan's rainy climate (see fig. 8.6). The tightly packed, easily uprooted trees are also subject to wind damage, at Fuji as elsewhere.[86]

Efforts have been made on the national and local levels to reverse the decline of Japanese forests and forestry. The government's Forestry Agency has been working with local partners to consolidate forest management entities,

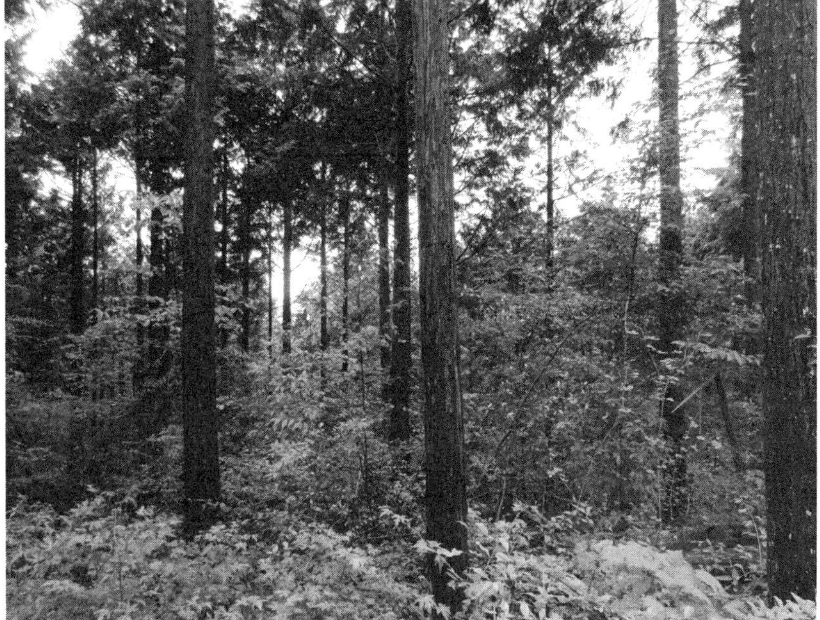

FIGURE 8.7. Test plot where trees have been thinned and selectively cut. Photo by author.

develop new technologies, build new forestry roads, and encourage the use of wood in public buildings, among other initiatives.[87] Some progress has been made since 2002, when domestic wood production hit a low, but the government is fighting an uphill battle (both literally and figuratively) as the stock of harvest-ready timber on and around Japanese mountains continues to expand.[88] On Fuji's southern skirt, forestry expert Watanabe Sadamoto has established test plots to show that a strategic combination of row thinning and selective cutting rather than clear-cutting can lead to both greater forest health and economic gain (see fig. 8.7). His methods have yet to be widely adopted, so whether they will make a large-scale difference remains to be seen.

Guns and Grass

Climbers making their way up Fuji are sometimes accompanied by artillery fire booming in the distance. Most think of the mountain as a glorious wonder of nature, yet the US and Japanese militaries currently use about 13,400 hectares (over 33,000 acres) of its lower slopes for barracks, training grounds, and target ranges, which are grouped into two sites, the North Fuji Maneuver Area and the East Fuji Maneuver Area (see map 8.1). The militarization of Fuji took shape in the years leading up to World War II and during the Cold War that immediately followed. On one level, this is a familiar story of those with more power imposing their will on those with less. But while Fuji was drafted to serve the nation—and after World War II, the American-led battle against communism—villagers living at the base of the volcano also worked to incorporate the Japanese and then US militaries into an ecosystem shaped by traditional common land practices. The agency of these villagers demonstrates how militarized landscapes are often fashioned not only by the goals of military and bureaucratic elites but also by the needs and desires of those who live in or nearby them.

There are two main reasons why the Japanese and US militaries find the area attractive. One, it is located only a little over an hour's drive from Tokyo. Two, it consists for the most part of grasslands, with the Nashigahara grassland constituting most of the northern maneuver area and the Ōnohara grassland the eastern one. Together with the adjoining woods, these grasslands—which are the largest in the Japanese Islands outside Hokkaido—offer a terrain well suited for military exercises.

This terrain would not exist were it not for centuries-old common land practices that developed in response to the local geology and climate. Relatively cold and lacking a steady supply of surface water, much of it is unsuitable for intensive farming, so for hundreds of years locals used the land communally for a variety of other purposes. In the coppiced woodlands they hunted small animals and gathered construction materials and fuel—including firewood to heat silkworm houses and boil water during the silk-spinning process. Like their counterparts in other rural communities in Japan and elsewhere around the world, they also maintained grasslands by burning and/or cutting them. Vast grasslands circled almost the entirety of Fuji's lower skirt into the early twentieth century, providing edible plants, fertilizer for agricultural fields, material for thatched roofs, and fodder to feed horses.[89] During the Tokugawa period, locals raised horses for both agricultural work and the lucrative packhorse trade. Horses continued to prove their worth in the late nineteenth and early twentieth centuries by pulling carriages on horse-drawn railroads in addition to working on farms.[90]

Access to the common lands at Fuji and elsewhere in Tokugawa Japan was determined by locally negotiated common land use rights, in Japanese called *iriai-ken*. But while the shogunate recognized them to be legally binding, these rights were thrown into question in the years following the Meiji Restoration. One of the new regime's first orders of business was to implement reforms concerning the taxation and registration of land—the aim being a market-based system of land ownership and exchange that generated predictable tax revenues for the state. In the process, the government made it difficult for communities to formally register their iriai lands as lands *owned* and not just *used* in common. Officials also chose to register as state property those common lands for which adequate documentation of ownership and iriai rights was unavailable. Of course, it was the government that determined what documentation was adequate, so in the 1870s and early 1880s, it expropriated a vast amount of iriai land.[91] In Yamanashi Prefecture, nearly all iriai land was declared government property.[92]

While the Meiji government made it difficult for communities to establish common ownership of iriai lands, it nevertheless recognized their customary entry into and use of them. In fact, the Meiji Civil Code (in effect from 1898) explicitly stated that iriai rights had legal standing and could function according to local custom, meaning that groups of traditional iriai rights-holders could, in theory, continue to access their common lands regardless of who owned them.[93] Trying to implement modern property rights while retaining customary use rights created situations ripe for conflict, and the government worked to undermine the iriai rights it claimed to respect. In 1915 Japan's Supreme Court dealt an especially significant blow to iriai groups who claimed usufruct (use rights) on state lands, ruling that any common lands the early Meiji government had registered as state property no longer had iriai rights attached to them (the ruling was overturned in 1973).[94] Nevertheless, if iriai groups put up enough of a fight, or if iriai practices on land registered as state property did not interfere with state goals, officials often tried to reach some kind of accommodation—all the while maintaining that they were not formally recognizing iriai rights, but simply showing respect for iriai customs.

People living around Fuji were zealous in controlling access to their commons, as the Japanese military discovered when it began establishing itself in the region early in the twentieth century. During the 1890s the army trained occasionally in what is now the East Fuji Maneuver Area, but when it tried to build its first permanent barracks in 1908, it encountered stiff resistance from local residents, many of whom not only claimed iriai rights but were also landowners. They were won over only after the army gave them the right to collect spent bullets and artillery shells, which could be sold for scrap metal, and the manure produced by army horses, which could be used for fertilizer. In 1909 villagers also entered an agreement with the army stipulating that military exercises would not hinder visitors from Tokyo and elsewhere from climbing Fuji during the summer months and that soldiers would keep out of cultivated fields and avoid damaging stands of trees.[95] Despite these precautions, it was impossible to keep artillery and gun fire from inflicting a certain amount of damage on trees and fields, so in 1912 the army signed another agreement with local village officials, property owners, and iriai rights-holders to pay compensation for harm done to land in the area, whether public or private.[96] In effect, the army rented its own, limited form of iriai rights, making it a fellow, although temporary, stakeholder in the region.

Of course, the army was not just any stakeholder, but at no point could it simply ignore the wishes of landowners and iriai groups—even in the years before World War II, when the military was at its most powerful. Looking to find more space to test new equipment, in the mid-1930s the army set its sights on the Nashigahara grassland in what is now the North Fuji Maneuver Area. Most of the grassland and the surrounding woods were under the management of the Onshirin Kumiai, an association representing the iriai rights-holders of eleven formerly separate villages (today amalgamated into Fujiyoshida City, Yamanakako Village, and Oshino Village). It was named the Onshirin Kumiai, the "Imperial Gift Forest Association," because the wooded areas managed by the association had been designated imperial household forest in 1889 but were "gifted" to Yamanashi Prefecture in 1911.[97] The Onshirin Kumiai was established to manage the transferred iriai lands and continues in that role today.[98] When the army approached the association in 1936 to purchase access to the commons, its members resisted. In response, the military police barged into the homes of association leaders, threatening to send young men conscripted from their villages to the most dangerous front lines in China. Under the gun, the members eventually agreed to transfer their land to the army, but only after stipulating in writing that association members could continue to use the resources of their common lands and that they, like residents near the East Fuji Maneuver Area, had the right to collect spent bullets, artillery shells, and horse manure. The army agreed, promising to respect iriai customs so long as they did not interfere with military operations.[99]

The deals made between locals and the military lasted until the end of World War II, but after Japan's defeat, the Japanese government announced that these arrangements were now defunct. For a brief time, communities near the maneuver areas thought the land would return to peaceful purposes under their control. They quickly learned that Occupation forces had other plans. The US military took over both maneuver areas and expanded them while restricting entry far more than the imperial army ever had.[100] Use of the maneuver areas intensified after the outbreak of the Korean War (1950–1953), with exercises occurring day and night.[101]

The impact on surrounding communities was profound. Some locals were wounded and even killed. Farmers lost the right to gather grass, crops, timber, mulberry leaves (to feed silkworms), and brushwood (to heat homes and make charcoal); and farmers around the East Fuji Maneuver Area were no longer able to harvest turf that had been sold for years to maintain airstrips around Japan. Meanwhile, American tanks and artillery destroyed fields, woodlands, and streams, leading to floods and destruction of the water supply for surrounding communities.[102] When locals complained to officials at the prefectural level or in Tokyo, they were told that the Americans were calling the shots, so nothing could be done.[103]

Once Japan regained its sovereignty in 1952, the bargaining position of landowners and iriai rights-holders improved somewhat. The US-Japan Security Treaty, which went into effect that year, guaranteed American forces uninterrupted use of facilities in Japan, including the Fuji maneuver areas, so the government was able to use its treaty obligations to the United States to deflect local demands. But in response to pressure from local groups, it did arrange for greater access of rights-holders to the maneuver areas to cut grass, collect wood, and even cultivate certain plots of land. It also reached agreements with different communities to provide limited compensation for damage to their commons as well as their private property.[104]

At the end of the 1950s, the bulk of American forces in the Fuji region were relocated to Okinawa,[105] but US marines stationed at Camp Fuji in Gotenba have continued to use the maneuver areas on a regular basis. So have troops based in Okinawa and elsewhere in the Pacific region. The official website of the US Marine Corps makes today's military operations at Fuji seem only natural by situating them in a tradition going back centuries. We learn from the site that "the ground adjacent to Camp Fuji was used for training samurai warriors long before the Marines arrived. As far back as 1198 AD, the Kamakura Feudal Government trained more than 30,000 Samurai warriors on the same ground where Marines and other U.S. forces train today."[106]

From the 1950s, Japan's Self-Defense Forces (SDF) have used the maneuver areas as well, generating conflicts and compromises with local communities. Article 9 of the 1946 constitution written by Occupation authorities explicitly states that "the Japanese people forever renounce war as a sovereign right of the nation and the threat or use of force as a means of settling international disputes," and that "land, sea, and air forces, as well as other war potential, will never be maintained."[107] Japan's armed forces were, however, reconstituted amid the pressure of the Cold War—and were named the "Self-Defense" Forces so as to preserve the constitutional provision in name if not in spirit.[108] No longer able to assign the blame to the Americans, the Japanese government felt increased pressure from locals who argued that it was illegitimate for the SDF to use the maneuver areas without entering into formal contracts; and in 1959, farmers around the East Fuji Maneuver Area filed a lawsuit arguing that the military had no right to use the training grounds. At the time, the government faced fierce opposition to the US military presence throughout Japan, with large protests taking place in 1959 and 1960 to block (unsuccessfully) the ratification of a revised US-Japan Security Treaty. It was in this context that the beleaguered government decided to settle the lawsuit, reaching a long-term agreement with the East Fuji Maneuver Area Regional Farmers' Rehabilitation League (Higashi Fuji Enshūjō Chiiki Nōmin Saiken Renmei) that allowed locals to access the maneuver area on designated days and provided rental fees and compensation for damage. The government also created a plan to develop horticulture and rice and livestock farming in the region.[109]

In the years that followed, the Japanese government reached temporary deals with groups of iriai rights-holders around the North Fuji Maneuver Area—consisting mainly of common lands registered in the name of Yamanashi Prefecture or the state—but conflict was ongoing and sometimes erupted in dramatic protests. One high-profile demonstration took place on October 2, 1965, when a US artillery battalion fired a Little John rocket from the East Fuji Maneuver Area to the North Fuji Maneuver Area. Hundreds of locals occupied both the launch site and the impact zone to stop the testing. The police managed to clear the launch site of protestors but were unable to remove them from the impact zone. To everyone's astonishment, the commander of the battalion refused to back down. The rocket test went forward and through sheer luck no one was injured.[110] The artillery battalion scheduled another test for October 7, when it planned to fire the rocket from the North Fuji Maneuver Area to the East Fuji Maneuver Area. This time locals succeeded in stopping the test. On October 6, about two hundred farmers, along with thirty bulldozers, occupied the impact zone; and they were joined by hundreds more protestors in the early hours of the seventh. Police were mobilized to clear the protestors, but in the face of fierce resistance, the launch was called off.[111]

Highlighting the fact that the Little John rockets could carry nuclear pay-loads, left-wing activists around Japan used the conflict to advance their na-tionwide battle against the US-Japan Security Treaty; and during the Vietnam War and the decades that followed, they were joined by local residents who shared their antiwar sentiments. A prime example was a group of women from the hamlet of Shibokusa (part of Oshino Village) called Haha-no-kai (Mothers' Association) that repeatedly used guerrilla tactics to obstruct mili-tary exercises in the North Fuji Maneuver Area and were repeatedly arrested as a consequence. In a meeting with peace activist Leonie Caldecott in 1981, Haha-no-kai member Amano Mie said, "As we carried on with our campaign, we realised that the whole phenomenon of militarism is violence against the land, wherever it takes place. So we are really a part of the wider anti-war movement. You see, Mount Fuji is the symbol of Japan. If they are preparing war on her flanks, how can they say Japan desires peace?"[112]

Members of the Haha-no-kai viewed their struggle as part of the larger peace movement, but others living around the North Fuji Maneuver Area did not seek an end to the military exercises. Instead they wanted higher pay-ments from the government to compensate for interference with common land practices.[113] They succeeded in 1973, when government officials and the Onshirin Kumiai concluded a five-year renewable agreement with the SDF similar to the one signed with communities next to the East Fuji Maneuver Area in 1959. The North Fuji Maneuver Area contract required the govern-ment only to respect iriai "customs" rather than "rights," but it differed from prior agreements in that it also required the government to provide billions of yen for "initiatives to stabilize the livelihoods" of local iriai rights-holders and the "protection of natural resources."[114] Since 1973 Japan's military has renewed its agreements with iriai rights-holders, represented by the Onshirin Kumiai, on similar terms, providing funds for job-creating public works proj-ects that have supported the local economy.[115]

Meanwhile, radical changes to the economic and material conditions of the Japanese countryside rendered many age-old iriai practices either peripheral or obsolete. These included the maintenance of grasslands. As horses gave way to cars and tractors, and roof thatch was replaced with tiles and tin, rural com-munities stopped burning or mowing their grasslands, which subsequently turned into forest, both planned and unplanned, or were put to other uses such as golf courses. It is estimated that 14 percent of Japan's total land area at the start of the twentieth century consisted of grasslands.[116] Today this figure is less than 1 percent.[117] The Fuji region was no exception to this trend. Most of the grasslands that surrounded the mountain vanished in the mid- to late twentieth century, replaced mainly with forest.

Yet the grasslands in the North Fuji Maneuver Area and East Fuji Maneuver Area have survived because they benefit both local communities and the Japanese and American armed forces. Since the middle of the twentieth century, thousands of residents around Fuji have gathered each spring to burn the grasslands (see fig. 8.8). They do this not to feed horses or provide thatch for roofs but to assert common land claims that in turn perpetuate government payments. The SDF and US military are pleased that locals maintain the grasslands because doing so produces the terrain they need for their training (see fig. 8.9). There are still some local residents who oppose the military's use of their lands, but there is now wide acceptance of a status quo that has persisted for decades and will likely continue well into the future.

This relationship between the military and people who live at the base of Fuji also benefits plants, insects, birds, and other wildlife that depend on the grasslands for their survival. Among them are butterfly species like Reverdin's Blue (*miyamashijimi*) that have disappeared in other parts of Japan due to habitat loss. They would vanish around Fuji too were it not for the maintenance of the grassland vegetation that they need to survive.[118] This is not to say the militarization of Fuji has generated more positive ecological consequences than negative ones. But it does show, as scholars have noted in studies of warfare and the environment, that dedicating a landscape to military purposes is not necessarily bad for all species. It can indirectly benefit or even save them; and in the case of the Fuji maneuver areas, unlike places such as Colorado's Rocky Flats, this has happened not because civilians have been excluded from the landscape, but because they have worked to maintain their place in it.[119]

THE HUMAN IMPACT on Fuji has compounded over time and is obvious to any visitor. As Philip Brown notes, "Japan's 'controlling urge' has a history that extends well before nineteenth-century industrialization,"[120] but the technologies created by industrialization empowered that urge to an unprecedented degree. Today, roads built in the middle of the twentieth century enable buses and cars to reach the fifth stations on the northern, eastern, and southern sides of the mountain. During the summer months, a bulldozer even trundles all the way to the summit. First used in the 1960s to carry materials to construct a radar dome that detected typhoons far out to sea (rendered obsolete at the turn of the millennium by weather satellites, the

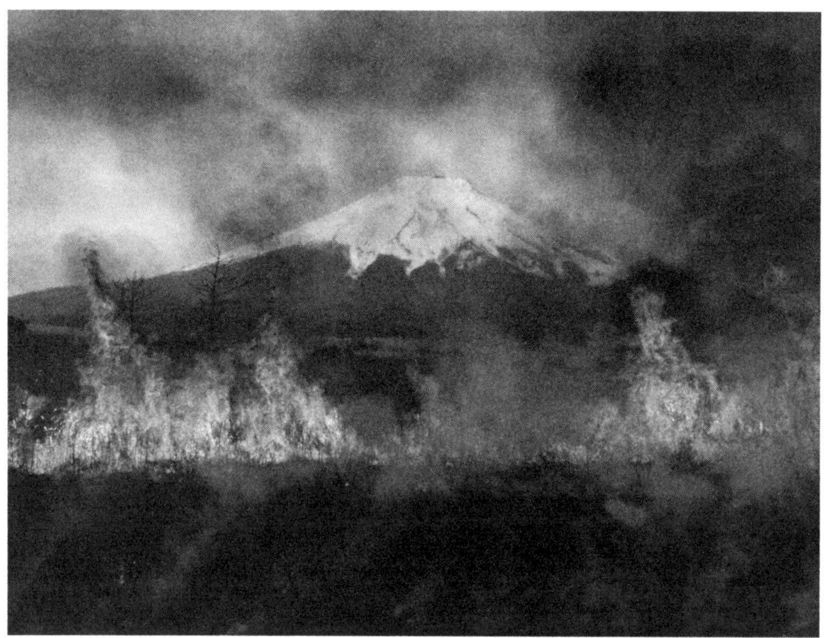

FIGURE 8.8. Spring burning of Nashigahara grassland. Photo by Watanabe Michihito.

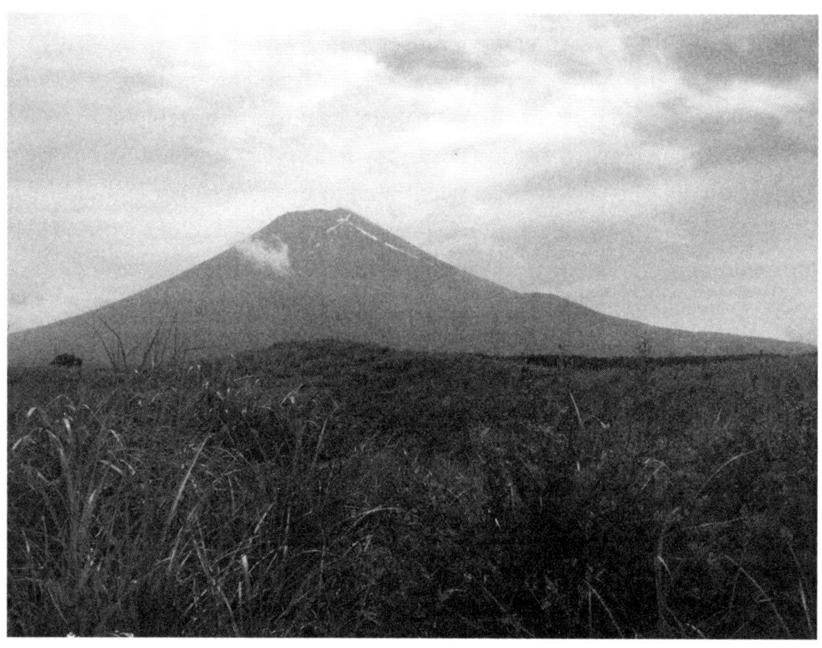

FIGURE 8.9. Nashigahara grassland in July. Photo by author.

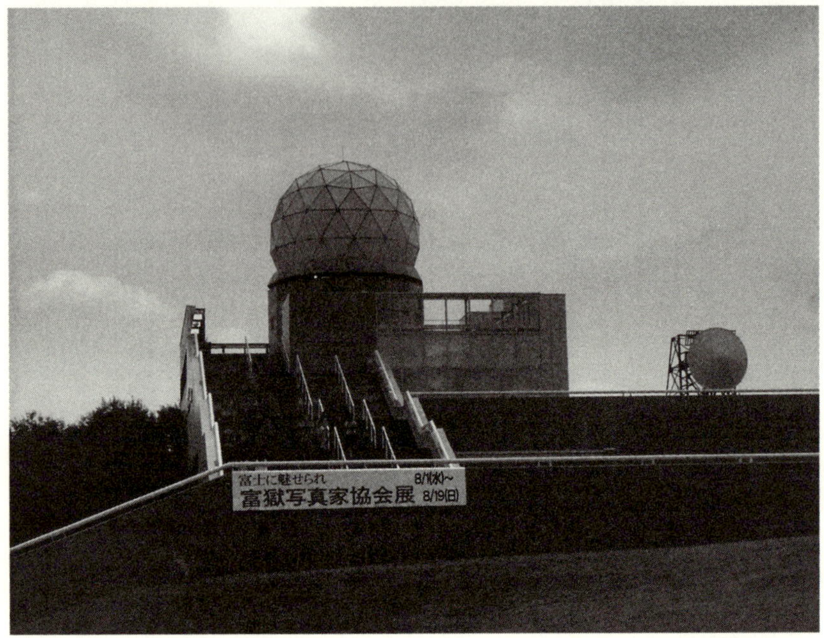

FIGURE 8.10. Radar dome museum in Fujiyoshida. Photo by author.

dome is now the centerpiece of a museum in Fujiyoshida; see fig. 8.10),[121] each summer the bulldozer delivers water, food, and other supplies to huts along the paths and carries mail down from the post office at the summit (see fig. 8.11). There are also erosion control works (*sabō*) located up and down the mountain. The largest are concentrated in the alluvial fan at the base of the Ōsawa Failure (see fig. 8.12). Consisting of channels, dams, groundsills, and sediment basins that descend like a gradual staircase for over four kilometers, the *sabō* at the base of the failure are designed to manage debris and mud flows, like the ones in 1972 that led to massive flooding in the Urui River Valley and deposited sediment as far as Tagonoura Port. The debris that accumulates in the sediment basins behind the groundsills is now removed on a regular basis and deployed in construction projects such as reinforcing beaches and building roads.[122]

Fuji shows the unintended effects of long-distance air pollution and global warming as well. In recent years, atmospheric chemists have tracked in the air at Fuji's summit fluctuating levels of mercury and volatile organic compounds like chloroform and toluene that originate in China and other parts of Asia.[123] The atmosphere at the summit has also warmed considerably since 1970, with the average maximum temperature between June and September rising from

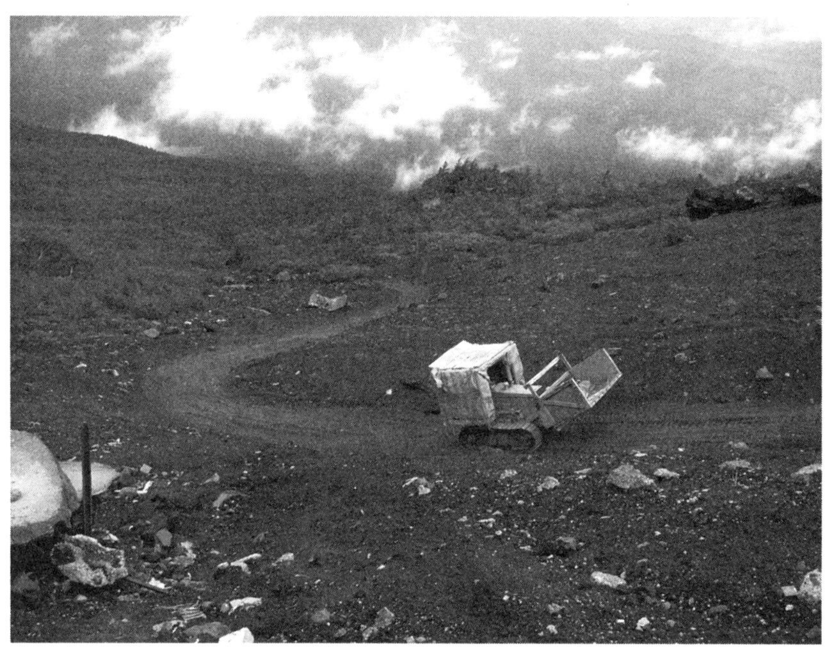

FIGURE 8.11. Bulldozer climbing Fuji. Photo by author.

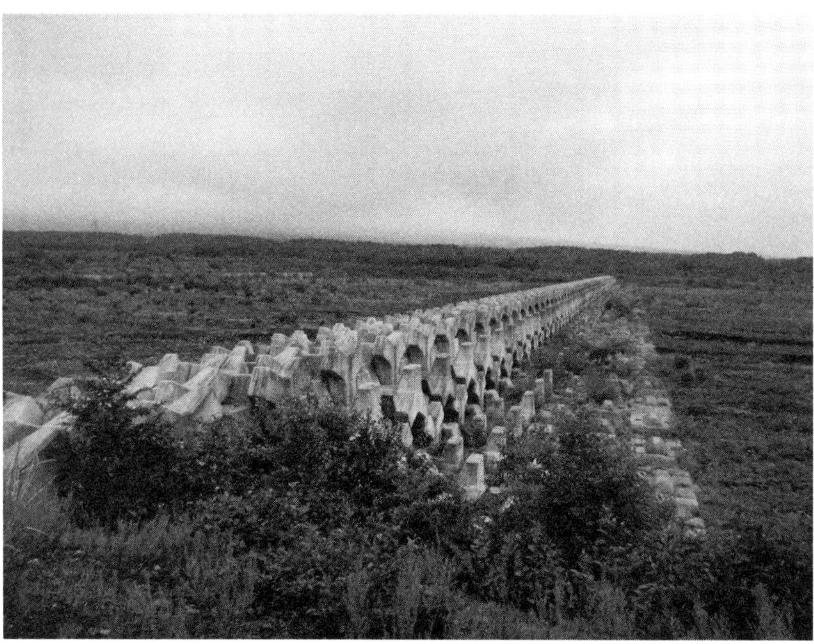

FIGURE 8.12. One of the groundsills that form sediment basins where debris from the Ōsawa Failure is captured. Photo by author.

about 6°C to almost 8°C. According to research conducted from 1978 to 2018, the timberline has climbed rapidly along with temperatures and concentrations of carbon dioxide, with the makeup and size of trees at the timberline changing as well.[124] The impact of climate change on Fuji made headlines in the fall of 2024, when media around the world reported that, for the first time on record, no measurable snow had fallen on the mountain in the month of October.[125]

So while Fuji has propelled the forces of the Anthropocene—by supplying water for tea plantations and paper mills, for example—it has been changed by those forces in ways large and small, both intended and unexpected. How to lessen the negative impact of such forces on ecosystems of every sort is one of the most urgent questions facing humanity. But as the connection between butterflies and bombs on the flanks of Fuji demonstrates, there are no simple answers. Many plants and animals benefit when human activities are kept at bay, but once a landscape has been transformed by human efforts, it is no easy matter to remove them from the picture. "You break it, you own it," as the saying goes. As the struggle with increased threats from floods, fires, pandemics, and other disasters of the Anthropocene continues, this simple truth is worth keeping firmly in mind.

9

World Heritage Fuji

ON JUNE 22, 2013, UNESCO's World Heritage Committee inscribed Fuji as Japan's seventeenth—and most famous—World Heritage Site.[1] This was a bit like conferring an honorary doctorate on a scholar who had already enjoyed a long and illustrious academic career. Heritage status can lift sites out of obscurity, but Fuji's inscription, which was widely celebrated in the Japanese press and reported overseas, did not introduce the mountain to the world so much as confirm its iconic presence in it. It also reinforced a Japanese vision of Fuji as an enduring symbol of national unity and continuity with little space for any complexities of material or ideological conflict.

Despite its fame, Fuji's two-decade journey to World Heritage status proved neither smooth nor straight. After Japan signed the Convention Concerning the Protection of the World Cultural and Natural Heritage in 1992 (the convention dates to 1972), environmental groups in Shizuoka and Yamanashi prefectures spearheaded a campaign to make Fuji a World Heritage Site.[2] They had three categories to choose from: natural, cultural, and mixed. The last category is reserved for "cultural landscapes" that meet the criteria for both natural and cultural heritage sites. Because Fuji is a mountain, the groups thought it only fitting to pursue "natural" World Heritage status. Central to their effort was a nationwide signature-gathering campaign, backed by the local newspaper *Shizuoka shinbun*, which yielded 2.46 million signatures in the spring of 1994.[3] But despite popular support and the blessing of politicians and bureaucrats on the local and national levels, the campaign failed when, in 2003, an investigative body established by the Ministry of the Environment and the Forestry Agency (part of the Ministry of Agriculture, Forestry, and Fisheries) opposed the idea of putting Fuji on Japan's list of World Heritage candidates because it was not sufficiently unique, at least in geological terms.[4] According to UNESCO, World Heritage Sites must be of "outstanding universal value," meaning they are of "cultural and/or natural significance which is so exceptional as to transcend national boundaries and to be of common importance for present and future generations of all humanity."[5] The problem with Fuji was that similar stratovolcanoes could be found in other parts of the world. One of the character compounds used to write Fuji means "peerless" (不二), but it turned out that the mountain had too many peers. Although this reason alone would have stopped the Japanese government from nominating Fuji as a natural World Heritage Site, concerns about overdevelopment and garbage on and around the mountain did not help the case.[6]

The failure to nominate Fuji for its natural features opened the door for a new campaign to nominate it as a cultural heritage site ("mixed" status was not an option since Fuji failed to meet the criteria for natural heritage). The focus on culture did not spring out of nowhere. In 1995 the idea of pursuing cultural status had come up at an international forum on making Mount Fuji a World Heritage Site, which included officials from the World Heritage Centre, the UNESCO body that coordinates the nomination process.[7] In addition, the Mount Fuji Charter promulgated by Shizuoka and Yamanashi prefectures in 1998 highlights the mountain's cultural, as well as natural, importance. The charter refers to "culture" multiple times and ends with this charge: "Let us pass down the nature, scenery, history, and culture of Mount Fuji to succeeding generations far into the future."[8] It was only after the rejection of the bid to nominate Fuji as a natural World Heritage Site, however, that a systematic effort to add it to the cultural heritage list gathered steam.

At the heart of the new campaign was a group formed in 2005 called the National Council on Fujisan World Heritage (Fujisan Sekai Isan Kokumin Kaigi). The movement to inscribe Fuji as a natural World Heritage Site had been driven largely by citizens' groups in Shizuoka and Yamanashi prefectures, but the National Council on Fujisan World Heritage, chaired by former prime minister Nakasone Yasuhiro (1918–2019), was orchestrated by political and business elites, many of them based in Tokyo.[9] They played a key role in adding Fuji to Japan's list of World Heritage candidates, although it still took years of work to win the support of local stakeholders. The collection of signatures from several hundred leisure boat businesses around Lake Yamanaka and Lake Kawaguchi, for example, held up the nomination by a year. The signatures were necessary to add the lakes to Japan's list of Important Cultural Properties, a government prerequisite for the nomination of a site for World Heritage status.[10]

After the Japanese government submitted Fuji's nomination file in early 2012, the International Council on Monuments and Sites (ICOMOS), an advisory body to the World Heritage Committee, performed on-site inspections in the same year to decide whether to recommend making Fuji a World Heritage Site. ICOMOS gave its stamp of approval in March 2013 after concluding that Fuji met two criteria for cultural World Heritage status: one, to "bear a unique or at least exceptional testimony to a cultural tradition or to a civilization which is living or which has disappeared"; and two, to "be directly or tangibly associated with events or living traditions, with ideas, or with beliefs, with artistic and literary works of outstanding significance."[11] Several months later, at its annual meeting, the World Heritage Committee voted to inscribe Fuji as a cultural World Heritage Site, with the official title "Fujisan, sacred place and source of artistic inspiration," making clear that Fuji's placement on the World Heritage List depended not on its intrinsic qualities but on the ways it was viewed and experienced by people ranging from Fujikō pilgrims who worshipped Fuji as the center of the cosmos to Western artists inspired by Japanese woodblock prints.[12]

To say that Fuji is a World Heritage Site, or "property" in UNESCO parlance, may give the impression of a single unit with uninterrupted borders, but the site actually consists of twenty-five "component parts," some contiguous and some not. These parts include geographical features as well as human structures, which is why the World Heritage Committee used the term "cultural landscape" in its decision even though the mountain was officially inscribed as a cultural property, not a mixed one.[13] The central part is the "mountain area," which covers the main climbing trails, mountain worship sites, and Kitaguchi Hongū Fuji Sengen Shrine (in Fujiyoshida). It also includes lakes Motosu, Shōji, and Sai. The other twenty-four parts are six Sengen shrines (including Fujisan Hongū Sengen Taisha, the shrine in Fujinomiya that owns the summit), two oshi lodgings for Fujikō pilgrims, lakes Yamanaka and Kawaguchi, the eight Oshino springs, Shiraito Falls, two sets of lava tree molds (the "womb caves" visited by pilgrims), the Hitoana Cave (where Kakugyō performed his austerities), and the Miho-no-matsubara pine grove (the famous viewpoint and setting for *Hagoromo*, the Nō play in which a fisherman steals the feathered robe of a heavenly maiden).[14] Surrounding the component parts is a protective "buffer zone" that conspicuously—and unsurprisingly—leaves out the military's two maneuver areas, exempting them from the regulations that obtain within the designated zone (see map 9.1).

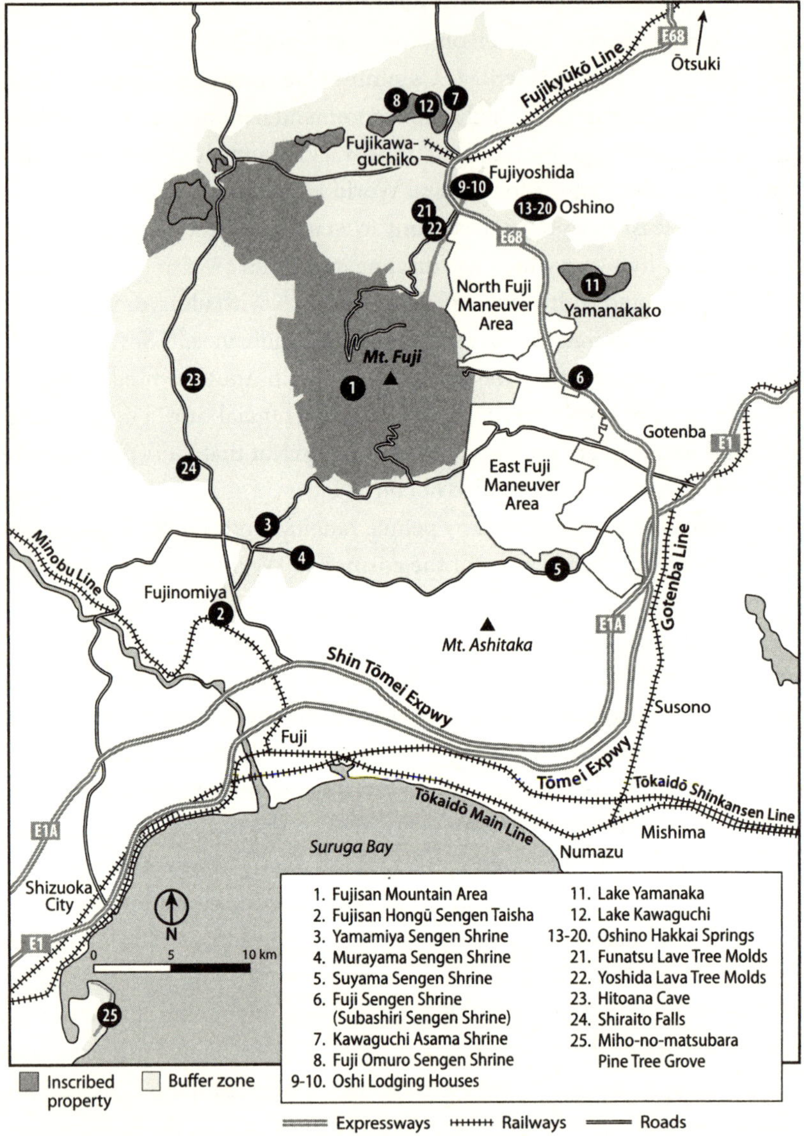

MAP 9.1. Component parts of Mount Fuji World Heritage Site. Based on maps in Government of Japan, *Nomination of Fujisan*, Executive Summary.

The list of component parts was not foreordained. The pine grove, for example, almost did not make the cut. Views of Fuji from and including Miho-no-matsubara are popular motifs in Japanese art. The grove appears, for instance, at the bottom of the famous painting of Fuji, Miho, and Seikenji Temple attributed to Sesshū and in Kanō Motonobu's Fuji mandala (see fig. 5.1

and plate 1). But ICOMOS recommended against including it because of its distance from the mountain (about forty-five kilometers). The group's report states, "ICOMOS considers that long distance views of Fujisan have been an extremely significant part of its development and are still valued. Those that remain need to be protected to help promote an understanding of the property but they cannot be said to be an integral part of the spiritual and inspirational mountain."[15] The report also notes the unsightliness of concrete breakwaters along the Miho-no-matsubara shoreline.[16]

It is not uncommon for the World Heritage Committee—which consists of representatives of twenty-one nations elected for terms of up to six years—to overrule recommendations by ICOMOS. Thanks to last-minute personal lobbying by Kondō Seiichi, the commissioner of Japan's Agency for Cultural Affairs, that is exactly what committee members decided to do in the case of Miho-no-matsubara. In his book describing the process and significance of inscribing Fuji as a World Heritage Site, Kondō celebrates the inclusion of the pine grove as a triumph of cultural diplomacy. He emphasizes that he did not strong-arm committee members with political pressure but instead worked to convince them that Fuji and Miho-no-matsubara were intangibly connected "through the spirit (*kokoro*) of the Japanese people" (although it was probably not irrelevant that Japan was the second-largest financial contributor to UNESCO at the time).[17]

Yet if Miho-no-matsubara could be considered part of the Fujisan World Heritage Site, why not any number of places integral to Fuji's history as a "sacred place and source of artistic inspiration" but not nominated as component parts? What about, for example, the Fuji Takasago Sake Brewery in Fujinomiya that houses Buddhist icons rescued from the hands of radical Shintoists just after the Meiji Restoration? Or the Fujizuka (mini-Fujis) located in Tokyo and elsewhere? As Kati Lindström points out, one can make the case that some Fujizuka should be included "for demarcating the widespread cultural identification with the mountain." She also notes that Miho-no-matsubara, the "only component that stands for the Japanese admiration of Mt. Fuji from afar," was "only included in the inscription with references to art, that is, its national and global value, and regional non-administrative community identity remained invisible."[18] The case of Miho-no-matsubara reveals not only the contingent and subjective nature of defining the parameters of a World Heritage Site but also how the focus on "outstanding universal value" can obscure local interpretations and experiences.

Japanese news outlets hailed Fuji's inscription as a World Heritage Site, but not everyone was happy about it. An especially vocal critic was the celebrity mountaineer Noguchi Ken, who for years had led high-profile campaigns to clean up trash on Mount Everest and Fuji. The nomination file submitted in 2012 repeatedly associates worship of Fuji with the "coexistence" of humans and nature; and in his book, Kondō celebrates the inscription as an important step in teaching the world about the Japanese people's "feeling of reverence" toward the natural world.[19] Noguchi's perspective could not have been more different. In the same month that Kondō's book was published (June 2014), his own book hit the shelves with the blunt title *Made into a World Heritage Site, Fuji is Weeping* (*Sekai isan ni sarete Fujisan wa naite iru*).[20] Noguchi argues in his book that Fuji should not have become a heritage site unless such environmental problems as illegal dumping, vehicle pollution, insufficient toilets, and a lack of crowd control on a mountain summited by several hundred thousand people each summer were adequately addressed.[21] From a distance, people only see the beautiful "A side" of Fuji, he says, but in the shadow of that beauty is a problem-filled "B side" that reflects the larger "B side" of Japan, a nation where foreign visitors admire the cleanliness of its cities but where "it is common to toss garbage out in nature and illegal dumping is widespread."[22]

This is not to say things have not improved over the years. Noguchi recalls that, when he first climbed Fuji in 2000, he came across large amounts of garbage and "white rivers" of toilet paper, making it the dirtiest mountain he had ever seen.[23] He therefore decided to work with the Fujisan Club, an environmental group with offices in both Yamanashi and Shizuoka prefectures, and the *Mainichi shinbun*, a major national newspaper, to conduct cleanup efforts and educational campaigns to stop climbers from littering.[24] The Fujisan Club took a leading role in the movement to install environmentally friendly "biotoilets" along climbing trails and at the summit.[25] As Noguchi acknowledges, it is thanks to the work of the Fujisan Club and other organizations and individuals that the upper reaches of Fuji are much cleaner than they once were.[26]

The reason Noguchi wrote his book was to stress not what had been accomplished but everything that still had to be done. The trails above the fifth station had been cleaned up, but illegal dumping along roads running through forested areas at the base of the mountain, including the Aokigahara "sea of trees," remained a serious problem. Noguchi reports that he came across as many as eighteen hundred discarded tires at one location in 2013.[27] To this day, the Fujisan Club coordinates efforts to remove trash ranging from batteries and plastic bottles to televisions and refrigerators.[28] The movements to make Fuji a World Heritage Site—first "natural" and then "cultural"—called attention to dumping and other environmental impacts, and thus played a role in rallying citizens' groups, businesses, government officials, and individual volunteers to deal with them. But to Noguchi's frustration, the work of cleaning up Fuji was too often viewed as a means to designate the mountain a World Heritage Site, not as an end in itself.[29] When Fuji's inscription was announced, he was both surprised and dismayed—surprised that Fuji was made a World Heritage Site in its current condition, and dismayed that, amid the hoopla, people were ignoring the work that still remained to be done.[30]

Many share this skepticism about the benefits of World Heritage status. In some places, economically struggling communities support the inscription of nearby sites to increase tourism, but the numbers of new visitors turn out to be far fewer than expected. In others, an influx of tourists strains local resources and creates more problems than it solves.[31] And while inscription might encourage the protection of individual sites, it can also exacerbate, or at least draw attention away from, negative environmental effects in surrounding areas. Mark McGuire argues that this is the case for the pilgrimage routes and sacred sites of Japan's Kii Mountain Range, which were collectively inscribed as a World Heritage Site in 2004. In the course of McGuire's field research, a Shugendō priest gave him a tour of degraded landscapes, including "sites where former sacred peaks have been smashed into gravel by road construction outfits," and repeatedly asked, "Is *this* World Heritage?"[32] McGuire observes that, "rather than suggest remediation of degraded sites prior to or following nomination, unnamed ICOMOS officials opted instead to simply leave them out of the designation."[33]

In the case of Fuji, World Heritage status was awarded on the condition that the Japanese government submit a conservation report to the World Heritage Centre by February 1, 2016. Drawing on recommendations made by ICOMOS, the World Heritage Committee asked that the report "provide an update on the progress with the development of an overall vision for the property, a tourism strategy, a conservation approach for the access routes, an interpretation strategy, [and] a risk management strategy with the overall revision of the management plan to reflect a cultural landscape approach."[34] Here, once again, we see Fuji described as a cultural landscape even though it was officially designated a "cultural," not a "mixed," heritage site.

Noguchi and others were concerned that no one had the vision or will to develop a coherent plan for managing Fuji in the face of clashing interest groups—including government officials, landowners, tourist operators, and environmental organizations—and that Fuji might become an "endangered" heritage site and even be removed from the World Heritage List.[35] Complicating matters was that, although Fuji was part of the Fuji-Hakone-Izu National Park (established in 1936), there was no single authority comparable to the US National Park Service that could mandate land use on and around the mountain. The Ministry of the Environment is in charge of administering Japan's national parks, but it owns a mere 0.2 percent of the land in them, with the Forestry Agency controlling 62 percent and the rest owned by prefectural and local governments and private landholders. Meanwhile, the well-funded Ministry of Land, Infrastructure, and Transport wields authority over large roads and rivers, as well as the construction and management of infrastructure—including the erosion control works (*sabō*) on and near Fuji—to mitigate disasters like floods and rockslides. In the case of the Fuji region of the Fuji-Hakone-Izu National Park, the Ministry of the Environment owns outright only two branch offices staffed by a couple of rangers, the trail that circles the crater, and some public toilets on the mountain. The Forestry Agency owns 21 percent of the land, Shizuoka and Yamanashi prefectures 41 percent, and private landholders the remaining 38 percent.[36] This does not mean the ministry is powerless, especially when it comes to the tightly regulated "special protection zone" that covers the upper reaches of the mountain and the Aokigahara forest, a zone that largely overlaps with the "mountain area" component of the Fujisan World Heritage Site. But the "special protection zone" comprises only 4,642 hectares (11,471 acres) out of the 60,645 hectares (149,857 acres) that make up the Fuji region of the national park, with most of the region divided into zones with much looser oversight.[37]

It was challenging to get Fuji's different local and national stakeholders on the same page, but in 2016, Japan submitted a conservation report that not only passed muster in the eyes of the World Heritage Committee but was praised as a model for managing cultural landscapes in other countries.[38] The report includes a detailed "Vision and Strategies" document developed by the Fujisan World Cultural Heritage Council (Fujisan Sekai Bunka Isan Kyōgikai)—a group established in 2012 with representatives from local municipalities, the governments of Yamanashi and Shizuoka prefectures, and the national government[39]—as well as a lengthy "Comprehensive Preservation and Management Plan" to encourage better tourism management, conservation and renovation of component parts, "visual harmonization" of buildings and signs, historical and archaeological research, and visitor education. Central to the goal of supporting research and educating visitors was the construction of two World Heritage centers, one in Fujikawaguchiko (opened in 2016) and the other, designed by renowned architect Ban Shigeru, in Fujinomiya (opened in 2017; see plate 15). Other measures in the planning documents include standardizing signs, relocating shops from the base of Shiraito Falls, removing utility poles and burying cables to open up views of Fuji, regulating designs for structures ranging from convenience stores to mountain huts, and replacing concrete breakwaters at Miho-no-matsubara with more visually pleasing and effective structures.[40]

The report describes ways to monitor and manage the approximately three million people who go to the fifth station each year as well as the several hundred thousand who make their way to the summit during the summer climbing season. To provide funding for safety and conservation measures such as repairing signs and maintaining toilets, a one-thousand-yen voluntary donation fee was put into place in 2013 for visitors climbing to the summit.[41] Other measures listed in the planning documents include GPS climber movement studies, increased trail maintenance, beefed-up patrols, private vehicle restrictions, and guidance to climbers in multiple languages on an official Mount Fuji climbing website managed by an organization established by the Ministry of the Environment and Yamanashi and Shizuoka prefectures.[42] The website makes a special point of dissuading climbers from making nighttime "bullet climbs" from the fifth station to the summit without adequate rest, explaining that they create a much greater risk of injury and altitude sickness.[43]

These concrete measures won over Noguchi, who said in a newspaper interview in 2022 that he had been wrong to worry that World Heritage status would do more harm than good. He said he was happy with the level of cooperation among local governments, the expanded restrictions on driving private cars to the fifth station, and the increase in both the number of rangers (supplied by Yamanashi Prefecture) and the number of people willing to pay the voluntary conservation fee. He was also relieved, he said, that there had not been a sharp increase in the number of people climbing Fuji.[44] In 2013, the year of Fuji's inscription, the number of climbers reaching the eighth station was 310,700, a slight drop from the year earlier, when it was 318,565. After 2013 the number fell below 300,000 and stayed there until 2020, when the climbing routes were closed due to Covid restrictions.[45]

While the number of climbers was still relatively modest when Noguchi gave his interview in 2022 (it had recovered to 78,500 in 2021 and 160,100 in 2022), the situation dramatically changed the following summer, when post-Covid "revenge tourism" led to a surge in people making their way to the summit. The total of 221,322 climbers in 2023 still fell short of pre-Covid levels, but the capacity of mountain huts to accommodate climbers was far lower than it had been before the pandemic.[46] Despite the conservation and safety efforts made over the previous decade, the challenge of managing crowds on Fuji's slopes was once again a pressing issue. In response, Yamanashi Prefecture for the first time instituted a policy of temporarily stopping climbers from continuing up the Yoshida trail when it became too congested.[47] During the 2024 season it went several steps further by setting a daily limit of four thousand climbers above the fifth station, establishing a mandatory two-thousand-yen hiking fee, and barring those without a mountain hut reservation from climbing between 4:00 p.m. and 3:00 a.m. (to discourage "bullet climbs").[48] These measures could explain why the number of people climbing to the summit in 2024 dropped 7.7 percent (to 204,316) from the year before.[49]

The large number of visitors presents a problem at the base of the mountain too. In the spring of 2024, for example, the town of Fujikawaguchiko erected a black mesh barrier across the street from a Lawson's convenience store. It did this to deter tourists who were crowding the sidewalk to take their own version of a photo popularized on social media that showed the store with Fuji in the background.

Climbing restrictions and barriers aside, there is also only so much that can be done to shape the experience of the throngs of people who come to the Fuji region each year. Early in the twentieth century, first railways, then buses and cars, lured a growing number of visitors to the mountain and its environs, which saw the development of hotels, second homes, and leisure activities.[50] Tourism around Fuji especially surged from the 1960s. In 1964, the year Tokyo hosted the Olympics, the opening of the Subaru Line allowed visitors to drive up to the fifth station on the northern side of the mountain. From 1970 drivers could take the Skyline Drive to the fifth station on the southern side as well. The opening of the Skyline Drive came a year after the completion of the Chūō Expressway's Kawaguchiko Interchange and the full opening of the Tōmei Expressway on the Shizuoka side of the mountain, making it far more convenient for people to get to the Fuji region by car and bus.[51] As a consequence, the number of people coming to Fuji-Hakone-Izu National Park on a yearly basis had topped one hundred million by the mid-1980s, with the majority of them going to the Fuji region of the park.[52]

As the number of visitors increased, so too did recreational options, thanks in large part to Fujikyūkō, a transportation and tourism company that operates an electric railway between Ōtsuki and Kawaguchiko in addition to regional and long-distance bus lines. Starting as early as the 1920s, but particularly from the 1960s, the company has been a key player in developing tourist accommodations and facilities around Fuji, the most famous being the Fuji-Q Highland amusement park that sits on top of the Kenmarubi lava flow in Fujikawaguchiko. It started as a skating center in 1961 and grew in stages to be one of the most popular amusement parks in Japan.[53]

Other tourist attractions proliferated around Fuji during and after the 1960s, including golf courses, the Fuji Speedway (in Oyama), and Fuji Safari Park (in Susono), where tourists can take photos of giraffes, elephants, and other animals with the mountain rising behind them. As these and numerous other facilities appeared, so too did convenience stores, shopping centers, and roadside motels, severely limiting the degree to which the goal of "visual harmonization" around Fuji can now be achieved, despite good-faith efforts by local businesses and governments. A massive tourist infrastructure focused on leisure instead of worship also presents a challenge to those trying to cultivate an appreciation for Fuji's history as a pilgrimage site where worship went hand in hand with leisure during the Tokugawa period. Efforts have been made to revitalize trails that Fujikō pilgrims trod from the base of the mountain to the fifth station, but it is unlikely that more than a small percentage of visitors to Fuji will ever walk them. It is also hard to imagine how the Hitoana Cave or an oshi inn can compete with the thrill of Fuji-Q Highland's Fujiyama roller coaster.

A one-minute video uploaded to UNESCO's official web page for the Fujisan World Heritage Site excludes the roller coaster or anything else that would interfere with the message that Fuji is a beautiful and sacred mountain. As stirring music plays, the video shows the sun rising over Fuji with a sea of clouds below, followed by another view of the mountain with blossoming cherry trees in the foreground. It also depicts local residents parading their Fuji-shaped *mikoshi* (portable shrine) through the streets of Fujiyoshida during the annual Fire Festival (see plate 14) and ends with an aerial shot of the Kitaguchi Hongū Fuji Sengen Shrine that pans out to a view of the mountain rising above the forest. Nowhere do we see paper mills or artillery fire.[54] A gallery of photos posted to the same web page shows a bird's-eye view of the northern side of the mountain in which the Nashigahara grassland is visible in the distance, but there is no way to tell that it is part of a military maneuver area. Another photo shows Fuji looming above a tea plantation, but how many viewers realize that those machine-sculpted rows of tea plants are grown with fertilizers that pollute the groundwater?[55] It is also unlikely many know that much of the woodland on Fuji's skirt consists of abandoned plantation forests.

The UNESCO web page does include language that points to development around the mountain as a problem, but only in that it presents an obstacle to visitors' aesthetic and spiritual experience of Fuji. The section titled "Integrity" states that "because of development in the lower part of the mountain, the relationship between pilgrims' routes and supporting shrines and lodging houses cannot readily be appreciated," while the one titled "Management and

Protection Requirements" refers to "conflicting needs between access and recreation on the one hand and maintaining spiritual and aesthetic qualities on the other hand."[56] Ecological impacts go unmentioned.

The web page also leaves out Fuji's history as a site of competition, conflict, and destruction. In the "Brief Synthesis" of the mountain's "Outstanding Universal Value," both Buddhism and Shinto are mentioned, but not the modern and contested development of Shinto as a "native religion" or the nativist movement that destroyed much of Fuji's Buddhist heritage. And while the web page refers to the "Shinto deity Asama no Okami," it says nothing about Great Bodhisattva Sengen (Sengen Daibosatsu), Dainichi Buddha, and other deities associated with Fuji over the centuries. Nor does it acknowledge that Tokugawa officials viewed Fujikō adherents as troublesome upstarts who had to be regulated and even punished. The Hōei eruption is omitted, and although Fuji's volcanic activity is recognized, it is only in its contribution to a sense of awe that "transformed into religious practices that linked Shintoism and Buddhism, people and nature, and symbolic death and re-birth."[57]

The Fujisan World Cultural Heritage Council's web page on the "Outstanding Universal Value of Mt. Fuji" similarly displays no interest in the complex and contested history of the mountain. Like the nomination file, it extols a nationalized and dehistoricized relationship with nature. It states that Fuji has exerted a "large influence" on "the Japanese people's view of nature and on Japanese culture" and that "one large characteristic that distinguishes Fujisan is the way in which people and nature coexist through the realms of religious belief and art."[58] There is no inkling of any human impact that does not support the idea of "coexistence" with the natural world.

It is also misleading to suggest a singular "view of nature" that endured across time. In the Tokugawa period, for example, *honzōgaku* scholars ("naturalists") studied, and in the process created, a disenchanted world of nonhuman objects, often with the goal of resource extraction.[59] Fujikō members, in contrast, viewed their sacred mountain not as a natural feature to be categorized, analyzed, and exploited but as the Original Father and Mother (Moto no Chichihaha) that sustained all beings, both human and nonhuman. Imagined in the form of a human body, Fuji was at the center of a cosmos in which the ethical was inextricable from the physical, with no room for "nature" as a realm separate from that of humans (or "culture"). In fact, the English word "nature"—a notoriously ambiguous word that can signify everything from "that which is necessarily so" to "that which is not culture" to "the essence of what makes us human" (human nature)[60]—did not have a Japanese equivalent (or, at least, something close to it) until the repurposing of the word *shizen*—which had been used to indicate "idleness," "purposelessness," or, in the form of an adjective, "spontaneous"—toward the end of the nineteenth century.[61] Before that time, people in Japan may have valued relationships with particular animals, plants, landscapes, or other features of what we today call the "natural world," but they would not have imagined themselves "coexisting with nature" in either a universal or national sense. And just as they did not view nonhumans through the lens of "nature," they did not group worship, art, and other human activities into the modern category of "culture" (*bunka*).

Which returns us to Fuji's inscription as a "cultural" versus a "natural" or "mixed" World Heritage Site. To state the obvious, Fuji would not have failed as a natural World Heritage Site if there were no concept of nature as something separable from humans to begin with. Only by removing humans from the picture could Fuji be treated as one stratovolcano in a world of other stratovolcanoes (if a bit worse for wear). It was therefore up to "culture" to come to the rescue. Fuji may not have been geologically unique, but there was no denying its importance in terms of worship and art. Of course, that does not change the fact that Fuji is a mountain whose physical existence owes nothing to humans. People would not have worshipped, climbed, and depicted Fuji were it not for the tectonic forces that generated it in the first place.

Even though Fuji is not a mixed World Heritage Site, use of the term "cultural landscape" in official documents acknowledges that it is at once a geological feature and a product of the imagination. Fuji is celebrated as a cultural landscape because of its value as a "sacred place and source of artistic inspiration" over the past thousand or so years, but in truth, any landscape occupied by humans might be described as "cultural." Once people think and talk about a waterfall, a mountain, or any other such feature, they turn it into a cultural phenomenon with stories attached to it. In this sense, Fuji has been as much "cultural" as "natural" since Paleolithic hunter-gatherers first saw the mountain tens of thousands of years ago, even if the stories they might have told about it are lost.

Wherever human populations have left their mark, what we call "nature" has never truly been separated from "culture" in physical terms either. People have always transformed the world not just in their heads but through their activities and technologies. These range from the fires set by hunter-gatherers to the computer used to write this book. In the epoch of the Anthropocene, the entanglements of humans and nonhumans have intensified and multiplied to such a degree that the nature/culture divide seems even more illusory—and distracting—than ever. Consider the Miho-no-matsubara pine grove. The justification for including it as a component part of the Fujisan World Heritage Site is that, together with Fuji, its natural beauty has inspired works of art. But while the trees survive through photosynthesis, a process that has supported life on Earth for billions of years, the city of Shizuoka has resorted to pesticides to save them from pine sawyer beetles, making them dependent on human actions as well.[62]

Just as the pine grove might be considered "cultural" in the ways it is aesthetically represented and physically maintained, so might the pesticides be considered "natural" in that they are created from elements like nitrogen and chlorine and are the result of the "natural" propensity of humans to make tools and alter their environments. This is not to say that all human actions can be justified by saying they are "natural." Some are clearly more destructive—to both humans and nonhumans—than others. But this does suggest that the separation between humans and those things we call "natural" exists more in our heads than in reality. This is why some have argued for abandoning the totalizing word "nature"—and its imagined opposition to culture, society, and politics—entirely.[63] Others do not discard "nature" altogether, but advocate more thoughtful ways of understanding and describing our relationships to it. William Cronon, for example, criticizes the American tendency to view "wilderness" as nature in its most pure and authentic form, arguing that the "flight from history that is very nearly the core of wilderness represents the false hope of an escape from responsibility," and "to the extent that we celebrate wilderness as the measure with which we judge civilization, we reproduce the dualism that sets humanity and nature at opposite poles. We thereby leave ourselves little hope of discovering what an ethical, sustainable, *honorable* human place in nature might actually look like."[64] In his history of the Columbia River, Richard White strikes a similar chord: "To call for a return to nature is posturing. It is a religious ritual in which the recantation of our sins and a pledge to sin no more promises to restore purity. Some people believe sins go away. History does not go away."[65]

In Japan, as in much of the world, there is less focus on untrammeled wilderness than on a romantic vision of humans living in harmony with nature—and each other—in a countryside free from both the ills of industrialization and social strife. Here, too, is a "flight from history," but in this case it is to a mythical realm where it is in the nature of the Japanese people to harmoniously coexist with the nature of their homeland. Any evidence to the contrary (such as the pollution of Suruga Bay or the meltdown of three nuclear reactors in 2011) can be either ignored or dismissed as a deviation from a cultural, if not a historical, norm.

Fuji has long occupied a central place in this national vision of both the Japanese people and the natural world, a vision bolstered by the celebration of Fuji in literature and art from the poems of the eighth-century *Collection of Ten Thousand Leaves* to the countless photographs taken of it today. This long-standing celebration of the mountain, which expanded to a national scale in the Tokugawa period, is a key part of its biography. But what about the conflicts that erupted among those looking to make money from pilgrims? Or the long-running struggle over the ownership of Fuji's summit? Or the militarization of its lower flanks? It is only by accounting for the good and the bad, the beautiful and the ugly, that we can do historical justice to the multiple lives of Fuji and, one could argue, even more fully celebrate this famous but in many ways hidden peak.

In seeing Fuji whole, we must remember that it is not only a mountain but a volcano, as the Jōgan and Hōei eruptions explosively attest. The fact that it is a living volcano has received fresh attention in recent years, as volcanologists work with government officials to develop hazard maps and evacuation plans based on the history of past eruptions.[66] Concern about a future eruption has grown not only in communities immediately around Fuji but also in Tokyo, where the municipal fire department has begun developing plans to deal with the impact of falling ash on waterways, electrical and communication systems, and mass transit.[67] Of course, such preparations can only go so far when dealing with something as powerful as a volcanic eruption.

In short, while humans have played a major role in the making of Fuji, they certainly do not control it. Fuji will act according to its own devices whether we like it or not. The same holds true for the planet as a whole. As we continue to travel ever further into the Anthropocene, we must accept responsibility in a spirit of humility. We are only human after all.

NOTES

Introduction

1. Here I follow the lead of Mark Cioc, who uses the word "biography" to characterize his history of the Rhine. As he puts it, "rivers seem alive to us—restless, temperamental, fickle, sometimes raging, sometimes calm." The same can be said of volcanoes like Fuji. See Cioc, *The Rhine: An Eco-Biography, 1815–2000*, p. 5. Journalist and mountain enthusiast Jon Bell also employs "biography" to describe his book about Oregon's Mount Hood, *On Mount Hood: A Biography of Oregon's Perilous Peak.*

2. Sigurdsson, "Introduction," pp. 8, 2.

3. Endō, *Volcano*, p. 27.

4. Dating based on Miyaji, "Kako ichi man sen nenkan no Fuji kazan," pp. 79–95. Some argue that the distinction between New and Old Fuji is negligible, but the two are widely considered to be separate volcanoes because mineral ratios differ from one to the other. See Aramaki, "Fujisan no tokushoku," pp. 157–58; and Togashi and Takahashi, "Fujisan no maguma," pp. 219–31.

5. Yamanashi Nichinichi Shinbunsha, *Narusawa sonshi*, vol. 1, pp. 626–29.

6. Fujiyoshida Shishi Hensan Iinkai, *Fujiyoshida shishi tsūshihen*, vol. 2, pp. 261, 316–17.

7. Miyaji et al., "High-Resolution Reconstruction," p. 114.

8. Hearn, *Exotics and Retrospectives*, p. 11.

9. Hearn, *Exotics and Retrospectives*, p. 14.

10. I am grateful to Fujisan Club leader Aoki Naoko for coordinating the program logistics.

11. Sato, "Trap-Pit Hunting," pp. 395–96.

Chapter One: Active Volcano, Lofty Mountain

1. The "Sea of Se" (Se-no-umi) refers to a lake at the northern foot of Fuji that was split in two by lava flows from the Jōgan eruption (864–866), resulting in what are now Lake Shōji and Lake Sai.

2. *Man'yōshū*, book 3, poem no. 319. Takahashi Mushimaro, "Poem about Mount Fuji," in Levy, *The Ten Thousand Leaves*, p. 179. Original appears in Takagi et al., *Nihon koten bungaku taikei*, vol. 4, pp. 169–70.

3. A note appears after one of the accompanying "envoy" poems (poem no. 321) specifying that "the poem to the right" is Mushimaro's. Some argue that it does not extend to the main poem (no. 319), but I provisionally accept that Mushimaro is the author. To learn about the debate

surrounding the authorship of poem no. 319, see Sugano, "Mushimaro no Fuji no yama no uta," pp. 166–80.

4. Takagi et al., *Nihon koten bungaku taikei*. The two-character compound now most commonly used for Fuji is 富士, which can be translated as "prosperous retainer" or alternatively as "abundance of retainers." A number of theories have been proposed to explain the phonetic origins of Fuji, with some suggesting the name might derive from one of two Ainu words, *pushi* (eruption) and *funchi* (mountain of fire), and others that it stems from *puji*, the Malay word for magnificent. Another theory locates its origins in the Japanese word for wisteria, which is also pronounced *fuji*, while yet another traces its origins to *niji*, the Japanese word for rainbow. These and other speculations are exactly that, however, speculations, and it does not seem likely any consensus about the linguistic origins of the word "Fuji" will ever be reached. Tani, *Fujisan wa naze Fujisan ka*, pp. 10–11, 282–92.

5. Takagi et al., *Nihon koten bungaku taikei*.

6. Easy-to-read explanations of plate tectonics and volcanism appear in Marshak, *Earth: Portrait of a Planet*.

7. Aramaki, "Fujisan no tokushoku," pp. 154–55.

8. Barnes, *Tectonic Archaeology*, pp. 104–6; Hirata et al., "Puroto Izu-Mariana tōko," pp. 1125, 1129.

9. Aoki, Tsunematsu, and Yoshimoto, "Recent Progress of Geophysical and Geological Studies," p. 264.

10. Dating based on Miyaji, "Kako ichi man sen nenkan no Fuji kazan," pp. 79–95.

11. Aramaki, "Fujisan no tokushoku," pp. 157–58; Togashi and Takahashi, "Fujisan no maguma no kagaku sosei to gansekigakuteki tokuchō," pp. 219–31.

12. Nakada, Yoshimoto, and Fujii, "Sen Fuji kazangun," pp. 69–77.

13. Miyaji, "Kako ichi man sen nenkan no Fuji kazan," pp. 79–95.

14. Anma, "Fujisan de hassei suru rahāru," pp. 285–301.

15. Watanabe, "The Nature and Scenery of Mt. Fuji Area," p. 204.

16. Watanabe, "The Nature and Scenery of Mt. Fuji Area," p. 207.

17. Sato, "Trap-Pit Hunting," p. 400.

18. The very oldest, dating from thirty-five to thirty-eight thousand years ago, are located about one thousand kilometers to the southwest on the island of Tanegashima. Sato, "Trap-Pit Hunting," pp. 395–400.

19. Kuzmin and Glascock, "Introduction," pp. 1–2.

20. Ikeya, "Marine Transport of Obsidian," pp. 362, 369–72; Tsutsumi, "Prehistoric Procurement of Obsidian," pp. 30–47; Shimada, "Pioneer Phase of Obsidian Use," pp. 133–39.

21. Tsutsumi, "Prehistoric Procurement of Obsidian," pp. 30–36.

22. Considering the mobile nature of Paleolithic humans, groups possessing obsidian sourced up to two hundred kilometers away from where it was used could have acquired it directly—whether in the regular course of seasonal migrations or by dispatching procurement bands with that specific goal. The other possibility is that they gained it indirectly through exchange networks. On the lower flanks of Ashitaka, excavations have revealed that obsidian was introduced from a variety of sources at any given point in time, suggesting that not just one but a number of different groups were involved. See Tsutsumi, "Prehistoric Procurement of Obsidian," pp. 33, 39–40, 48.

23. Marshak, *Earth: Portrait of a Planet*, pp. 145, 157.

24. Marshak, *Earth: Portrait of a Planet*, pp. 248–63.

25. Different ideas have been proposed for why Fuji mainly erupts basalt, but the reason is still not well understood. One hypothesis is that a crack in the Philippine Plate allows direct access to mantle-derived basaltic lava located in a magma chamber far deeper than those feeding the largely andesitic volcanoes nearby. The reason why Fuji has erupted inordinately vast quantities of basaltic magma is difficult to determine, however. See Fujii, "Fuji kazan no magumagaku," pp. 233–42; and Aizawa, Yoshimura, and Oshiman, "Splitting of the Philippine Sea Plate," pp. 1–4.

26. For a concise meteorological explanation of Fuji's "cap clouds," see Yamakawa, "Fujisan no kasagumo," pp. 109–13.

27. Although he does not single out Fuji, Ikeya Nobuyuki notes the important role that "weather lore" would have played in navigating the waters between Honshu and Kōzushima. Ikeya, "Marine Transport of Obsidian," p. 373.

28. "Clouds over Mount Fuji Predict the Weather," *Nipponia*, no. 35 (December 2005), http://web-japan.org/nipponia/nipponia35/en/feature/feature05.html.

29. There are different theories about where the immigrants originated, but the distribution, dating, and composition of stone tools throughout Japan indicate that the earliest settlers crossed the Tsushima Strait from the Korean Peninsula. Ono and Yamada, "Upper Palaeolithic," p. 227; Tsutsumi, "MIS3 Edge-Ground Axes," p. 77; Shimada, "Pioneer Phase of Obsidian Use," p. 142.

30. Kudo and Kumon, "Paleolithic Cultures of MIS 3 to MIS 1," p. 29; Clark et al., "Last Glacial Maximum," pp. 710–14.

31. Shizuoka-ken, *Shizuoka kenshi tsūshihen*, vol. 1, p. 40.

32. Kudo, "Absolute Chronology of Archaeological and Paleoenvironmental Records," pp. 13–14; Izuho et al., "The Upper Paleolithic of Hokkaido," p. 111.

33. Adachi, Shinoda, and Izuho, "Further Analyses of Hokkaido Jomon Mitochondrial DNA," pp. 406–13; Kuznetsov, "Origin of Microblade Industries," pp. 87–88; Izuho et al., "The Upper Paleolithic of Hokkaido," pp. 188, 120; Habu, *Ancient Jomon of Japan*, p. 34.

34. Many argue that overhunting was the main culprit, but recent research suggests that climate change may have been primarily to blame. Kawamura and Nakagawa, "Terrestrial Mammal Faunas"; Takahashi and Izuho, "Formative History of Terrestrial Fauna"; Iwase, Takahashi, and Izuho, "Late Pleistocene Megafaunal Extinction."

35. Bleed and Matsui, "Why Didn't Agriculture Develop in Japan?," pp. 364–66.

36. A soil monolith from Fujiyoshida now stored at the Soil Museum of Japan's Institute for Agro-Environmental Sciences shows evidence of extensive burning that, over the course of thousands of years, helped transform volcanic ejecta into soils known as andosols (also called andisols). It is likely that humans were responsible for much of that burning, although exactly how much is a subject of debate. Maejima Yuji (Senior Researcher at the Institute for Agro-Environmental Sciences' Soil Inventory Unit), lecture given on tour of museum's soil monolith collection for Lewis and Clark Mount Fuji Study Abroad Program, July 3, 2017.

37. Tsuchi, "Fujisan no chikasui, yūsui," pp. 383–84; Tosaki and Asai, "Fujisan no chikasui nendai."

38. Tsuchi, "Fujisan no chikasui, yūsui," pp. 376, 381.

39. Yamamoto et al., "Fujisan hokuroku, Kawaguchiko no kotei"; Hayashi, "Fujisan hokubu ni okeru chikasui."

40. Fuji Shishi Hensan Iinkai, *Fuji shishi*, vol. 1, p. 156; Oyama Chōshi Hensan Senmon Iinkai, *Oyama chōshi*, vol. 6, pp. 149–50; Fujinomiya Shishi Hensan Iinkai, *Fujinomiya shishi*, vol. 1, pp. 160–62.

41. Fuji Shishi Hensan Iinkai, *Fuji shishi*, vol. 1, p. 156; Oyama Chōshi Hensan Senmon Iinkai, *Oyama chōshi*, vol. 6, pp. 149–50; Fujinomiya Shishi Hensan Iinkai, *Fujinomiya shishi*, vol. 1, pp. 160–62.

42. Habu, *Ancient Jomon of Japan*, pp. 7–14, 62–63.

43. Shizuoka-ken, *Shizuoka kenshi tsūshihen*, vol. 1, pp. 80–81.

44. Tsutsumi, "Prehistoric Procurement of Obsidian," pp. 49–51; Habu, *Ancient Jomon of Japan*, pp. 221–24; Shizuoka-ken, *Shizuoka kenshi tsūshihen*, vol. 1, pp. 106–7. A number of other items traded over substantial distances in Jōmon Japan included jade, shells, pottery, cinnabar (used to produce red lacquer), and asphalt (used in making tools and repairing pottery). Habu, *Ancient Jomon of Japan*, pp. 224–33.

45. Fuji Shishi Hensan Iinkai, *Fuji shishi*, vol. 1, pp. 154, 158–59; Takahashi and Kanayama, "Fujisan ni tai suru Jōmonjin."

46. Habu, *Ancient Jomon of Japan*, pp. 182–87.

47. The stones at Tenmanzawa form a triangle, with Fuji occupying the center of one side of the triangle when viewed from a large stone placed at the corner directly opposite to it. This suggests that the mountain may have been the object of some sort of worship and is in keeping with stone arrangements elsewhere in Japan that seem to be oriented toward mountains. See Fuji Shishi Hensan Iinkai, *Fuji shishi*, vol. 1, p. 154; and Pearson, "New Perspectives on Jomon Society," pp. 65–66.

48. Fujiyoshida Shishi Hensan Iinkai, *Fujiyoshida shishi tsūshihen*, vol. 1, pp. 112–14.

49. Habu, *Ancient Jomon of Japan*, pp. 142–51; Fujinomiya Shishi Hensan Iinkai, *Fujinomiya shishi*, vol. 1, pp. 193–94; Susono Shishi Hensan Senmon Iinkai, *Susono shishi*, vol. 8, p. 49.

50. Habu, *Ancient Jomon of Japan*, p. 151; Fujinomiya Shishi Hensan Iinkai, *Fujinomiya shishi*, vol. 1, p. 148.

51. Pearson, "Debating Jomon Social Complexity," p. 363.

52. Koyama's estimates are regularly employed by scholars writing on the Jōmon period. See Habu, *Ancient Jomon of Japan*, pp. 46–50 for an explanation of Koyama's methodology.

53. Habu, *Ancient Jomon of Japan*, pp. 254–55.

54. Fujiyoshida Shishi Hensan Iinkai, *Fujiyoshida shishi tsūshihen*, vol. 1, pp. 110, 112.

55. Sugiyama and Kaneko, "Jōmon jidai kōhanki no Izu, Hakone, Fujisan."

56. Miyaji, Togashi, and Chiba, "Fuji kazan tōshamen," pp. 237–42; Oyama Chōshi Hensan Senmon Iinkai, *Oyama chōshi*, vol. 6, pp. 101–2.

57. Totman, *Japan: An Environmental History*, p. 34.

58. Shizuoka-ken, *Shizuoka kenshi tsūshihen*, vol. 1, pp. 106–7, 124, 126; Fujiyoshida Shishi Hensan Iinkai, *Fujiyoshida shishi tsūshihen*, vol. 1, pp. 105, 126.

59. Barnes, *Archaeology of East Asia*, pp. 270–71; Mizoguchi, *The Archaeology of Japan*, pp. 26–27, 33–36.

60. Barnes, *Archaeology of East Asia*, pp. 275–78; Nakaoka et al., "HLA idenshi takei," pp. 37–44.

61. Bleed and Matsui, "Why Didn't Agriculture Develop in Japan?," pp. 366–67.

62. Mizoguchi, *Archaeology of Japan*, pp. 167, 185, 200, 208.

63. Mizoguchi, *Archaeology of Japan*, pp. 105–7, 141–42.

64. Mizoguchi, *Archaeology of Japan*, p. 142.

65. For a detailed explanation of andosols in the Japanese context, see Barnes, *Tectonic Archaeology*, pp. 175–203.

66. Ono, "Tokugawa Tsunayoshi to dojō hiryō gaku."

67. Fuji Shishi Hensan Iinkai, *Fuji shishi*, vol. 1, p. 207; Watai, "Fuji sanroku ni okeru kodai," p. 66.

68. Fuji Shiritsu Hakubutsukan, *Ukishimanuma to kome zukuri*, pp. 1–2.

69. Fuji Shishi Hensan Iinkai, *Fuji shishi*, vol. 1, pp. 201–3, 207.

70. Watai, "Fuji sanroku ni okeru kodai," p. 66; Fujiyoshida Shishi Hensan Iinkai, *Fujiyoshida shishi tsūshihen*, vol. 1, pp. 140–45.

71. Fujiyoshida Shishi Hensan Iinkai, *Fujiyoshida shishi tsūshihen*, vol. 1, pp. 144–46.

72. Totman, *Japan: An Environmental History*, pp. 51–52; Fuji Shishi Hensan Iinkai, *Fuji shishi*, vol. 1, pp. 199–213, 215.

73. Piggott, *Emergence of Japanese Kingship*, pp. 28–37; Barnes, *State Formation in Japan*, pp. 104–49; Barnes, "A Hypothesis for Early Kofun Rulership"; Mizoguchi, *Archaeology of Japan*, pp. 214–96.

74. Some clusters of small tombs were, however, built in what is now Gotenba toward the end of the Tomb period. Oyama Chōshi Hensan Senmon Iinkai, *Oyama chōshi*, vol. 6, p. 154.

75. Watai, "Fuji sanroku ni okeru kodai," p. 67.

76. Watai, "Suruga ni okeru shuchōbo," p. 1.

77. Watai, "Suruga ni okeru shuchōbo," p. 3.

78. Watai, "Suruga ni okeru shuchōbo," pp. 3–5.

79. Watai, "Suruga ni okeru shuchōbo," pp. 6–8.

80. Watai, "Suruga ni okeru shuchōbo," p. 2.

81. Mizoguchi, *Archaeology of Japan*, pp. 290, 312–13.

82. Piggott, *Emergence of Japanese Kingship*, pp. 77–78.

83. Piggott, *Emergence of Japanese Kingship*, pp. 56–57, 71.

84. Piggott, *Emergence of Japanese Kingship*, pp. 145–46; Ooms, *Imperial Politics and Symbolics in Ancient Japan*, pp. 30, 46–47.

85. Piggott, *Emergence of Japanese Kingship*, pp. 143–44.

86. Piggott, *Emergence of Japanese Kingship*, pp. 167–235.

87. For example, the preface to the *Record of Ancient Matters* portrays the *tennō* Tenmu (who reigned from 673 to 686) as a sage ruler who had mastered the knowledge of yin and yang and the five elements. According to Chinese cosmology, yin and yang are universal forces existing in a polar relationship with one another. The former is passive and female, the latter active and male. An imbalance of the two can produce a host of problems, such as medical ailments or failed harvests. The five elements (or the five phases) are wood, fire, earth, metal, and water. Failing to understand and manipulate them properly can also create troubles ranging from the minor to the catastrophic. *Kojiki*, p. 41.

88. *Kojiki*, p. 240.

89. Revolts by the Emishi persisted into the early Heian period. Koyama, "East and West in the Late Classical Age," pp. 378–82.

90. *Nihongi*, vol. 1, p. 205.

91. *Nihongi*, vol. 2, p. 188.

92. "Hitachi Province Gazetteer (*Hitachi Fudoki*)," pp. 54–55.

93. The Tsukuba poems in book 9 of the *Collection of Ten Thousand Leaves* come from Mushimaro's poetry collection. It is possible that he did not compose these poems, but it is widely accepted that he did. English translations of Mushimaro's poems on Tsukuba (*Man'yōshū* book 9, poem nos. 1753–54, 1757–58, 1759–60) can be found in Nippon Gakujutsu Shinkōkai, *The Manyōshū*, pp. 220–22. Original versions appear in Takagi et al., *Nihon koten bungaku taikei*, vol. 5, pp. 390–94.

94. *Man'yōshū* book 9, poem no. 1753. Takahashi Mushimaro, "Composed when Lord Ōtomo, the revenue-officer, climbed Mount Tsukuba," in Nippon Gakujutsu Shinkōkai, *The Manyōshū*, p. 220. Original appears in Takagi et al., *Nihon koten bungaku taikei*, vol. 5, p. 390.

95. Plutschow, *Chaos and Cosmos*, p. 120.

96. *Man'yōshū* book 9, poem no. 1759. Takahashi Mushimaro, "Climbing Mount Tsukuba on the Day of *kagai*," in Nippon Gakujutsu Shinkōkai, *The Manyōshū*, p. 222. Original appears in Takagi et al., *Nihon koten bungaku taikei*, vol. 5, p. 394.

97. *Man'yōshū* book 3, poem no. 317. Yamabe Akahito, "Poem on viewing Mount Fuji," in Levy, *The Ten Thousand Leaves*, p. 178. Original appears in Takagi et al., *Nihon koten bungaku taikei*, vol. 4, p. 168.

98. *Man'yōshū* book 14, poem no. 3355. Anonymous and untitled. Author's translation. Original appears in Takagi et al., *Nihon koten bungaku taikei*, vol. 6, p. 410.

99. *Man'yōshū* book 11, poem no. 2695. Anonymous and untitled. Author's translation. Original appears in Takagi et al., *Nihon koten bungaku taikei*, vol. 6, p. 224.

100. *Man'yōshū* book 11, poem no. 2697. Anonymous and untitled. Author's translation. Original appears in Takagi et al., *Nihon koten bungaku taikei*, vol. 6, p. 224.

Chapter Two: From Angry God to Parent of the World

1. Takahashi et al., "Fujisan jōgan funka to Aokigahara yōgan," pp. 304–5.

2. The popularity of the Aokigahara jukai as a "suicide forest" is often attributed to the publication of Seichō Matsumoto's 1960 novel *Tower of Waves* (*Nami no tō*), whose lovelorn heroine enters the Aokigahara to die. In the decades since, books, films, and online media (both Japanese and foreign) have reinforced the Aokigahara forest's dark reputation. For an analysis of the forest's notorious place in popular culture, see Nelson, *Circulating Fear*, pp. 71–90.

3. Takahashi et al., "Fujisan jōgan funka to Aokigahara yōgan," p. 304.

4. Takahashi et al., "Fujisan jōgan funka to Aokigahara yōgan," pp. 304–305.

5. Seventeenth day of seventh month of Jōgan 6 (864), *Nihon sandai jitsuroku*, vol. 9. Takeda and Satō, *Kundoku Nihon sandai jitsuroku*, p. 246. After systematically examining records concerning Fuji's volcanic activity from the eighth to the eighteenth centuries, volcanologist Koyama Masato has judged the *Nihon sandai jitsuroku*'s account of the Jōgan eruption to be among the most reliable. Koyama, "Fujisan no rekishi funka sōran," pp. 122–23, 134.

6. Scapulimancy, which employs the scapulae of deer or oxen, was practiced in Japan as early as the Yayoi period and perhaps even earlier, but divination using tortoise shells eventually

became the preferred method of the Yamato court. Perhaps this was partly due to the fact that tortoises were associated with longevity, making the use of their shells in divination auspicious (though clearly not for the tortoises killed along the way). "Bokusen," *Encyclopedia of Shinto.*

7. Fifth day of eighth month of Jōgan 6 (864), *Nihon sandai jitsuroku*, vol. 9. Takeda and Satō, *Kundoku Nihon sandai jitsuroku*, p. 248.

8. Sixth day of seventh month of Ten'ō 1 (781), *Shoku Nihongi*, vol. 36. Fujiyoshida Shishi Hensan Iinkai, *Fujiyoshida shishi shiryōhen*, vol. 2, p. 20.

9. Sixth day of sixth month of Enryaku 19 (800), *Nihon kiryaku*, part 1, vol. 13. Fujiyoshida Shishi Hensan Iinkai, *Fujiyoshida shishi shiryōhen*, vol. 2, p. 22.

10. Koyama, "Fujisan no rekishi funka sōran," p. 121; Endō, "Fuji goko no nazo," pp. 134–40.

11. Yamamoto et al., "Eruptive History of Mt. Fuji," p. 5; Yamamoto et al., "Shinki Fuji kazan kōka kasaibutsu," pp. 574–76.

12. Nineteenth day of fifth month of Enryaku 21 (802), *Nihon kiraku*, part 1, vol. 13. Fujiyo-shida Shishi Hensan Iinkai, *Fujiyoshida shishi shiryōhen*, vol. 2, p. 23.

13. Imperial decree to Suruga and Sagami provinces, eighth day of first month of Enryaku 21 (802), *Nihon kiraku*, part 1, vol. 13. Fujiyoshida Shishi Hensan Iinkai, *Fujiyoshida shishi shiryōhen*, vol. 2, p. 22.

14. Actual distances varied in response to the terrain as well as the availability of water and grass for the horses stabled at them. Fuji Shishi Hensan Iinkai, *Fuji shishi*, vol. 1, p. 261.

15. Takeda, "Roads in the *Tennō*-Centered Polity," p. 150; Shizuoka-ken, *Shizuoka kenshi tsūshihen*, vol. 1, pp. 627–28.

16. Kanbara and Kashiwahara were established somewhere in what is now Fuji City. Kashi-wahara was located to the east of the Fuji River until it was closed not long after the Jōgan eruption of 864–866 (for reasons explained below), and Kanbara was originally located on the west side of the Fuji River but was moved to the east after Kashiwahara was eliminated. Nagakura occupied two different locations near the border between Suruga and Izu provinces (in or around today's Mishima City), and Yokohashiri was most likely located in Gotenba. Shizuoka-ken, *Shizuoka kenshi tsūshihen*, vol. 1, p. 632.

17. The pass also sat at the intersection with a road that ran north into the province of Kai. The Kai road had two stations along Fuji's northeastern skirt. One of them, Kakichi, sat on the northern side of the Kagosaka Pass near Lake Yamanaka. The other, Kawaguchi, was located by Lake Kawaguchi just to the south of the Misaka Mountain Range separating the northern base of Fuji from the Kōfu Basin. Their precise location is unclear, but archaeologists have unearthed coins, specialized tools, and other remains at a number of Heian-period sites between Lake Yamanaka and Lake Kawaguchi, all of them pointing to the presence of a state road that seems to have shifted over time in response to volcanic activity. Shizuoka-ken, *Shizuoka kenshi tsūshihen*, vol. 1, p. 639; Fujiyoshida Shishi Hensan Iinkai, *Fujiyoshida shishi tsūshihen*, vol. 1, pp. 240–41, 250–80.

18. Nineteenth day of fifth month of Enryaku 21 (802), *Nihon kiraku*, part 1, vol. 13. In Fu-jiyoshida Shishi Hensan Iinkai, *Fujiyoshida shishi shiryōhen*, vol. 2, p. 23.

19. Farris, *Japan's Medieval Population*, p. 9.

20. Grapard, "Economics of Ritual Power," p. 86; Ooms, *Imperial Politics and Symbolics in Ancient Japan*, pp. 235–36.

21. Tenth day of twelfth month of Jōgan 6 (864), *Nihon sandai jitsuroku*, vol. 9. Takeda and Satō, *Kundoku Nihon sandai jitsuroku*, p. 257; Shizuoka-ken, *Shizuoka kenshi tsūshihen*, vol. 1, p. 632.

22. Ooms, *Imperial Politics and Symbolics in Ancient Japan*, p. 89.

23. Grapard, "The Economics of Ritual Power," p. 72; "Bokusen," *Encyclopedia of Shinto*.

24. Sixth day of seventh month of Ten'ō 1 (781), *Shoku Nihongi*, vol. 36. Fujiyoshida Shishi Hensan Iinkai, *Fujiyoshida shishi shiryōhen*, vol. 2, p. 20.

25. Fifth day of fifth month of Jōgan 2 (860), *Nihon sandai jitsuroku*, vol. 4. Takeda and Satō, *Kundoku Nihon sandai jitsuroku*, vol. 9, p. 88.

26. Grapard, "The Economics of Ritual Power," p. 73.

27. See, for example, *The Last Testament of Lord Kujō* (*Kujōdono ikai*), part of which has been translated into English in Lu, *Japan: A Documentary History*, vol. 1, pp. 73–74.

28. Eighth day of first month of Enryaku 21 (802), *Nihon kiraku*, part 1, vol. 13. Fujiyoshida Shishi Hensan Iinkai, *Fujiyoshida shishi shiryōhen*, vol. 2, p. 22. It is unclear where the priests who performed these rituals were based, although it seems likely they worked at shrines attached to district (*gun*) offices and were responsible for a range of duties not limited to worship of Fuji. Watai Hideyo argues that the Suruga priests mentioned in the court's response to the Enryaku eruption were most likely attached to the district office of Fuji-gun. Watai, "Reimeiki no Fujisan shinkō," p. 45.

29. Tanaka, Morita, and Sano, "Yama no kashi'iki kara mita Fujisan," p. 76.

30. Tanaka, Morita, and Sano, "Yama no kashi'iki kara mita Fujisan," p. 76.

31. Twenty-fifth day of fifth month of Jōgan 6 (864), *Nihon sandai jitsuroku*, vol. 8. Takeda and Satō, *Kundoku Nihon sandai jitsuroku*, p. 243.

32. Endō, "Fujisan shinkō no hassei," p. 11.

33. Ninth day of twelfth month of Jōgan 7 (865). *Nihon sandai jitsuroku*, vol. 11. Takeda and Satō, *Kundoku Nihon sandai jitsuroku*, pp. 302–3.

34. Ooms, *Imperial Politics and Symbolics in Ancient Japan*, pp. 232–34; Stockdale, *Imagining Exile in Heian Japan*, pp. 65–68.

35. Watai, "Reimeiki no Fujisan shinkō," p. 45.

36. Watai, "Reimeiki no Fujisan shinkō," p. 46. Shrine legend holds that it was established during the Daidō era (806–809) instead, but there is no independent evidence to corroborate this claim. Endō, "Fujisan shinkō no hassei," p. 7.

37. Watai, "Kōkogakuteki na kenchi," pp. 25–26.

38. For English translations of the *Kokinshū*, see *Kokinshū*; and *Kokin Wakashū*.

39. *Kokin wakashū*, book 19, poem no. 1028. *Kokinshū*, p. 353.

40. The use of "pivot words" (*kakekotoba*) with multiple meanings is a popular device in classical Japanese poetry.

41. *Saigyō: Poems of a Mountain Home*, p. 210.

42. Kubota, *Fujisan no bungaku*, p. 65.

43. *Man'yōshū*, book 3, poem no. 317. Yamabe Akahito, "Poem on viewing Mount Fuji," in Levy, *The Ten Thousand Leaves*, p. 178. Original appears in Takagi et al., *Nihon koten bungaku taikei*, vol. 4, p. 170.

44. The fifteenth day of the sixth month fell in the middle of summer. *Man'yōshū*, book 3, poem no. 320. Envoy to Takahashi Mushimaro's "Poem about Mount Fuji," in Levy, *The Ten Thousand Leaves*, p. 180.

45. "The Tales of Ise," p. 189.

46. Kubota, *Fujisan no bungaku*, pp. 85–86.

47. Kubota, *Fujisan no bungaku*, pp. 100, 120.

48. Miyako no Yoshika, *Fujisanki*, pp. 414–15.

49. Chūjō, "Miyako no Yoshika saku 'Fujisanki' ni tsuite," pp. 39–40.

50. Miyako no Yoshika, *Fujisanki*, pp. 414–15; Amano, "Kodaijin no Fujisan kan," p. 47; Miyata, "Taketori monogatari to Fuji shinkō," p. 30.

51. For an English translation of the tale, see Kawabata, *The Tale of the Bamboo Cutter*.

52. The suitors were stand-ins for historical figures. Marra, *Aesthetics of Discontent*, p. 29.

53. Miyata, "Taketori monogatari to Fuji shinkō," p. 28.

54. Marra, *Aesthetics of Discontent*, pp. 17–18.

55. For an English translation of *Hagoromo*, see Waley, *The Nō Plays of Japan*, pp. 218–26.

56. Takeya, *Fujisan no seishinshi*, pp. 90–91.

57. Takeya, *Fujisan no seishinshi*, pp. 92–93.

58. Takeya, *Fujisan no seishinshi*, pp. 95–96.

59. Takeya, *Fujisan no seishinshi*, pp. 77–78.

60. Earhart, *Mount Fuji: Icon of Japan*, pp. 26–27.

61. The earliest existing account of this legend appears in the *Record of Miraculous Events in Japan* (*Nihon ryōiki*), a collection of miracle tales assembled toward the end of the eighth century by the Buddhist monk Kyōkai. *Miraculous Stories from the Japanese Buddhist Tradition*, pp. 140–41.

62. Kamigaito, *Fujisan: sei to bi no yama*, pp. 29–31; Naruse, "Fujisan no kaiga," pp. 151–54.

63. Endō, "Fujisan shinkō no hassei," p. 27.

64. Moerman, "Archeology of Anxiety," p. 266.

65. Endō, "Fuji shinkō no seiritsu to Murayama shugen," pp. 35–36.

66. This translation of *Shugendō* comes from Castiglioni, Rambelli, and Roth, *Defining Shugendō*, p. 1. Also see Suzuki, "A Critical History of the Study of Shugendō," pp. 43–46. For an in-depth study of the development of Shugendō at a Japanese sacred mountain, see Carter, *A Path into the Mountains*.

67. Endō, "Fuji shinkō no seiritsu to Murayama shugen," p. 38; Sawada, *Faith in Mount Fuji*, p. 28.

68. Koyama, "Fujisan no rekishi funka sōran," pp. 125, 127.

69. Mills, "*Soga Monogatari, Shintōshū*," pp. 51–53.

70. For an English translation of the story, see "The Tale of the Fuji Cave," pp. 197–216.

71. Endō, "Fuji shinkō no seiritsu to Murayama shugen," p. 39.

72. Andrei, "*Sankei Mandara*," p. 383.

73. Naruse, "Fujisan no kaiga," p. 155; Takeya, *Fujisan no seishinshi*, pp. 157–63.

74. Andrei, "*Sankei Mandara*," p. 388.

75. Andrei, "*Sankei Mandara*," p. 389.

76. Endō, "Fuji shinkō no seiritsu to Murayama shugen," pp. 56–67; Miyachi and Hirono, *Sengen jinja no rekishi*, pp. 213–14; Aoyagi, *Fugaku tabi hyakkei*, pp. 175–77.

77. Oyama Chōshi Hensan Senmon Iinkai, *Oyama chōshi*, vol. 7, pp. 510–11.

78. Tyler, "'The Book of the Great Practice,'" p. 287.

79. Tyler, "The Tokugawa Peace and Popular Religion," pp. 104–5.

80. The biography is called *The Book of the Great Practice* (*Gotaigyō no maki*) and appears to date from the late eighteenth or early nineteenth century. An English translation can be found in Tyler, "'The Book of the Great Practice,'" pp. 284–316.

81. Tyler, "'The Book of the Great Practice,'" p. 255.

82. For a close study of the contents and ritual uses of these "body extractions," see Sawada, *Faith in Mount Fuji*, pp. 52–91.

83. Sawada, *Faith in Mount Fuji*, pp. 69–83.

84. Gottardo, "On the Identity of Mt. Fuji's Deity," p. 51.

85. Hirano, *Fuji shinkō to Fujikō*, p. 130.

Chapter Three: Fallout of the Hōei Eruption

1. Koyama, "Fujisan no rekishi funka sōran," pp. 126–29, 134.

2. Esper et al., "European Summer Temperature Response"; Parker, *Global Crisis*, p. 587.

3. Miyaji et al., "High-Resolution Reconstruction of the Hoei Eruption," p. 120.

4. Miyaji et al., "High-Resolution Reconstruction of the Hoei Eruption," p. 114.

5. Inoue, "Fujisan Hōei funka (1707)," p. 427. Recovery took even longer for especially hard-hit villages like Ōmika, which had thirty-six households prior to the eruption but only fourteen as late as the early twentieth century. Nagahara, *Fujisan Hōei daibakuhatsu*, p. 258.

6. Miyaji et al., "High-Resolution Reconstruction of the Hoei Eruption," p. 113.

7. Ravina, *Land and Lordship*, pp. 23, 27.

8. Ravina, *Land and Lordship*, p. 16; Totman, *Early Modern Japan*, pp. 118–20.

9. Cooper, "Tokugawa Yoshimune versus Tokugawa Muneharu," p. 62.

10. Miyaji et al., "High-Resolution Reconstruction of the Hoei Eruption," p. 113.

11. Bodart-Bailey, *The Dog Shogun*, p. 260.

12. Kitahara et al., *Nihon rekishi saigai jiten*, pp. 211–15; Smits, *Seismic Japan*, pp. 11–12.

13. Bodart-Bailey, *The Dog Shogun*, pp. 257–59.

14. Kitahara et al., *Nihon rekishi saigai jiten*, p. 217; Smits, *Seismic Japan*, p. 12.

15. Kitahara et al., *Nihon rekishi saigai jiten*, p. 219; Smits, *Seismic Japan*, p. 12.

16. Koyama, *Fujisan funka to hazādo mappu*, pp. 112–13; Miyaji et al., "High-Resolution Reconstruction of the Hoei Eruption," p. 125.

17. Koyama, *Fujisan funka to hazādo mappu*, pp. 107–11; Hayashi and Koyama, "Hōei yonen Fujisan funka," pp. 127–32; Miyaji et al., "High-Resolution Reconstruction of the Hoei Eruption," p. 125.

18. Koyama, *Fujisan funka to hazādo mappu*, pp. 108–10. For a brief explanation of "volcano acoustics," see Department of Geosciences at Oregon State University, Volcano World, "Volcano Sounds Before Eruptions," http://volcano.oregonstate.edu/volcano-sounds-eruptions.

19. Koyama, *Fujisan funka to hazādo mappu*, p. 111.

20. Miyaji et al., "High-Resolution Reconstruction of the Hoei Eruption," p. 115.

21. Nagahara, *Fujisan Hōei daibakuhatsu*, pp. 34–35. It is thought that the author was a Buddhist priest living in the village of Ubusuna, located in today's Oyama-chō.

22. Arai Hakuseki, *Told Round a Brushwood Fire*, p. 90.

23. Smits, *Seismic Japan*, pp. 45–46.

24. Smits, *Seismic Japan*, p. 47.

25. Koyama, *Fujisan funka to hazādo mappu*, pp. 25–98; Miyaji et al., "High-Resolution Reconstruction of the Hoei Eruption," p. 119.

26. Miyaji et al., "High-Resolution Reconstruction of the Hoei Eruption," p. 121.

27. Miyaji et al., "High-Resolution Reconstruction of the Hoei Eruption," pp. 125–27.

28. Miyaji et al., "High-Resolution Reconstruction of the Hoei Eruption," p. 119.

29. Arai Hakuseki, *Told Round a Brushwood Fire*, p. 90.

30. Nagahara, *Fujisan Hōei daibakuhatsu*, p. 34.

31. Koyama, *Fujisan funka to hazādo mappu*, pp. 15–16.

32. Koyama, *Fujisan funka to hazādo mappu*, p. 27.

33. Some date the beginning of the Yayoi period as far back as 1,000 BCE.

34. Farris, *Japan's Medieval Population*, p. 262.

35. Ono, "Tokugawa Tsunayoshi to dojō hiryō gaku"; Barnes, *Tectonic Archaeology*, pp. 175–203.

36. Wilson, *Turbulent Streams*, pp. 55–60; Fuji Shiritsu Hakubutsukan, *Kashima kome to mizu*, pp. 1–14.

37. Susono Shishi Hensan Senmon Iinkai, *Susono shishi*, vol. 8, pp. 343–45.

38. Odawara-shi, *Odawara shishi tsūshihen*, vol. 2, pp. 195–96.

39. Susono Shishi Hensan Senmon Iinkai, *Susono shishi*, vol. 8, pp. 342–71.

40. Oyama Chōshi Hensan Senmon Iinkai, *Oyama chōshi*, vol. 7, pp. 87–93.

41. Thanks in large part to irrigation schemes, it is estimated that the amount of land under cultivation in Japan grew about 200 percent over the course of the seventeenth century. Kitahara, *Nihon shinsaishi*, p. 15.

42. Nagahara, *Fujisan Hōei daibakuhatsu*, p. 34.

43. Koyama, *Fujisan funka to hazādo mappu*, p. 34.

44. Fujiyoshida Shishi Hensan Iinkai, *Fujiyoshida shishi shiryōhen*, vol. 3, p. 408.

45. Kuroita and Kokushi Taikei Henshūkai, *Tokugawa jikki*, vol. 6, p. 676–77; Tōkyō Shiyakusho, *Tōkyō shishikō shūkyōhen*, vol. 2, p. 109.

46. Thanks to the efforts of historians like Nagahara Keiji, these documents have been assembled and published in several regional histories. Nagahara, *Fujisan Hōei daibakuhatsu*, p. 39.

47. Nagahara, *Fujisan Hōei daibakuhatsu*, p. 43.

48. Kitahara, *Nihon shinsaishi*, p. 138.

49. Nagahara, *Fujisan Hōei daibakuhatsu*, pp. 47–48.

50. Odawara-shi, *Odawara shishi shiryōhen, kinsei*, vol. 2, pp. 418–19.

51. Scott, *Seeing Like a State*, pp. 11–83.

52. Berry, *Japan in Print*, p. 42.

53. Nagahara, *Fujisan Hōei daibakuhatsu*, p. 47; Odawara-shi, *Odawara shishi shiryōhen, kinsei*, vol. 2, pp. 410–11.

54. Odawara-shi, *Odawara shishi shiryōhen, kinsei*, vol. 2, pp. 411–16.

55. The shogunate had provided assistance to the domain after the earthquake, but in the form of a loan, not a grant. Kitahara, *Nihon shinsaishi*, p. 78.

56. Odawara-shi, *Odawara shishi shiryōhen, kinsei*, vol. 2, pp. 419–20; Nagahara, *Fujisan Hōei daibakuhatsu*, pp. 50–51.

57. Odawara-shi, *Odawara shishi shiryōhen, kinsei*, vol. 2, p. 420.

58. Oyama Chōshi Hensan Senmon Iinkai, *Oyama chōshi*, vol. 2, pp. 873–74.

59. Odawara-shi, *Odawara shishi shiryōhen, kinsei*, vol. 2, pp. 420–21.

60. Bodart-Bailey, *The Dog Shogun*, pp. 103, 284.

61. Bodart-Bailey, *The Dog Shogun*, pp. 134.

62. Naitō Chisō, *Tokugawa jūgodaishi*, vol. 3, p. 1438. The original source of this account is unspecified.

63. Bodart-Bailey, *The Dog Shogun*, p. 203.

64. Bodart-Bailey, *The Dog Shogun*, p. 143.

65. Bodart-Bailey, *The Dog Shogun*, pp. 268, 287.

66. Odawara-shi, *Odawara shishi shiryōhen, kinsei*, vol. 2, pp. 421–22.

67. Odawara-shi, *Odawara shishi shiryōhen, kinsei*, vol. 2, pp. 422–23.

68. The value of one *ryō* fluctuated throughout the Tokugawa period, but a rule of thumb is that it could purchase one *koku* (about 180 liters) of rice, supposedly enough to feed an adult for an entire year.

69. Odawara-shi, *Odawara shishi shiryōhen, kinsei*, vol. 2, pp. 425–28.

70. Nagahara, *Fujisan Hōei daibakuhatsu*, p. 54.

71. Odawara-shi, *Odawara shishi shiryōhen, kinsei*, vol. 2, pp. 428–29.

72. Odawara-shi, *Odawara shishi shiryōhen, kinsei*, vol. 2, pp. 429–30.

73. Odawara-shi, *Odawara shishi shiryōhen, kinsei*, vol. 2, pp. 434–35.

74. Odawara-shi, *Odawara shishi shiryōhen, kinsei*, vol. 2, p. 435.

75. Ó Gráda, *Black '47 and Beyond*, pp. 6, 52, 66; quote at p. 6.

76. Tōkyō-shi, *Tōkyō shishikō shigaihen*, vol. 9, pp. 910–11; Nagahara, *Fujisan Hōei daibakuhatsu*, pp. 62–63. The plan was officially put into effect on the seventh day of the first intercalary month. Tōkyō-shi, *Tōkyō shishikō shigaihen*, vol. 9, p. 911.

77. Bodart-Bailey, *The Dog Shogun*, p. 192.

78. Nagahara, *Fujisan Hōei daibakuhatsu*, pp. 47, 62.

79. Nagahara, *Fujisan Hōei daibakuhatsu*, p. 63.

80. Kitahara, *Nihon shinsaishi*, pp. 314–15.

81. Bodart-Bailey, *The Dog Shogun*, p. 191.

82. Kitahara, *Nihon shinsaishi*, pp. 124, 315.

83. Kitahara, *Nihon shinsaishi*, pp. 82–103, 109–20.

84. Tōkyō-shi, *Tōkyō shishikō shigaihen*, vol. 9, p. 912; Nagahara, *Fujisan Hōei daibakuhatsu*, p. 65.

85. Kitahara, *Nihon shinsaishi*, pp. 145–46.

86. Nagahara, *Fujisan Hōei daibakuhatsu*, pp. 65–66. The exact figure includes some silver too.

87. Nagahara, *Fujisan Hōei daibakuhatsu*, pp. 66–67; Kitahara, *Nihon shinsaishi*, pp. 135–36.

88. Nagahara, *Fujisan Hōei daibakuhatsu*, p. 67.

89. Asahi, *Tekiroku Ōmu rōchūki*, vol. 2, p. 146.

90. Bodart-Bailey, *The Dog Shogun*, p. 274.

91. Nagahara, *Fujisan Hōei daibakuhatsu*, pp. 69–70.

92. Oyama Chōshi Hensan Senmon Iinkai, *Oyama chōshi*, vol. 2, p. 836.

93. Nagahara, *Fujisan Hōei daibakuhatsu*, p. 35.

94. Oyama Chōshi Hensan Senmon Iinkai, *Oyama chōshi*, vol. 2, p. 836.

95. Inoue, "Genroku jishin (1703) to Fujisan Hōei funka (1707)," pp. 251–54; Amano, "Yokonosannōbara iseki no hakkutsu chōsa," pp. 94–97.

96. Oyama Chōshi Hensan Senmon Iinkai, *Oyama chōshi*, vol. 2, pp. 838–50.

97. Odawara-shi, *Odawara shishi shiryōhen, kinsei*, vol. 2, pp. 446–60; Nagahara, *Fujisan Hōei daibakuhatsu*, pp. 77–80.

98. Kitahara, *Nihon shinsaishi*, pp. 109–25.

99. Nagahara, *Fujisan Hōei daibakuhatsu*, p. 116.

100. Nagahara, *Fujisan Hōei daibakuhatsu*, p. 117.

101. Oyama Chōshi Hensan Senmon Iinkai, *Oyama chōshi*, vol. 2, pp. 850–51, 856–57, 872–78, 886–87.

102. Oyama Chōshi Hensan Senmon Iinkai, *Oyama chōshi*, vol. 2, p. 857.

103. Oyama Chōshi Hensan Senmon Iinkai, *Oyama chōshi*, vol. 2, pp. 873–74.

104. Nagahara, *Fujisan Hōei daibakuhatsu*, p. 117.

105. Nagahara, *Fujisan Hōei daibakuhatsu*, pp. 108, 114, 118–19.

106. Oyama Chōshi Hensan Senmon Iinkai, *Oyama chōshi*, vol. 2, pp. 886–89.

107. Oyama Chōshi Hensan Senmon Iinkai, *Oyama chōshi*, vol. 2, p. 893.

108. Nagahara, *Fujisan Hōei daibakuhatsu*, pp. 131–32.

109. Nagahara, *Fujisan Hōei daibakuhatsu*, p. 256.

110. Oyama Chōshi Hensan Senmon Iinkai, *Oyama chōshi*, vol. 2, pp. 893–94.

111. Nagahara, *Fujisan Hōei daibakuhatsu*, p. 122.

112. Nagahara, *Fujisan Hōei daibakuhatsu*, p. 129; Bodart-Bailey, *The Dog Shogun*, pp. 272–73; Oyama Chōshi Hensan Senmon Iinkai, *Oyama chōshi*, vol. 2, pp. 896–900.

113. Oyama Chōshi Hensan Senmon Iinkai, *Oyama chōshi*, vol. 2, p. 896.

114. Oyama Chōshi Hensan Senmon Iinkai, *Oyama chōshi*, vol. 2, pp. 897–900.

115. Nagahara, *Fujisan Hōei daibakuhatsu*, pp. 137–40.

116. Nagahara, *Fujisan Hōei daibakuhatsu*, pp. 137–41.

117. Oyama Chōshi Hensan Senmon Iinkai, *Oyama chōshi*, vol. 2, pp. 901–2, 928.

118. Oyama Chōshi Hensan Senmon Iinkai, *Oyama chōshi*, vol. 2, p. 902.

119. Nagahara, *Fujisan Hōei daibakuhatsu*, pp. 147–50.

120. Oyama Chōshi Hensan Senmon Iinkai, *Oyama chōshi*, vol. 2, pp. 786–88.

121. Oyama Chōshi Hensan Senmon Iinkai, *Oyama chōshi*, vol. 2, pp. 789–91.

122. Oyama Chōshi Hensan Senmon Iinkai, *Oyama chōshi*, vol. 2, pp. 793–94.

123. Nagahara, *Fujisan Hōei daibakuhatsu*, pp. 154–55.

124. Nagahara, *Fujisan Hōei daibakuhatsu*, pp. 79, 162–65.

125. Nagahara, *Fujisan Hōei daibakuhatsu*, p. 167.

126. Nagahara, *Fujisan Hōei daibakuhatsu*, pp. 177, 189–90.

127. Nagahara, *Fujisan Hōei daibakuhatsu*, pp. 174–75.

128. Nagahara, *Fujisan Hōei daibakuhatsu*, pp. 63, 178–79, 192–93.

129. Nagahara, *Fujisan Hōei daibakuhatsu*, pp. 190–93.

130. Nagahara, *Fujisan Hōei daibakuhatsu*, pp. 194–97, 205.

131. Nagahara, *Fujisan Hōei daibakuhatsu*, p. 197.

132. Nagahara, *Fujisan Hōei daibakuhatsu*, pp. 197–204.

133. Nagahara, *Fujisan Hōei daibakuhatsu*, pp. 204–6, 216–17, 220.

134. Inoue, "Fujisan Hōei funka (1707)," p. 433.

135. Nagahara, *Fujisan Hōei daibakuhatsu*, pp. 239–40.

136. Nagahara, *Fujisan Hōei daibakuhatsu*, p. 258.

137. Baba, "Genroku dai jishin to Hōei Fujisan funka, sono 2."

138. Walthall, "Popular Movements in Early Modern Japan," p. 663.

139. Parker, *Global Crisis*, pp. 493–95.

140. Smits, *Seismic Japan*, p. 91.

141. Kitahara, *Nihon shinsaishi*, pp. 160, 171.

142. Vlastos, *Peasant Protests and Uprisings*, pp. 20, 70–72.

143. Totman, *Early Modern Japan*, pp. 238–40, 243–44, 343.

144. Ehlers, *Give and Take*, pp. 201–46.

145. Nagahara, *Fujisan Hōei daibakuhatsu*, pp. 242–46.

Chapter Four: Holy Fuji

1. Jikigyō, *Sanjūichinichi no otsutae*, in Iwashina, *Fujikō no rekishi*, p. 536.

2. Jikigyō, *Ichiji fusetsu no maki*, *Osoegaki no maki*, and *Sanjūichinichi no otsutae*, in Iwashina, *Fujikō no rekishi*, pp. 501, 527, 529, 547.

3. John Hall devised the phrase "container society" to describe the status system of Tokugawa Japan. Hall, "Rule by Status in Tokugawa Japan," p. 48.

4. Earhart, *Mount Fuji: Icon of Japan*, p. 48.

5. Tyler, "The Tokugawa Peace and Popular Religion," pp. 109, 111, 113 (n.76).

6. Iwashina, *Fujikō no rekishi*, p. 140.

7. Moerman, "Archeology of Anxiety," pp. 245–274.

8. Miyata, "Fuji shinkō to Miroku," pp. 292–99.

9. See, for example, Jikigyō's multiple references to *roku* in the sense of "proper/rectified" in the *Unspoken Word* (*Ichiji fusetsu no maki*) and the *Addendum* (*Osoegaki no maki*), in Iwashina, *Fujikō no rekishi*, pp. 502, 524, 528–29, 532.

10. Jikigyō, *Osoegaki no maki* and *Sanjūichinichi no otsutae*, in Iwashina, *Fujikō no rekishi*, pp. 523, 538, 541, 550.

11. Iwashina, *Fujikō no rekishi*, pp. 290–91; Sawada, *Faith in Mount Fuji*, pp. 128–31.

12. Sawada, *Faith in Mount Fuji*, p. 120.

13. Jikigyō, *Osoegaki no maki*, in Iwashina, *Fujikō no rekishi*, pp. 530–34.

14. Jikigyō, *Ichiji fusetsu no maki*, in Iwashina, *Fujikō no rekishi*, p. 504.

15. Totman, *Early Modern Japan*, pp. 236–37, 308.

16. Sawada, *Faith in Mount Fuji*, p. 146.

17. In Jikigyō's time this was considered the seventh station. He refers to the "seventh-station Eboshi rock" in the *Sanjūichinichi no otsutae*, in Iwashina, *Fujikō no rekishi*, p. 537.

18. Iwashina, *Fujikō no rekishi*, pp. 187–89; Sawada, *Faith in Mount Fuji*, p. 144.

19. Jikigyō, *Sanjūichinichi no otsutae*, in Iwashina, *Fujikō no rekishi*, pp. 535–51; Tyler, "The Tokugawa Peace and Popular Religion," p. 113.

20. There is another version of Jikigyō's final teachings called the *Thirty-One-Day Book* (*Sanjūichinichi no onmaki*), but it was the *Transmission* that circulated widely among Fujikō members and became a staple of their faith. An annotated version of the *Sanjūichinichi no onmaki* appears in Yasumaru and Itō, "Fujikō," pp. 426–51.

21. Castiglioni, "Devotion in Flesh and Bone."

22. Jikigyō, *Sanjūichinichi no otsutae*, in Iwashina, *Fujikō no rekishi*, p. 536.

23. Sawada, *Faith in Mount Fuji*, pp. 149–55.

24. Iwashina, *Fujikō no rekishi*, pp. 194–95.

25. Jikigyō, *Osoegaki no maki*, in Iwashina, *Fujikō no rekishi*, p. 527.

26. Jikigyō, *Ichiji fusetsu no maki* and *Sanjūichinichi no otsutae*, in Iwashina, *Fujikō no rekishi*, pp. 503, 546.

27. Jikigyō, *Sanjūichinichi no otsutae*, in Iwashina, *Fujikō no rekishi*, pp. 537, 540.

28. Jikigyō, *Sanjūichinichi no otsutae*, in Iwashina, *Fujikō no rekishi*, pp. 537–38, 540.

29. Jikigyō, *Sanjūichinichi no otsutae*, in Iwashina, *Fujikō no rekishi*, p. 539.

30. Jikigyō, *Sanjūichinichi no otsutae*, in Iwashina, *Fujikō no rekishi*, pp. 537, 547.

31. Jikigyō, *Ichiji fusetsu no maki*, in Iwashina, *Fujikō no rekishi*, pp. 500–501; Yamanashi Kenritsu Fujisan Sekai Isan Sentā, *Yōgan dōketsu o meguru shinkō*, pp. 2–6, 10–14.

32. Jikigyō, *Sanjūichinichi no otsutae*, in Iwashina, *Fujikō no rekishi*, p. 549.

33. Jikigyō, *Sanjūichinichi no otsutae*, in Iwashina, *Fujikō no rekishi*, p. 550.

34. Jikigyō, *Sanjūichinichi no otsutae*, in Iwashina, *Fujikō no rekishi*, pp. 537–39, 548–49.

35. Miyazaki, "Female Pilgrims and Mt. Fuji," pp. 341–48.

36. Takemi, "'Menstruation Sutra' Belief in Japan."

37. Jikigyō, *Osoegaki no maki* and *Sanjūichinichi no otsutae*, in Iwashina, *Fujikō no rekishi*, pp. 527, 538.

38. Jikigyō, *Sanjūichinichi no otsutae*, in Iwashina, *Fujikō no rekishi*, p. 542.

39. Sawada, *Faith in Mount Fuji*, pp. 131–33.

40. Miyazaki, "Female Pilgrims and Mt. Fuji," pp. 368–76.

41. Jikigyō, *Sanjūichinichi no otsutae*, in Iwashina, *Fujikō no rekishi*, pp. 539, 541, 543, 545, 548; quote at p. 539.

42. Jikigyō, *Sanjūichinichi no otsutae*, in Iwashina, *Fujikō no rekishi*, pp. 538–39, 544–47.

43. Jikigyō, *Sanjūichinichi no otsutae*, in Iwashina, *Fujikō no rekishi*, pp. 538–42.

44. Jikigyō, *Sanjūichinichi no otsutae*, in Iwashina, *Fujikō no rekishi*, p. 540.

45. Jikigyō, *Sanjūichinichi no otsutae*, in Iwashina, *Fujikō no rekishi*, p. 540.

46. Jikigyō, *Sanjūichinichi no otsutae*, in Iwashina, *Fujikō no rekishi*, p. 542.

47. Jikigyō, *Sanjūichinichi no otsutae*, in Iwashina, *Fujikō no rekishi*, p. 547.

48. Iwashina, *Fujikō no rekishi*, p. 234.

49. Tyler, "'The Book of the Great Practice,'" pp. 261–62.

50. Tyler, "'The Book of the Great Practice,'" pp. 262–63, 307–12.

51. Earhart, *Mount Fuji: Icon of Japan*, p. 91; Iwashina, *Fujikō no rekishi*, pp. 253–55.

52. Iwashina, *Fujikō no rekishi*, pp. 268–69.

53. Jikigyō, *Osoegaki no maki*, in Iwashina, *Fujikō no rekishi*, p. 527.

54. Iwashina, *Fujikō no rekishi*, p. 270.

55. Iwashina, *Fujikō no rekishi*, pp. 268–70; Kawai, "Fujikō kara mita seichi Fujisan no fūkei," p. 354.

56. Takeuchi, "Making Mountains," p. 44.

57. Takeuchi, "Making Mountains," p. 27.

58. One particularly popular pilgrimage for miniaturization is the famous circuit of eighty-eight temples on the island of Shikoku.

59. Kawai, "Fujikō kara mita seichi Fujisan no fūkei," pp. 352–53.

60. Iwashina, *Fujikō no rekishi*, pp. 270–71.

61. Takeuchi, "Making Mountains," p. 40.

62. Vaporis, *Breaking Barriers*, pp. 198–216, 236–49.

63. Okuwaki, "Fujisan no shinkō to 'ohachi meguri,'" p. 89.

64. Records show that, from the middle of the Tokugawa period, the number of pilgrims who went to Kamiyoshida equaled the number of those who went to all three Suruga routes combined. Hirano, *Fuji shinkō to Fujikō*, p. 125.

65. Hirano, *Fuji shinkō to Fujikō*, p. 143; Nishida, "Kawaguchi-mura ni okeru Fujisan oshi," p. 83.

66. Nagashima, *Fujisan shinkei no zu*, pp. 8–9.

67. To discourage nighttime bullet climbs, in the summer of 2024, the Yoshida trail was closed to climbers without hut reservations between 4:00 p.m. and 3:00 a.m. "Fuji Subaru Line 5th Station," Japan-guide.com, accessed August 10, 2024, https://www.japan-guide.com/e/e6922.html.

68. The identities of some lakes in this circuit shifted over time. Tyler, "'The Book of the Great Practice,'" pp. 277–79.

69. For a close analysis of this picture-map, see Miyazaki, "An Artist's Rendering of the Divine Mount Fuji," pp. 98–101.

70. Watanabe, "Nature and Scenery of Mt. Fuji Area," p. 207.

71. Tyler, "'The Book of the Great Practice,'" p. 259.

72. Fuji Sabo Work Office, *Kuzure (Collapse)*.

73. Aoyagi, *Fugaku tabi hyakkei*, pp. 152–54.

74. Okuwaki, "Fujisan no shinkō to 'ohachi meguri,'" pp. 76–80.

75. Nagashima, *Fujisan shinkei no zu*, pp. 61–73, 190–94.

76. Nagashima, *Fujisan shinkei no zu*, pp. 61, 66.

77. Ōtani, "Kinmeisui to Fujikō," pp. 47–51.

78. Nagashima's guidebook lists these specific items and how much they cost. *Fujisan shinkei no zu*, p. 192.

79. Aoyagi, *Fugaku tabi hyakkei*, p. 73.

80. Shinno, "Journeys, Pilgrimages, Excursions," pp. 468–69; Aoyagi, *Fugaku tabi hyakkei*, pp. 9–10.

81. Aoyagi, *Fugaku tabi hyakkei*, pp. 76–77.

82. Gottardo, "On the Identity of Mt. Fuji's Deity," pp. 50–51.

83. Ambros, *Emplacing a Pilgrimage*, pp. 164–65.

84. Vaporis, *Breaking Barriers*, p. 218; Shinno, "Journeys, Pilgrimages, Excursions," pp. 455–56, 463.

85. Aoyagi, *Fugaku tabi hyakkei*, pp. 52–53.

86. Oyama Chōshi Hensan Senmon Iinkai, *Oyama chōshi*, vol. 2, pp. 995–99.

87. Oyama Chōshi Hensan Senmon Iinkai, *Oyama chōshi*, vol. 2, pp. 1000–1003. The dispute between Subashiri and Sengen Shrine is covered in Aoyagi, *Fugaku tabi hyakkei*, pp. 186–89.

88. Oyama Chōshi Hensan Senmon Iinkai, *Oyama chōshi*, vol. 7, pp. 509–11. For an in-depth discussion of the background and outcome of this dispute, see Aoyagi, *Fugaku tabi hyakkei*, pp. 193–209.

89. Aoyagi, *Fugaku tabi hyakkei*, pp. 176–77.

90. Aoyagi, *Fugaku tabi hyakkei*, pp. 171–73.

91. Miyazaki, "Female Pilgrims and Mt. Fuji," p. 357.

92. Miyazaki, "Female Pilgrims and Mt. Fuji," p. 353; Sawada, "Sexual Relations as Religious Practice," p. 352.

93. Sawada, "Sexual Relations as Religious Practice," pp. 353–55.

94. Sawada, "Sexual Relations as Religious Practice," p. 358.

95. Miyazaki, "Networks of Believers in a New Religion," p. 150.

96. Miyazaki, "Networks of Believers in a New Religion," pp. 157–59.

97. Sawada, "Sexual Relations as Religious Practice," p. 352.

98. Miyazaki, "Female Pilgrims and Mt. Fuji," p. 360.

99. Miyazaki, "Female Pilgrims and Mt. Fuji," pp. 357–60.

100. Roberts, *Performing the Great Peace*, p. 3.

101. Roberts, *Performing the Great Peace*, pp. 6–7.

102. Miyazaki, "Female Pilgrims and Mt. Fuji," p. 361.

103. Miyazaki, "Female Pilgrims and Mt. Fuji," p. 362.

104. Miyazaki, "Female Pilgrims and Mt. Fuji," p. 362.

105. Miyazaki, "Female Pilgrims and Mt. Fuji," p. 363.

106. A series of lawsuits between Kawaguchi and Yoshida oshi can be found in Itō, *Fujisan oshi*, pp. 275–95.

107. Miyazaki, "Female Pilgrims and Mt. Fuji," pp. 363–64. The text of the lawsuit is located in Fujiyoshida Shishi Hensan Iinkai, *Fujiyoshida shishi shiryōhen*, vol. 5, pp. 655–59.

108. Miyazaki, "Female Pilgrims and Mt. Fuji," pp. 366–68.

109. Miyazaki, "Female Pilgrims and Mt. Fuji," pp. 368–72.

110. See Date, *Nihon shūkyō seido shiryō ruiju kō*, for a list of Tokugawa regulations for temples and shrines (organized chronologically and by subject).

111. Tamamuro, "Local Society and the Temple-Parishioner Relationship," p. 262.

112. Hardacre, *Shinto: A History*, pp. 240–41.

113. Tōkyō-shi, *Tōkyō shishikō sangyōhen*, vol. 16, pp. 112–13.

114. Kinsei Shiryō Kenkyūkai, *Edo machibure shūsei*, vol. 10, pp. 9–10.

115. The text of the petition appears in Sawato, "Fuji shinkō girei to Edo bakufu," p. 154.

116. A blow-by-blow account of the investigation appears in Sawato, "Fuji shinkō girei to Edo bakufu," pp. 151–59.

117. Kinsei Shiryō Kenkyūkai, *Edo machibure shūsei*, vol. 10, pp. 9–10.

118. Kinsei Shiryō Kenkyūkai, *Edo machibure shūsei*, vol. 10, p. 12.

119. Tōkyō-shi, *Tōkyō shishikō sangyōhen*, vol. 41, pp. 624–27.

120. Sawato, "Fuji shinkō girei to Edo bakufu," pp. 166–69.

121. Kinsei Shiryō Kenkyūkai, *Edo machibure shūsei*, vol. 11, pp. 48, 356–57; vol. 14, p. 29; vol. 15, pp. 445–49.

122. Iwashina, *Fujikō no rekishi*, p. 365.

123. Iwashina, *Fujikō no rekishi*, p. 353.

124. Kinsei Shiryō Kenkyūkai, *Edo machibure shūsei*, vol. 11, p. 357; vol. 15, p. 445.

125. Kinsei Shiryō Kenkyūkai, *Edo machibure shūsei*, vol. 14, pp. 20–21, 29–30; vol. 15, p. 266.

126. Kinsei Shiryō Kenkyūkai, *Edo machibure shūsei*, vol. 15, p. 445.

127. Kinsei Shiryō Kenkyūkai, *Edo machibure shūsei*, vol. 15, pp. 446–49; Iwashina, *Fujikō no rekishi*, p. 362.

128. Miyazaki, "Minshū no shūkyō undō," pp. 276–77.

129. Miyazaki, "Networks of Believers in a New Religion," p. 154.

130. Miyazaki, "Networks of Believers in a New Religion," p. 156.

131. Miyazaki, "Networks of Believers in a New Religion," p. 152.

132. Miura, *Agents of World Renewal*, p. 9.

133. Iwashina, *Fujikō no rekishi*, p. 362; Miyazaki, "Minshū no shūkyō undō," pp. 283–89.

134. Iwashina, *Fujikō no rekishi*, p. 363.

135. Fujiyoshida Shishi Hensan Iinkai, *Fujiyoshida shishi tsūshihen*, vol. 2, p. 840; Iwashina, *Fujikō no rekishi*, pp. 363–64; Miyazaki, "Daikyō senpu no jidai no shinshūkyō," p. 63.

136. Miyazaki, "Formation of Emperor Worship," p. 289.

137. Jikigyō, *Sanjūichinichi no otsutae*, in Iwashina, *Fujikō no rekishi*, p. 547.

Chapter Five: From Pride of Edo to Symbol of Japan

1. Toby, *Engaging the Other*, p. 289.

2. Totman, *Early Modern Japan*, pp. 66–67; McClain and Merriman, "Edo and Paris: Cities and Power," pp. 11–14.

3. Nenzi, *Excursions in Identity*, p. 22.

4. Lippit, *Painters of the Realm*, pp. 80–81.

5. Screech, *Tokyo Before Tokyo*, p. 183.

6. Matsushima, "Mount Fuji and the Tokugawa Shogunate," p. 84.

7. Matsushima, "Mount Fuji and the Tokugawa Shogunate," pp. 78–81.

8. Geographer Tanaka Kei has calculated that about a third of all Japanese today live within Fuji's viewshed. Tanaka, "Fujisan no kashi'iki," p. 358.

9. A partial list of places in Edo named "Fujimi" can be found in Inobe, *Fuji no rekishi*, p. 366.

10. Tanaka, "Fujisan no kashi'iki," pp. 360–61.

11. Takebayashi, "Nihon zenkoku no mitate Fuji," pp. 321–22, 325.

12. Inobe, *Fuji no rekishi*, pp. 376–403.

13. Inobe, *Fuji no rekishi*, pp. 404–5. A list of eighteen Fujis outside Japan appears in Takebayashi, "Nihon zenkoku no mitate Fuji," p. 320.

14. "The Tales of Ise," p. 189.

15. Nenzi, *Excursions in Identity*, pp. 126–28.

16. Nenzi, *Excursions in Identity*, p. 40.

17. Nenzi, *Excursions in Identity*, p. 131.

18. Ikku, *Shank's Mare*, p. 51.

19. Matsuo Basho, *Basho: The Complete Haiku*, pp. 42, 104.

20. "Matsuo Bashō, Two Haiku."

21. "Matsuo Bashō, Two Haiku."

22. Author's translation. Original found in Matsuo Basho, *Basho: The Complete Haiku*, p. 246.

23. Author's translation. Original found in Kobayashi Issa, *Issa haikushū*, p. 91.

24. Author's translation. Original found in Kubota, *Fujisan no bungaku*, p. 160.

25. Torii et al., *Nihon no kokoro*, pp. 97–120; Okamoto, *Fugaku roppyaku kei*, pp. 16–23, 168–71.

26. Clark, *100 Views of Mount Fuji*, pp. 18–19.

27. Clark, *100 Views of Mount Fuji*, p. 39.

28. Martini, "Katsushika Hokusai and Mount Fuji," pp. 82–84.

29. Arguments about when to date the publication of the third volume are summarized in Smith, *Hokusai: One Hundred Views of Mt. Fuji*, pp. 17–18.

30. Martini, "Utagawa Hiroshige and Mount Fuji," p. 101.

31. Smith, *Hiroshige: One Hundred Famous Views of Edo*.

32. Martini, "Utagawa Hiroshige and Mount Fuji," pp. 103, 107.

33. Smith, *Hokusai: One Hundred Views of Fuji*, p. 14; Clark, *100 Views of Mount Fuji*, p. 20.

34. Smith, *Hokusai: One Hundred Views of Fuji*, p. 14.

35. For a book-length study of the influence of "the Western scientific gaze" on Tokugawa popular culture, see Screech, *The Lens Within the Heart*.

36. Clark, *100 Views of Mount Fuji*, p. 16.

37. The painting was exhibited at Atagoyama Shrine from 1796 to 1811, "after which it was removed from display for fear that a work of such exotic style might bring calamity on the shrine." Clark, *100 Views of Mount Fuji*, p. 16.

38. Clark, *100 Views of Mount Fuji*, p. 16.

39. Martini, "Katsushika Hokusai and Mount Fuji," pp. 81–82.

40. Guth, *Hokusai's Great Wave*, p. 17.

41. Guth, *Hokusai's Great Wave*, p. 19.

42. For a close examination of the import of Berlin blue and its impact on Japanese printmaking, see Smith, "Hokusai and the Blue Revolution in Edo Prints," pp. 235–69. Charts showing changes in the amount and price of Berlin blue imports can be found on pp. 240, 241, and 244.

43. The preface was written by the same Ryūtei Tanehiko in whose book of stories the advertisement for *Thirty-Six Views of Mount Fuji* appeared. Smith, *Hokusai: One Hundred Views of Mt. Fuji*, p. 194.

44. Smith, *Hokusai: One Hundred Views of Mt. Fuji*, p. 19.

45. Smith, *Hokusai: One Hundred Views of Mt. Fuji*, pp. 14–15.

46. Smith, *Hokusai: One Hundred Views of Mt. Fuji*, p. 7.

47. Smith, *Hokusai: One Hundred Views of Mt. Fuji*, p. 19.

48. Martini, "Utagawa Hiroshige and Mount Fuji," p. 100.

49. Aoyagi, *Fugaku tabi hyakkei*, pp. 82–83.

50. For an overview of different kinds of Tokugawa-period maps and their uses, see Traganou, *The Tōkaidō Road*, pp. 25–47.

51. Horiuchi, "'Fujimi jūsanshū yochi no zenzu' ihan saikō," pp. 9–10.

52. Chamberlain, *Japanese Things*, p. 309. Chamberlain's book was originally published in 1890 under the title *Things Japanese*.

53. Horiuchi, "'Fujimi jūsanshū yochi no zenzu' ihan saikō," p. 15.

54. Many historians refrain from using the term "nation" to describe Tokugawa Japan, given its significant differences—economic, social, and political—from the centralized nation-state that Japan would later become. As Mary Elizabeth Berry argues, however, texts produced in the Tokugawa period did generate a sense of nationhood defined by "an integral conception of territory, an assumption of political union under a paramount state, and a prevailing agreement

about the cultural knowledge and social intercourse that bound 'our people.' They also indulged in chauvinistic pieties about 'our country,' a land rich in history, harvests, enterprise, taste." Berry, *Japan in Print*, p. 248.

55. Takeya, *Fujisan no seishinshi*, pp. 77–78.

56. Takeya, *Fujisan no seishinshi*, pp. 94–96.

57. Toby, *Engaging the Other*, p. 258.

58. Toby, *Engaging the Other*, pp. 282–92.

59. Smith, *Hokusai: One Hundred Views of Fuji*, pp. 66–67, 213–14.

60. Toby, *Engaging the Other*, pp. 262–69, 297–304.

61. Ronald Toby suggests that the two figures might be a captured Korean prince and his servant. Toby, *Engaging the Other*, p. 301.

62. Toby, *Engaging the Other*, p. 274.

63. For an English-language summary and close analysis of the entire tale, see Yonemoto, *Mapping Early Modern Japan*, pp. 111–24. The original is found in Hiraga Gennai, *Fūrai sanjin shū*, pp. 153–224.

64. Hiraga Gennai, *Fūrai sanjin shū*, pp. 205–8.

65. Yonemoto, *Mapping Early Modern Japan*, p. 123.

66. Suzuki, "Seeking Accuracy," pp. 129–32.

67. Toby, *Engaging the Other*, p. 281.

68. Fujiyoshida Shishi Hensan Iinkai, *Fujiyoshida shishi tsūshihen*, vol. 2, pp. 232–33.

69. Guth, *Hokusai's Great Wave*, p. 32.

Chapter Six: Native Mountain, Foreign Threat

1. Horiuchi, "'Fujimi jūsanshū yochi no zenzu' ihan saikō," p. 9.

2. See Date, *Nihon shūkyō seido shiryō ruiju kō*, for a list of Tokugawa regulations for temples and shrines (organized chronologically and by subject).

3. Hardacre, *Shinto: A History*, p. 240.

4. Hardacre, *Shinto: A History*, pp. 240–43.

5. Hardacre, *Shinto: A History*, pp. 216–18. Quote at p. 216.

6. Hardacre, *Shinto: A History*, p. 216.

7. Hardacre, *Shinto: A History*, pp. 212–15.

8. Hardacre, *Shinto: A History*, p. 229.

9. Fujiyoshida Shishi Hensan Iinkai, *Fujiyoshida shishi shiryōhen*, vol. 4, p. 726; Hirano, *Fuji shinkō to Fujikō*, p. 132.

10. Fujiyoshida Shishi Hensan Iinkai, *Fujiyoshida shishi shiryōhen*, vol. 4, pp. 725–26.

11. Fujiyoshida Shishi Hensan Iinkai, *Fujiyoshida shishi shiryōhen*, vol. 4, p. 726.

12. Fujiyoshida Shishi Hensan Iinkai, *Fujiyoshida shishi shiryōhen*, vol. 4, pp. 728–30.

13. Fujiyoshida Shishi Hensan Iinkai, *Fujiyoshida shishi shiryōhen*, vol. 4, pp. 731–33.

14. Fujiyoshida Shishi Hensan Iinkai, *Fujiyoshida shishi shiryōhen*, vol. 4, p. 735.

15. Hardacre, *Shinto: A History*, p. 345; Nishida, "Kawaguchi-mura ni okeru Fujisan oshi no seiritsu," p. 96.

16. Fujiyoshida Shishi Hensan Iinkai, *Fujiyoshida shishi shiryōhen*, vol. 4, p. 738.

17. Fujiyoshida-shi Kyōiku Iinkai Rekishi Bunkaka, *Yoshida no himatsuri*, p. 143.

18. Fujiyoshida Shishi Hensan Iinkai, *Fujiyoshida shishi shiryōhen*, vol. 4, p. 738.

19. Fujiyoshida-shi Kyōiku Iinkai Rekishi Bunkaka, *Yoshida no himatsuri*, pp. 102–4.

20. Fujiyoshida-shi Kyōiku Iinkai Rekishi Bunkaka, *Yoshida no himatsuri*, pp. 130–32.

21. Fujiyoshida Shishi Hensan Iinkai, *Fujiyoshida shishi shiryōhen*, vol. 4, pp. 738–42.

22. Nishida, "Kawaguchi-mura ni okeru Fujisan oshi no seiritsu," p. 95.

23. Nishida, "Kawaguchi-mura ni okeru Fujisan oshi no seiritsu," pp. 91–95.

24. Nishida, "Kawaguchi-mura ni okeru Fujisan oshi no seiritsu," pp. 96–97.

25. Nishida, "Kawaguchi-mura ni okeru Fujisan oshi no seiritsu," pp. 97, 101–3.

26. Fujiyoshida Shishi Hensan Iinkai, *Fujiyoshida shishi tsūshihen*, vol. 2, p. 840.

27. Hardacre, *Shinto: A History*, pp. 256, 260.

28. Teeuwen, "Kokugaku vs. Nativism," p. 227.

29. *Nihongi*, vol. 1, pp. 70–72.

30. Takeya, *Fujisan no seishinshi*, pp. 57–58; Uematsu, "Fuji shinkō to kami Kaguya-hime," p. 7.

31. Takeya, *Fujisan no seishinshi*, pp. 125–26.

32. Smith, *Hokusai: One Hundred Views of Mt. Fuji*, p. 188.

33. Uematsu, "Fuji shinkō to kami Kaguya-hime," pp. 7–8.

34. Nishida, "Kawaguchi-mura ni okeru Fujisan oshi no seiritsu," p. 103.

35. Sawada, "Sexual Relations as Religious Practice," p. 362.

36. Hirano, *Fuji shinkō to Fujikō*, p. 139; Fujiyoshida Shishi Hensan Iinkai, *Fujiyoshida shishi tsūshihen*, vol. 2, pp. 844–45.

37. For more on the pressure applied by Westerners on Tokugawa Japan and the shogunate's response, see Totman, *Early Modern Japan*, pp. 482–503, 528–29; and Ravina, *To Stand with the Nations of the World*, pp. 56–69.

38. Guth, "Hokusai's Great Waves," pp. 473, 483.

39. For a book-length treatment of bakufu efforts to bolster coastal defenses, see Wilson, *Defensive Positions*.

40. Lu, *Japan: A Documentary History*, vol. 2, pp. 290–91.

41. Harris, *The Complete Journal of Townsend Harris*, p. 416. It is worth noting that, like many others, Harris thought Fuji lost its luster when viewed up close. At one point in his journal he observes, "Like many other things in this world, the nearer approach does not add to its beauty or grandeur." Harris, *The Complete Journal of Townsend Harris*, pp. 418–19.

42. Heusken, *Japan Journal: 1855–1861*, pp. 124–25.

43. For a detailed examination of Heusken's assassination and what it reveals about the *shishi* who killed him, see Hesselink, "The Assassination of Henry Heusken."

44. Perry and Hawks, *Narrative of the Expedition of an American Squadron*, p. 212.

45. Debarbieux and Rudaz, *The Mountain: A Political History*, p. 69.

46. Alcock, *The Capital of the Tycoon*, vol. 1, p. 396.

47. Alcock, *The Capital of the Tycoon*, vol. 1, pp. 396, 400.

48. Alcock, *The Capital of the Tycoon*, vol. 1, p. 401.

49. Miyazaki, "Gaikokujin no Fuji tozan," pp. 19–20.

50. Tani, *Kurofune Fujisan ni noboru!*, p. 109.

51. Alcock, *The Capital of the Tycoon*, vol. 1, p. 396.

52. Miyazaki, "Gaikokujin no Fuji tozan," p. 20.

53. Alcock, *The Capital of the Tycoon*, vol. 1, pp. 235–39.

54. Ravina, *To Stand with the Nations of the World*, p. 103.

55. Alcock, *The Capital of the Tycoon*, vol. 1, p. 399.

56. Alcock, *The Capital of the Tycoon*, vol. 1, p. 397.

57. Virginia Mills, "From Chelsea to Mount Fuji, the Legacy of Veitch Nurseries," Royal Botanic Gardens, Kew, accessed November 5, 2024, https://www.kew.org/read-and-watch /from-chelsea-to-mount-fuji-legacy-of-veitch-nurseries.

58. Alcock, *The Capital of the Tycoon*, vol. 1, pp. 412–13.

59. Alcock, *The Capital of the Tycoon*, vol. 1, p. 426.

60. Miyazaki, "Gaikokujin no Fuji tozan," p. 17; Tani, *Kurofune Fujisan ni noboru!*, pp. 85–86.

61. Miyazaki, "Gaikokujin no Fuji tozan," p. 21.

62. Toby, *Engaging the Other*, p. 309.

63. Miyazaki, "Gaikokujin no Fuji tozan," pp. 22–23.

64. Miyazaki, "Gaikokujin no Fuji tozan," pp. 23–25.

65. Miyazaki, "Gaikokujin no Fuji tozan," p. 26.

66. Toby, *Engaging the Other*, p. 310.

67. Matsushima, "Man'en gannen no deipuromateikku gifuto."

68. Matsushima, "Man'en gannen no deipuromateikku gifuto," p. 12.

69. Miyazaki, "Gaikokujin no Fuji tozan," pp. 15–16. John Black, a prominent English journalist living in Japan, wrote in his memoir that, after Alcock's climb, "no other attempts had been made to obtain permission" to climb Fuji, explaining, "It certainly would not have been granted; and during the entire period would have been altogether too hazardous." Black, *Young Japan: Yokohama and Yedo*, vol. 2, p. 11.

70. Miyazaki, "Gaikokujin no Fuji tozan," p. 29.

71. Suzuki, *Nyōnin kinsei*, p. 9.

72. Ravina, *To Stand with the Nations of the World*, pp. 85, 97.

73. Ravina, *To Stand with the Nations of the World*, p. 2.

74. For more on the nationwide movements to "separate kami and buddhas" and "destroy the buddhas, eliminate Shākyamuni," see Ketelaar, *Of Heretics and Martyrs in Meiji Japan*.

75. Iwashina, *Fujikō no rekishi*, p. 369; Suzuki, "Meiji ishin to shugendō."

76. Fujiyoshida Shishi Hensan Iinkai, *Fujiyoshida shishi tsūshihen*, vol. 3, pp. 211–14.

77. Iwashina, *Fujikō no rekishi*, pp. 367, 369.

78. Iwashina, *Fujikō no rekishi*, p. 368. For a short biography of Shishino in English, see "Shishino Nakaba," *Encyclopedia of Shinto*.

79. For a more thorough explanation of Meiji policies toward shrine and temple lands as well as their particular effect on the Ōmiya Sengen Shrine (today's Fujisan Hongū Sengen Taisha), see Bernstein, "Whose Fuji?," pp. 64–67.

80. The short- and long-term consequences of the decision to allow monks and nuns to marry is the subject of Jaffe's *Neither Monk nor Layman*.

81. For a summary of early Meiji policies toward Shinto shrines and their priests, see Hardacre, *Shinto: A History*, pp. 373–76, 387–400. In her history of Shikoku's Konpira/Kotohira Shrine, Sarah Thal provides a detailed look at the impact of early Meiji policies on a major pilgrimage site. Thal, *Rearranging the Landscape of the Gods*, pp. 127–76.

82. Iwashina, *Fujikō no rekishi*, p. 370.

83. Iwashina, *Fujikō no rekishi*, pp. 370–72.

84. Sawada, "Mind and Morality in Nineteenth-Century Japanese Religions," p. 117.

85. Sawada, "Mind and Morality in Nineteenth-Century Japanese Religions," pp. 119–20; Iwashina, *Fujikō no rekishi*, pp. 372–73.

86. Sawada, "Mind and Morality in Nineteenth-Century Japanese Religions," p. 120.

87. "Maruyamakyō," *Encyclopedia of Shinto*. Byron Earhart provides an ethnographic account of Maruyamakyō's annual pilgrimage to Fuji in the late 1980s in *Mount Fuji: Icon of Japan*, pp. 145–50.

88. Miyazaki, "The Formation of Emperor Worship," p. 290.

89. Miyazaki, "The Formation of Emperor Worship," p. 295.

90. Miyazaki, "The Formation of Emperor Worship," p. 303.

91. Hardacre, *Shinto: A History*, p. 376.

92. Sawada, "Mind and Morality in Nineteenth-Century Japanese Religions," p. 118; Miyazaki, "Daikyō senpu no jidai no shinshūkyō," p. 86; Hardacre, *Shinto: A History*, p. 376.

93. Hardacre, *Shinto: A History*, pp. 378–79.

94. Bernstein, *Modern Passings*, pp. 67–90.

95. Bernstein, *Modern Passings*, pp. 95–96.

96. Nitta, "Shinto as a 'Non-religion,'" pp. 266–67.

Chapter Seven: A Nation and Mountain in Flux

1. Winichakul, *Siam Mapped*, pp. 16–17.

2. Kim, "In the Shadow of Mount Fuji," pp. 39–65.

3. Much of this section on Nonaka Itaru and his wife Chiyoko draws on Bernstein, "Weathering Fuji," pp. 152–74.

4. Marcon, *The Knowledge of Nature*, p. 7.

5. Despite the early Meiji trend to dismiss the institutions and practices of a supposedly benighted Tokugawa past, the development of natural sciences after the Restoration in fact showed numerous continuities with the work of naturalists in the preceding decades. See Marcon, *The Knowledge of Nature*, pp. 300–305.

6. Clancey, *Earthquake Nation*, pp. 160–61; Kotō, "The Scope of the Vulcanological Survey of Japan," pp. 89–103.

7. Nonaka Itaru, *Fuji annai*, p. 58.

8. Janković, *Reading the Skies*, p. 11.

9. The Tokyo-based system of collecting and interpreting weather data was mostly based on methods worked out in the United States by the Smithsonian Institute, which had, during the mid-nineteenth century, set up a far-reaching network of volunteers and professionals to process numerous observations into coherent sets of data. Arakawa, *Nihon kishōgakushi*, p. 62. See also Fleming, *Meteorology in America*, for an in-depth history of the development of meteorology in the United States in the nineteenth century.

10. Janković, *Reading the Skies*, p. 3.

11. Nonaka Itaru, *Fuji annai*, p. 74.

12. Nonaka Itaru, *Fuji annai*, pp. 65–86.

13. Nonaka Itaru, *Fuji annai*, p. 79.

14. Chiyoko was accompanied by the Satōs' son Gorō, the boss of a station shelter named Jūsaburō, and Itaru's younger brother Kiyoshi, who had decided to lend his support. Nonaka Chiyoko, *Fuyō nikki*, pp. 194–95.

15. Nonaka Chiyoko, *Fuyō nikki,* pp. 180–81.

16. Nonaka Chiyoko, *Fuyō nikki,* p. 204. Author's translation.

17. Walthall, *The Weak Body of a Useless Woman,* pp. 227–31.

18. Nonaka Chiyoko, *Fuyō nikki,* p. 193.

19. Nonaka Chiyoko, *Fuyō nikki,* p. 196.

20. Nonaka Chiyoko, *Fuyō nikki,* pp. 199–200.

21. Nonaka Chiyoko, *Fuyō nikki,* p. 205.

22. Nonaka Chiyoko, *Fuyō nikki,* pp. 198, 201, 205–9; Nonaka Itaru, *Fuji annai,* pp. 99–105, 137–40.

23. "Nonaka Itaru-shi, fusai gesan no moyō," *Yomiuri shinbun,* December 26, 1895.

24. Nitta, *Fuyō no hito,* pp. 143–48.

25. Nonaka Itaru, *Fuji annai,* p. 102.

26. Nonaka Chiyoko, *Fuyō nikki,* pp. 209–19; Nonaka Itaru, *Fuji annai,* pp. 104–5.

27. "Nonaka Itaru-shi, fusai gesan no moyō"; Nonaka Chiyoko, *Fuyō nikki,* pp. 220–23; Nonaka Itaru, *Fuji annai,* pp. 106–8.

28. Nonaka Itaru, *Fuji annai,* pp. 141–43.

29. Nonaka Itaru, *Fuji annai,* p. 140.

30. Ōmori, *Fuji annai, Fuyō nikki,* p. 253.

31. Wada, "Nonaka Itaru-shi no Fujisan kansokujo," p. 200.

32. Nonaka Chiyoko, *Fuyō nikki,* p. 218.

33. In the 1890s, a truly nationalized (and anthologized) canon of Japanese literature took shape; and since much of it consisted of women's writing, especially that of the Heian period (794–1185), there was a receptive audience for a product like Chiyoko's, which told her story through culturally familiar and critically validated norms. For more on the process of literary canonization in Japan, see Shirane and Suzuki, *Inventing the Classics.*

34. Nonaka Chiyoko, *Fuyō nikki,* p. 202. Author's translation.

35. Ochiai, *Takane no yuki.*

36. Starr, *Fujiyama: The Sacred Mountain of Japan,* pp. 18–19.

37. Hashimoto, *Fuji sanchō.*

38. Nitta, *Fuyō no hito,* p. 239.

39. Ōmori, *Fuji annai, Fuyō nikki.*

40. Kamogawa Masashi, "The Mount Fuji Research Station: A Scientific Treasure Trove," *Nippon.com,* March 30, 2020, https://www.nippon.com/en/japan-topics/g00822/the-mount -fuji-research-station-a-scientific-treasure-trove.html.

41. Okada, "'Landscape' and the Nation-State," p. 96.

42. Thomas, *Reconfiguring Modernity,* p. 176.

43. Shiga, *Nihon fūkeiron,* pp. 85–97.

44. Shiga, *Nihon fūkeiron,* p. 96. For an in-depth study of the development of Japanese seismology, including the role played by Milne, see Clancey, *Earthquake Nation.*

45. Shiga, *Nihon fūkeiron,* p. 97.

46. Clancey, *Earthquake Nation,* pp. 146–47.

47. Wigen, "Discovering the Japanese Alps," pp. 24, 22.

48. Weston, *Mountaineering and Exploration in the Japanese Alps.*

49. Karatani, *Origins of Modern Japanese Literature,* p. 29.

50. For an in-depth treatment of Kojima's navigation—and production—of these boundaries, see Jasny, "In Praise of the Peaks."

51. Kojima, *Fujisan*, in Shimada et al., *Kojima Usui zenshū*, vol. 4, pp. 223–365; Fujioka, "Vision or Creation?," p. 289. The original Japanese can be found in Kojima, *Kikōbun ron*, in Shimada et al., *Kojima Usui zenshū*, vol. 5, p. 381.

52. Fujioka, "Vision or Creation?," p. 289.

53. Wigen, "Discovering the Japanese Alps," p. 4.

54. Shiga, *Nihon fūkeiron*, pp. 203–4.

55. "Fujin no Fuji tozan," *Yamanashi nichi nichi shinbun*, August 4, 1905; "Joshi to Fuji tozan," *Yamanashi nichi nichi shinbun*, June 29, 1906; "Onna no Fuji tozan," *Yamanashi nichi nichi shinbun*, July 6, 1906; "Jogakusei Fuji tozan," *Yamanashi nichi nichi shinbun*, July 9, 1907.

56. "Fujin no Fuji tozan," *Yamanashi nichi nichi shinbun*, August 4, 1905.

57. Sugiura Yagaibō, "Tōgaku nikki," *Fūzoku gahō*, August 15, 1905, pp. 9–10, 12.

58. "Gakusei Fuji tozankai," *Yomiuri shinbun*, June 25, 1902.

59. "Dai ni kai, Gakusei Fuji Tozankai," *Yomiuri shinbun*, July 19, 1903.

60. Fujiyoshida Shishi Hensan Iinkai, *Fujiyoshida shishi tsūshihen*, vol. 3, pp. 336–38.

61. "Fujisan yūbinkyoku kaishi," *Yomiuri shinbun*, July 31, 1906.

62. "Fujisan denwa," *Yamanashi nichi nichi shinbun*, July 27, 1907.

63. "Tonari no uwasa," *Yomiuri shinbun*, August 4, 1907; "Fujisan no denwa," *Yomiuri shinbun*, August 11, 1908.

64. Shizaki, *Fujisan sokkōjo monogatari*, p. 7.

65. "Fujisan hogoron," *Yomiuri shinbun*, July 23–26, 1909.

66. "Fujisan hogoron," *Yomiuri shinbun*, July 25 and July 23, 1909.

67. "Fujisan dōro shūzen," *Yamanashi nichi nichi shinbun*, June 18, 1915; "Yoshida guchi dōro kaishū," *Yamanashi nichi nichi shinbun*, June 28, 1915.

68. Weston, *The Playground of the Far East*, p. 64.

69. "Kōen no Fuji," *Yomiuri shinbun*, October 28, 1887; "Asakusa Fuji no dentōryō to hason," *Yomiuri shinbun*, August 1, 1888.

70. "Asakusa no Fuji," *Yomiuri shinbun*, November 1, 1887.

71. "Fujisan mokei no chichinsai," *Yomiuri shinbun*, August 31, 1916.

72. "Sukii Fuji tozan," *Yamanashi nichi nichi shinbun*, December 15, 1912; "Sukii tozantai," *Yamanashi nichi nichi shinbun*, January 3, 1913; "Fuji sanchō de yakyū," *Yamanashi nichi nichi shinbun*, July 24, 1917.

73. Yokoi, "Sekaiteki kōen to shite no Fuji," p. 44; Yokoi, "Fuji tozan annai," pp. 100–111.

74. Maruyama, *Kindai Nihon kōenshi no kenkyū*, p. 365.

75. Figal, *Civilization and Monsters*, p. 5.

76. Iwashina, *Fujikō no rekishi*, p. 349.

77. In 1871 the nascent Ministry of Education described the goal of universal education as follows: "For enlightenment to flourish and civilization to advance, for people to be secure in their livelihood and preserve their families, each person's talents and skills (*sainō gigei*) must be developed. For this purpose schools are to be established in order that the people will learn." Gluck, *Japan's Modern Myths*, p. 104. For school attendance rates, see Abe, "Kindai Nihon no kyōkasho to Fujisan," p. 64.

78. Gluck, *Japan's Modern Myths*, p. 108.

79. Gluck, *Japan's Modern Myths*, pp. 112–13, 128.

80. Abe, "Kindai Nihon no kyōkasho to Fujisan," p. 60.

81. Kaigo, *Nihon kyōkasho taikei kindaihen*, vol. 25, p. 16.

82. Maruyama, *Kindai Nihon kōenshi no kenkyū*, p. 341.

83. Vincent van Gogh was a huge fan of Japanese prints (especially those by Hokusai and Hiroshige), in one letter enthusing to his brother Theo that "the more you see things with a Japanese eye, the more subtly you perceive colour." In another letter he confessed, "Japanese woodcuts, when once you love them, you never regret it." He even saw fit to include a depiction of Fuji in his *Portrait of Père Tanguy*. Berger, *Japonisme in Western Painting from Whistler to Matisse*, pp. 125–26, 132–33.

84. Guth, *Hokusai's Great Wave*.

85. Earhart, *Mount Fuji: Icon of Japan*, p. 114.

86. Kaigo, *Nihon kyōkasho taikei kindaihen*, vol. 6, p. 211.

87. Kaigo, *Nihon kyōkasho taikei kindaihen*, vol. 6, p. 262.

88. Kaigo, *Nihon kyōkasho taikei kindaihen*, vol. 6, pp. 26, 434.

89. Kaigo, *Nihon kyōkasho taikei kindaihen*, vol. 6, p. 434.

90. For examples of Fuji templates, see Kaigo, *Nihon kyōkasho taikei kindaihen*, vol. 26, pp. 89, 126, 179, 190, 192, 203, 209, 227, 278.

91. For a detailed look at the development and popularization of the three-peaked Fuji, see Takeya, *Fujisan no seishinshi*, pp. 1–42, 157–78.

92. Kaigo, *Nihon kyōkasho taikei kindaihen*, vol. 5, pp. 515–16.

93. Kaigo, *Nihon kyōkasho taikei kindaihen*, vol. 6, pp. 356–57.

94. Kaigo, *Nihon kyōkasho taikei kindaihen*, vol. 6, pp. 542–43.

95. Kaigo, *Nihon kyōkasho taikei kindaihen*, vol. 6, p. 543. Author's translation.

96. For a brief summary of Iwaya Sazanami's work, see Miyake, "Iwaya Sazanami."

97. Kaigo, *Nihon kyōkasho taikei kindaihen*, vol. 7, pp. 58–59. Author's translation.

98. Abe, "Kindai Nihon no kyōkasho to Fujisan," p. 73.

99. The original quote is: "All animals are equal, but some animals are more equal than others." Orwell, *Animal Farm and 1984*, p. 80.

100. Kaigo, *Nihon kyōkasho taikei kindaihen*, vol. 7, p. 170.

101. Kaigo, *Nihon kyōkasho taikei kindaihen*, vol. 7, pp. 354–55.

102. Kaigo, *Nihon kyōkasho taikei kindaihen*, vol. 7, p. 563.

103. Kaigo, *Nihon kyōkasho taikei kindaihen*, vol. 8, p. 405; Kaigo, *Nihon kyōkasho taikei kindaihen*, vol. 25, p. 430.

104. Kaigo, *Nihon kyōkasho taikei kindaihen*, vol. 8, p. 405.

105. Abe, "Kindai Nihon no kyōkasho to Fujisan," p. 85.

106. Gluck, *Japan's Modern Myths*, p. 14.

107. Rosenfield, "Nihonga and Its Resistance," p. 176.

108. Rosenfield, "Nihonga and Its Resistance," p. 176.

109. "Reihō no sanchō takaku dainisshōki hirugaeru," *Yomiuri shinbun*, evening edition, August 13, 1934.

110. "Sanchō de seiju banzai: Fuji mezashi sanzen mei," *Yamanashi nichi nichi shinbun*, July 8, 1940. For a book-length study of the commemoration of the empire's two thousand six-hundredth anniversary, see Ruoff, *Imperial Japan at its Zenith*.

111. Tange, "Dai Tōa kensetsu kinen eizō keikaku," pp. 963–74.

112. Yoneyama, *Hiroshima Traces*, pp. 1–3.

113. "Sanchō de seiju banzai: Fuji mezashi sanzen mei," *Yamanashi nichi nichi shinbun*, July 8, 1940; "Nengan jūman nin naru: kotoshi no Fuji tozan subete kessan," *Yamanashi nichi nichi shinbun*, August 28, 1940.

114. "Basu wa unkyū, yūbinkyoku mo haishi," *Yamanashi nichi nichi shinbun*, July 5, 1943.

115. Earhart, *Mount Fuji: Icon of Japan*, p. 175.

116. Earhart, *Mount Fuji: Icon of Japan*, pp. 178–79.

117. Hirano, *Mr. Smith Goes to Tokyo*, pp. 52–53.

118. The new constitution written by the Americans in 1946 demoted the emperor from absolute sovereign to "the symbol of the State and of the unity of the people, deriving his position from the will of the people with whom resides sovereign power." Prime Minister of Japan and His Cabinet, *The Constitution of Japan*, https://japan.kantei.go.jp/constitution_and _government_of_japan/constitution_e.html.

119. "Fujisan" (part 4 in a 5-part series), *Asahi shinbun*, January 7, 1953.

120. Kashiwazaki, "The Politics of Legal Status," pp. 18–23.

121. Kim, "In the Shadow of Mount Fuji," p. 54.

122. Kim, "In the Shadow of Mount Fuji," p. 65.

123. Yi, "Fujisan," pp. 624–25.

124. The following examination of this conflict draws on Bernstein, "Whose Fuji?" For a critical assessment of Occupation policies concerning religion and "religious freedom" in particular, see Thomas, *Faking Liberties*.

125. The origins and implementation of the directive are covered in Woodard, *The Allied Occupation of Japan*, pp. 54–74. The full text of the directive can be found on pp. 295–98.

126. Woodard, *The Allied Occupation of Japan*, p. 297.

127. Prime Minister of Japan and His Cabinet, *The Constitution of Japan*.

128. Woodard, *The Allied Occupation of Japan*, p. 120.

129. Ōkurashō Kanzaikyoku, *Shaji keidaichi shobunshi*, pp. 228–31, 233, 266.

130. "Fuji zenbu koya Sengen keidai to no setsu," *Yamanashi nichi nichi shinbun*, July 1, 1917.

131. Aoyagi, *Fugaku tabi hyakkei*, pp. 176–77; Oyama Chōshi Hensan Senmon Iinkai, *Oyama chōshi*, vol. 7, pp. 510–11.

132. Fujisan Hongū Sengen Taisha, *Kokuyū keidaichi jōyo shinsei*, pp. 275–76.

133. Bernstein, "Sacred Space on Postwar Fuji," pp. 199–202.

134. Ōkurashō Kanzaikyoku, *Shaji keidaichi shobunshi*, pp. 256–57.

135. "Chotto Fujisan made kēburu de hitonobori," *Yamanashi nichi nichi shinbun*, November 21, 1946.

136. Ōkurashō Kanzaikyoku, *Shaji keidaichi shobunshi*, pp. 260–62.

137. "Fuji sanchō shiyūka hantai kokumin taikai," *Yamanashi nichi nichi shinbun*, February 6, 1953.

138. Fuji Sanchō Shiyūka Hantai Dōmei, *Fuji sanchō haraisage mondai no shinsō*.

139. Kokkai (National Diet of Japan), *Dai jūkyū kai Kokkai Shūgiin Gyōsei Kansatsu*, December 17, 1953, p. 10.

140. Kokkai, *Dai jūkyū kai Kokkai Shūgiin Gyōsei Kansatsu*, December 17, 1953, p. 29.

141. Kokkai, *Dai jūkyū kai Kokkai Shūgiin Gyōsei Kansatsu*, December 18, 1953, pp. 13–14.

142. Kokkai, *Dai jūkyū kai Kokkai Shūgiin Gyōsei Kansatsu*, December 17, 1953, pp. 18, 33; December 18, 1953, p. 14.

143. Kokkai, *Dai jūkyū kai Kokkai Shūgiin Gyōsei Kansatsu*, December 17, 1953, pp. 16, 18–19, 27, 35; December 18, 1953, p. 12.

144. Kokkai, *Dai jūkyū kai Kokkai Shūgiin Gyōsei Kansatsu*, December 18, 1953, p. 13.

145. Kokkai, *Dai jūkyū kai Kokkai Shūgiin Gyōsei Kansatsu*, December 18, 1953, p. 4.

146. Kokkai, *Dai jūkyū kai Kokkai Shūgiin Gyōsei Kansatsu*, December 18, 1953, p. 8.

147. Kokkai, *Dai jūkyū kai Kokkai Shūgiin Gyōsei Kansatsu*, December 17, 1953, p. 12.

148. "Fujisan, kokuyū to ketsuron," *Jinja shinpō*, February 15, 1954.

149. "Fuji sanchō mondai, shin dankai e," *Jinja shinpō*, February 16, 1957.

150. For an in-depth examination of the shrine's lawsuit and its progress through the court system, see Bernstein, "Whose Fuji?," pp. 85–88.

151. "Fuji sanchō Shizuoka no jinja ni," *Yamanashi nichi nichi shinbun*, December 18, 2004.

152. Natsume Sōseki, *Sanshirō*, p. 15.

Chapter Eight: A Global Mountain on a Human Planet

1. Paul Crutzen and Eugene Stoermer independently devised and began publicizing the term several decades ago. Thomas, Williams, and Zalasiewicz, *The Anthropocene*, p. 46.

2. In 2023 the International Commission on Stratigraphy's Anthropocene Working Group (AWG), established in 2009, officially proposed that the Anthropocene be added to the Geological Time Scale. The group chose a start date in the mid-twentieth century, when nuclear testing, the production of plastics, the rapid increase in the burning of fossil fuels, and other human activities left their mark in the geological record across the planet. In 2024 a majority of the Subcommission of Quaternary Stratigraphy (SQS) voted against the proposal, however. A major concern was the brevity of the proposed epoch and the shallowness of the geological record to determine it. It was also argued that major anthropogenic changes reached back centuries—even millennia—making it difficult to date the start of the epoch to a specific moment in time. The International Union of Geological Sciences, the parent body of the International Commission on Stratigraphy, backed the SQS vote while countering criticisms about "procedural irregularities" in the voting process. Its formal statement defending the decision nevertheless states that "the Anthropocene as a concept will continue to be widely used not only by Earth and environmental scientists, but also by social scientists, politicians, and economists, as well as by the public at large. As such, it will remain an invaluable descriptor in human-environment interactions." Thomas, Williams, and Zalasiewicz, *The Anthropocene*, pp. 54–67; Raymond Zhong, "Geologists Make it Official: We're Not in an 'Anthropocene' Epoch," *New York Times*, March 20, 2024; Witze, "Geologists Reject the Anthropocene as Earth's New Epoch"; International Union of Geological Sciences Executive Committee, "The Anthropocene: IUGS-ICS Statement," March 20, 2024, https://www.iugs.org/_files/ugd /f1fc07_40d1a7ed58de458c9f8f24de5e739663.pdf?index=true.

3. Fujiyoshida Shishi Hensan Iinkai, *Fujiyoshida shishi tsūshihen*, vol. 2, pp. 261, 316–17.

4. When it was first constructed in the seventeenth century, it soon collapsed but was successfully rebuilt in the 1860s. Fujiyoshida Shishi Hensan Iinkai, *Fujiyoshida shishi tsūshihen*, vol. 2, pp. 258–90.

5. Susono Shishi Hensan Senmon Iinkai, *Susono shishi*, vol. 8, pp. 342–71; Fuji Shishi Hensan Iinkai, *Yoshiwara shishi*, vol. 2, pp. 273–76.

6. Fujiyoshida Shishi Hensan Iinkai, *Fujiyoshida shishi tsūshihen*, vol. 2, p. 290.

7. Fujiyoshida Shishi Hensan Iinkai, *Fujiyoshida shishi tsūshihen*, vol. 2, pp. 366, 370.

8. Fujiyoshida Shishi Hensan Iinkai, *Fujiyoshida shishi tsūshihen*, vol. 2, p. 382.

9. Fujiyoshida Shishi Hensan Iinkai, *Fujiyoshida shishi tsūshihen*, vol. 2, pp. 379–81, 423.

10. Fujiyoshida Shishi Hensan Iinkai, *Fujiyoshida shishi tsūshihen*, vol. 2, pp. 371–74, 423.

11. Fujiyoshida Shishi Hensan Iinkai, *Fujiyoshida shishi tsūshihen*, vol. 2, p. 385.

12. Makimura, *Yokohama and the Silk Trade*, pp. xiii, 102–3.

13. Fujiyoshida Shishi Hensan Iinkai, *Fujiyoshida shishi tsūshihen*, vol. 3, p. 120.

14. Yamanashi Nichinichi Shinbunsha, *Narusawa sonshi*, vol. 1, pp. 626–29.

15. Fujiyoshida Shishi Hensan Iinkai, *Fujiyoshida shishi tsūshihen*, vol. 3, pp. 149.

16. Fujiyoshida Shishi Hensan Iinkai, *Fujiyoshida shishi tsūshihen*, vol. 3, pp. 455, 466–67, 469, 475.

17. Fujiyoshida Shishi Hensan Iinkai, *Fujiyoshida shishi tsūshihen*, vol. 3, p. 467.

18. Fujiyoshida Shishi Hensan Iinkai, *Fujiyoshida shishi tsūshihen*, vol. 3, pp. 472–75.

19. Fujiyoshida Shishi Hensan Iinkai, *Fujiyoshida shishi tsūshihen*, vol. 3, pp. 682–83.

20. Fujiyoshida Shishi Hensan Iinkai, *Fujiyoshida shishi shiryōhen*, vol. 7, pp. 265–66, 277, 299.

21. Fujiyoshida Shishi Hensan Iinkai, *Fujiyoshida shishi shiryōhen*, vol. 7, p. 336.

22. Fujiyoshida Shishi Hensan Iinkai, *Fujiyoshida shishi tsūshihen*, vol. 3, pp. 467, 475, 681.

23. Fujiyoshida Shishi Hensan Iinkai, *Fujiyoshida shishi tsūshihen*, vol. 3, p. 685.

24. Fujiyoshida Shishi Hensan Iinkai, *Fujiyoshida shishi tsūshihen*, vol. 3, pp. 690–92.

25. Fujiyoshida Shishi Hensan Iinkai, *Fujiyoshida shishi tsūshihen*, vol. 3, pp. 695, 711.

26. Watanabe Michihito (Director of Mount Fuji Biodiversity Lab), interview by author, Kawaguchiko, July 28, 2014.

27. Farris, *A Bowl for a Coin*, p. 104; Fuji Shiritsu Hakubutsukan, *Fuji no cha*, p. 4.

28. Farris, *A Bowl for a Coin*, pp. 127–29.

29. Fuji Shiritsu Hakubutsukan, *Fuji no cha*, p. 10.

30. For a book-length study of America's thirst for Japanese tea, see Hellyer, *Green with Milk and Sugar*.

31. Fuji Shiritsu Hakubutsukan, *Fuji no cha*, p. 18.

32. Fuji Shiritsu Hakubutsukan, *Fuji no cha*, pp. 22, 32–33; Farris, *A Bowl for a Coin*, pp. 152–55.

33. "Jugetsudo's Exquisite Teas from Mount Fuji Now Pouring at the Portland Japanese Garden" (Sponsored Content), *Portland Monthly*, October 27, 2017, https://www.pdxmonthly .com/sponsored/2017/10/jugetsudo-s-exquisite-teas-from-mount-fuji-now-pouring-at-the -japanese-garden.

34. Shikazono, Arakawa, and Nakano, "Fujisan nanroku no chikasui suiritsu."

35. Fuji Shiritsu Hakubutsukan, *Fuji-shi no seishigyō*, p. 1.

36. Fuji Shiritsu Hakubutsukan, *Fuji-shi no seishigyō*, p. 1.

37. Fuji Shiritsu Hakubutsukan, *Fuji-shi no seishigyō*, pp. 1–3.

38. Fuji Shiritsu Hakubutsukan, *Fuji-shi no seishigyō*, p. 37.

39. Fuji Shiritsu Hakubutsukan, *Fuji-shi no seishigyō*, p. 40.

40. Fuji Shiritsu Hakubutsukan, *Fuji-shi no seishigyō*, p. 67; Hayase, "Manila Hemp," pp. 176–78, 186.

41. Fuji Shiritsu Hakubutsukan, *Fuji-shi no seishigyō*, pp. 4–7.

42. Fuji Shiritsu Hakubutsukan, *Fuji-shi no seishigyō*, pp. 10–19.

43. Fuji Shiritsu Hakubutsukan, *Fuji-shi no seishigyō*, p. 23.

44. Nonaka Itaru, *Fuji annai*, p. 24.

45. Fuji Shiritsu Hakubutsukan, *Fuji-shi no seishigyō*, p. 25.

46. Fuji Shiritsu Hakubutsukan, *Fuji-shi no seishigyō*, p. 28.

47. Kurosawa and Hashino, "From the Non-European Tradition," p. 147.

48. Fuji Shiritsu Hakubutsukan, *Fuji-shi no seishigyō*, p. 34.

49. Fuji Shiritsu Hakubutsukan, *Fuji-shi no seishigyō*, pp. 51–55; Fedman, *Seeds of Control*, pp. 43–44.

50. Fedman, *Seeds of Control*, pp. 130–32, 145.

51. Kurosawa and Hashino, "From the Non-European Tradition," p. 147; Fuji Shiritsu Hakubutsukan, *Fuji-shi no seishigyō*, pp. 69–71.

52. Kurosawa and Hashino, "From the Non-European Tradition," p. 147.

53. Kurosawa and Hashino, "From the Non-European Tradition," p. 150; Fuji Shiritsu Hakubutsukan, *Fuji-shi no seishigyō*, p. 79.

54. Fujiyoshida Shishi Hensan Iinkai, *Fujiyoshida shishi tsūshihen*, vol. 3, p. 709; Tsutsui, "Landscapes in the Dark Valley," p. 203.

55. Kurosawa and Hashino, "From the Non-European Tradition," p. 151.

56. Fuji Shishi Hensan Iinkai, *Yoshiwara shishi*, vol. 2, pp. 273–310; Fuji Shiritsu Hakubutsukan, *Ukishimanuma to kome zukuri*, pp. 2–3, 8–17.

57. Fuji Shishi Hensan Iinkai, *Yoshiwara shishi*, vol. 2, pp. 625–32; Fuji Shiritsu Hakubutsukan, *Ukishimanuma to kome zukuri*, p. 18. For a handy timeline of development in the Ukishimagahara from the 1840s to the 1960s, see Fuji Shiritsu Hakubutsukan, *Ukishimanuma to kome zukuri*, p. 20.

58. Davis, "Paperpower," pp. 12–15; Kurosawa and Hashino, "From the Non-European Tradition," pp. 153–55.

59. Fuji Shishi Hensan Iinkai, *Yoshiwara shishi*, vol. 3, pp. 678–81; *Fuji shishi*, vol. 2, pp. 1001–11.

60. Fuji Shiritsu Hakubutsukan, *Ukishimanuma to kome zukuri*, p. 19.

61. Kumar and Dutt, "A Comparative Study of Conventional Chemical Deinking."

62. Fuji Shishi Hensan Iinkai, *Yoshiwara shishi*, vol. 3, pp. 866–72.

63. Walker, *Toxic Archipelago*. Walker's book presents a vivid study of the "hybrid causations" that generated severe cases of environmental pollution—and bodily pain—as Japan rapidly industrialized in the nineteenth and twentieth centuries. A close look at the litigation that resulted in the so-called "Big Four" pollution verdicts of the early 1970s (the ones that awarded damages to those who suffered from "it hurts, it hurts disease" in Toyama, mercury poisoning in both Minamata and Niigata, and fatal levels of asthma from air pollution in Yokkaichi) is provided in McKean, *Environmental Protest and Citizen Politics in Japan*, pp. 33–79. See George, *Minamata*, for a book-length treatment of the politics surrounding Minamata disease.

64. Donald Kirk, "Students in the Elementary School Grow Up Suffering from Asthma. Plants Wither and Die. The Birds Around Mount Fuji Are Decreasing in Number. They No Longer Visit the Town," *New York Times*, March 26, 1972.

65. *Gojira tai hedora*, directed by Banno Yoshimitsu (Tōhō Kabushiki Kaisha, 1971).

66. For a close examination of the dynamics of anti-pollution citizen's movements in Japan in the 1960s and '70s, see McKean, *Environmental Protest and Citizen Politics in Japan*; and Krauss and Simcock, "Citizens' Movements," pp. 187–227.

67. Fuji Shishi Hensan Iinkai, *Yoshiwara shishi*, vol. 3, p. 866. A clip of the flotilla is preserved in the archives of NHK (Japan Broadcasting Corporation). "Tagonoura nado, osen shinkokuka," NHK ākaibusu, https://www2.nhk.or.jp/archives/tv60bin/detail/index.cgi?das_id =D0009030100_00000.

68. The comprehensive nature of its activities and goals made environmental activism in Fuji City the "complete prototype movement" among environmental movements, as Margaret McKean puts it in *Environmental Protest and Citizen Politics in Japan*, pp. 30–31.

69. Fuji Shishi Hensan Iinkai, *Yoshiwara shishi*, vol. 3, pp. 879–85; McKean, *Environmental Protest and Citizen Politics in Japan*, p. 76.

70. Walker, *Toxic Archipelago*, pp. 219–21.

71. Fuji Shishi Hensan Iinkai, *Yoshiwara shishi*, vol. 3, pp. 681–96.

72. Lewis, "Civic Protest in Mishima," pp. 274–313; McKean, *Environmental Protest and Citizen Politics in Japan*, pp. 29–31.

73. Groundwork Mishima's activities include cleaning trash from rivers and ponds, monitoring the quality of spring water, and reintroducing *baikamo* (*Ranunculus nipponicus var. submersus*), a rare flowering plant that thrives only in cold, uncontaminated water. The organization provides up-to-date information about its work on its website: Gurandowāku Mishima, http://www.gwmishima.jp/.

74. Shizuoka Chiri Kyōiku Kenkyūkai, *Fujisan sekai isan e no michi*, pp. 15–23.

75. Takahashi et al., "Contamination and Specific Accumulation of Organochlorine and Butyltin Compounds."

76. Jamieson et al., "Bioaccumulation of Persistent Organic Pollutants," p. 2.

77. Nakajima et al., "Occurrence and Levels of Polybrominated Diphenyl Ethers (PBDEs)."

78. Davis, "Paperpower," pp. 7–9; Kijima, Sakurai, and Otsuka, "Iriaichi," pp. 869–71.

79. Davis, "Paperpower," pp. 9–13; Kurosawa and Hashino, "From the Non-European Tradition," pp. 153–54.

80. Forestry Agency, *Annual Report on Forest and Forestry in Japan*, p. 41.

81. Dauvergne, *Shadows in the Forest*, pp. 3–5.

82. Seo and Taylor, "Forest Resource Trade," pp. 93–97; Forestry Agency, *Annual Report on Forest and Forestry in Japan*, p. 23; Masuyama Toshimasa (Deputy Director of the International Forestry Cooperation Office, Forest Planning Division, Forestry Agency, Ministry of Agriculture, Forestry, and Fisheries), "Forests and Forestry in Japan" (lecture given on the Lewis and Clark Mount Fuji Study Abroad Program, July 6, 2017).

83. Forestry Agency, *Annual Report on Forest and Forestry in Japan*, p. 41.

84. Peter Dauvergne uses the terms "ecological shadow" and "shadow ecology" to describe Japan's destructive impact on the forests of Southeast Asia. Dauvergne, *Shadows in the Forest*, pp. 1–19.

85. Dauvergne, *Shadows in the Forest*; Davis, "Paperpower," pp. 19–23; Seo and Taylor, "Forest Resource Trade," pp. 92, 95–97.

86. Seo and Taylor, "Forest Resource Trade," pp. 94–95; Shizuoka Chiri Kyōiku Kenkyūkai, *Fujisan sekai isan e no michi*, pp. 3–5; Watanabe Sadamoto (Director, Forest Environmental

Institute), "Degradation of Forests and Their Restoration Through Sustainable Forest Management" (lecture given for Lewis and Clark Mount Fuji Study Abroad Program, August 12, 2017).

87. Forestry Agency, *Annual Report on Forest and Forestry in Japan*, pp. 15–16, 31.

88. Forestry Agency, *Annual Report on Forest and Forestry in Japan*, p. 14.

89. Pan-Pacific Science Congress, *Guide-book Excursion C-4*, p. 5; Batdorff, "Mapping a Mountain," pp. 16, 19; Saitō, "Seigyō no ba to shite no Fujisan," pp. 174–78.

90. Butchōji, *Fuji ni tatsu hito*, pp. 8–9; Fujiyoshida Shishi Hensan Iinkai, *Fujiyoshida shishi tsūshihen*, vol. 3, pp. 332–41. For more on traditional common land practices on the north side of Fuji, see McKean, "Management of Traditional Common Lands (*Iriaichi*) in Japan," pp. 63–98.

91. Hōjō, *Yamanashi-ken iriai tōsōshi*, pp. i–ii, 42–59; McKean, "Defining and Dividing Property Rights in the Japanese Commons," pp. 14–17.

92. Totman, *Japan's Imperial Forest*, p. 45.

93. Hōjō, *Yamanashi-ken iriai tōsōshi*, pp. i, 60–61; Fujiyoshida-shi Hoka Nikason Onshi Ken'yū Zaisan Hogo Kumiai, *Onshirin kumiaishi*, vol. 2, pp. 699–709.

94. Hōjō, *Yamanashi-ken iriai tōsōshi*, pp. 144–49, 257–60.

95. Gotenba Shishi Hensan Iinkai, *Gotenba shishi*, vol. 9, pp. 143–47.

96. Gotenba Shishi Hensan Iinkai, *Gotenba shishi*, vol. 9, pp. 181–82.

97. For a book-length study of imperial household forests, see Totman, *Japan's Imperial Forest*.

98. Fujiyoshida-shi Hoka Nikason Onshi Ken'yū Zaisan Hogo Kumiai, *Onshirin kumiaishi*, vol. 2, pp. 748–49, 789–91; vol. 3, pp. 7–20, 97–168. The association maintains a regularly updated website that provides information on its initiatives and events: https://www.onshirin.jp/.

99. Hōjō, *Yamanashi-ken iriai tōsōshi*, pp. 166–69.

100. Fujiyoshida Shishi Hensan Iinkai, *Fujiyoshida shishi shiryōhen*, vol. 7, p. 405; Gotenba Shishi Hensan Iinkai, *Gotenba shishi*, vol. 9, pp. 604–10, 637–38; Hōjō, *Yamanashi-ken iriai tōsōshi*, pp. 170–72.

101. Susono Shishi Hensan Senmon Iinkai, *Susono shishi*, vol. 9, p. 181.

102. Shizuoka Chiri Kyōiku Kenkyūkai, *Fujisan sekai isan e no michi*, pp. 98–101, 115–16; Hōjō, *Yamanashi-ken iriai tōsōshi*, pp. 170–72.

103. Gotenba Shishi Hensan Iinkai, *Gotenba shishi*, vol. 9, p. 612.

104. Butchōji, *Fuji ni tatsu hito*, p. 146; Gotenba Shishi Hensan Iinkai, *Gotenba shishi*, vol. 9, pp. 642–43, 657.

105. Shizuoka Chiri Kyōiku Kenkyūkai, *Fujisan sekai isan e no michi*, p. 102.

106. US Marines, "Combined Arms Training Center Camp Fuji," accessed September 20, 2024, https://www.fuji.marines.mil/About/History/.

107. Prime Minister of Japan and His Cabinet, *The Constitution of Japan*, https://japan.kantei.go.jp/constitution_and_government_of_japan/constitution_e.html.

108. For an in-depth study of Japan's postwar military, see Skabelund, *Inglorious, Illegal Bastards*.

109. Gotenba Shishi Hensan Iinkai, *Gotenba shishi*, vol. 9, pp. 661–66; Shizuoka Chiri Kyōiku Kenkyūkai, *Fujisan sekai isan e no michi*, pp. 102–4, 111–13.

110. Butchōji, *Fuji ni tatsu hito*, p. 160; Shizuoka Chiri Kyōiku Kenkyūkai, *Fujisan sekai isan e no michi*, pp. 106–7.

111. Shizuoka Chiri Kyōiku Kenkyūkai, *Fujisan sekai isan e no michi*, pp. 107–8.

112. Caldecott, "At the Foot of the Mountain," p. 105.

113. Butchōji, *Fuji ni tatsu hito*, p. 163; Robert Trumbull, "U.S. Warned That Japanese Left Wing Seeks to Exploit Missile Tests on Mt. Fuji," *New York Times*, October 6, 1965.

114. Butchōji, *Fuji ni tatsu hito*, pp. 312–21.

115. Onshirin Kumiai information session, Lewis and Clark Mount Fuji Study Abroad Program, July 29, 2014.

116. Some communities in Japan have recently made efforts to save grassland environments with the help of local governments, nonprofits, and outside volunteers. See Shimada, "Multi-level Natural Resources Governance"; and Miyanaga and Shimada, "'The Tragedy of the Commons' by Underuse," pp. 340–41.

117. Ministry of Agriculture, Forestry and Fisheries (Japan), "Census of Agriculture and Forestry 2020, Forest and Grass Area," e-Stat: Statistics of Japan, https://www.e-stat.go.jp/en /stat-search/files?stat_infid=000032172328.

118. Fieldwork with ecologist Watanabe Michihito (Director of Mount Fuji Biodiversity Lab), Lewis and Clark Mount Fuji Study Abroad Program, July 2014 and July 2017. For descriptions and photographs of threatened (and nonthreatened) species in the grasslands around Fuji, see Watanabe, *The Natural Features of Mount Fuji*, vol. 2, pp. 8–35.

119. For a book-length study of militarized landscapes and wildlife conservation, see Havlick, *Bombs Away*.

120. Brown, "Constructing Nature," p. 90.

121. Shizaki, *Fujisan sokkōjo monogatari*, pp. 124–25.

122. Yoshida, "Fujisan no sabō jigyō," pp. 17–20; Mount Fuji Sabō Office information session, Lewis and Clark Mount Fuji Study Abroad Program, August 13, 2014.

123. Ogawa et al., "Fuji santai o riyō shita"; Yamawaki et al., "Fuji santai o mochiita."

124. Sakio and Masuzawa, "Advancing Timberline on Mt. Fuji between 1978 and 2018," p. 1537.

125. See, for example, Julia Jacobo, "Japan's Mount Fuji Breaks Record for No Snow in October," ABC News, October 30, 2024, https://abcnews.go.com/International/japans-mount -fuji-breaks-record-snow-october/story?id=115301410.

Chapter Nine: World Heritage Fuji

1. Agency for Cultural Affairs, Government of Japan, List of World Heritage Sites in Japan, https://www.bunka.go.jp/english/policy/cultural_properties/introduction/world_heritage /list/.

2. United Nations Educational, Scientific, and Cultural Organization (UNESCO), Japan, Conventions Ratified, https://www.unesco.org/en/countries/jp/conventions; Shizuoka Chiri Kyōiku Kenkyūkai, *Fujisan sekai isan e no michi*, pp. 215–16.

3. Shizuoka Chiri Kyōiku Kenkyūkai, *Fujisan sekai isan e no michi*, p. 216.

4. Suzuki Kōshirō, "Yunesuko no tsuika kankoku ni miru Fujisan," p. 1001.

5. UNESCO, World Heritage Convention, Policy Compendium, https://whc.unesco.org /en/compendium/action=list&id_faq_themes=962.

6. Noguchi, *Sekai isan ni sarete Fujisan wa naite iru*, pp. 74–78.

7. Kondō, *Fujisan: Sekai isan e no michi*, p. 51; Shizuoka Chiri Kyōiku Kenkyūkai, *Fujisan sekai isan e no michi*, pp. 217–19; Suzuki, "Yunesuko no tsuika kankoku ni miru Fujisan," p. 1001.

8. Shizuoka Chiri Kyōiku Kenkyūkai, *Fujisan sekai isan e no michi*, p. 219.

9. For a list of the organization's officers and board of directors soon after it was formed, see Oda, *Fujisan ga sekai isan ni naru hi*, p. 288.

10. Lindström, "Universal Heritage Value," p. 282.

11. International Council of Monuments and Sites, *Fujisan (Japan), no. 1417*, p. 131. For a full list of criteria for world heritage sites, see UNESCO, World Heritage Convention, "The Criteria for Selection," https://whc.unesco.org/en/criteria/.

12. World Heritage Committee, Decision 37 COM 8B.29, "Fujisan, Sacred Place and Source of Artistic Inspiration (Japan)," UNESCO World Heritage Convention, https://whc.unesco.org/en/decisions/5157.

13. World Heritage Committee, Decision 37 COM 8B.29.

14. Government of Japan, *Nomination of Fujisan*, Executive Summary.

15. International Council of Monuments and Sites, *Fujisan (Japan), no. 1417*, pp. 131, 136.

16. International Council of Monuments and Sites, *Fujisan (Japan), no. 1417*, pp. 127–28.

17. Kondō, *Fujisan: Sekai isan e no michi*, pp. 24, 26; UNESCO, *UNESCO 2013*, p. 127.

18. Lindström, "Universal Heritage Value," p. 286.

19. Government of Japan, *Nomination of Fujisan*; Kondō, *Fujisan: Sekai isan e no michi*, p. 185.

20. Noguchi, *Sekai isan ni sarete Fujisan wa naite iru*.

21. Noguchi, *Sekai isan ni sarete Fujisan wa naite iru*, pp. 30–34.

22. Noguchi, *Sekai isan ni sarete Fujisan wa naite iru*, pp. 67–69.

23. Noguchi, *Sekai isan ni sarete Fujisan wa naite iru*, pp. 5, 20–21.

24. Noguchi, *Sekai isan ni sarete Fujisan wa naite iru*, pp. 5, 111–12.

25. In the summers of 2000, 2001, and 2002, the Fujisan Club set up biotoilets on the mountain to demonstrate their effectiveness. Fujisan Club, "Nenpyō, dai ikki," accessed September 20, 2024, https://www.fujisan.or.jp/About/history/chronology_01.html.

26. Noguchi, *Sekai isan ni sarete Fujisan wa naite iru*, pp. 6, 20, 24.

27. Noguchi, *Sekai isan ni sarete Fujisan wa naite iru*, pp. 22, 30–31.

28. Fujisan Club, "Gomi mondai wa ima," accessed August 4, 2024, https://www.fujisan.or.jp/Action/think/index.html.

29. Noguchi, *Sekai isan ni sarete Fujisan wa naite iru*, pp. 112–14.

30. Noguchi, *Sekai isan ni sarete Fujisan wa naite iru*, pp. 4, 7, 107.

31. Jones, Yang, and Yamamoto, "Assessing the Recreational Value of World Heritage Site Inscription," pp. 68–69; Ginanne Brownell, "A World Heritage Designation Can Be a Blessing, or a Curse," *New York Times*, April 25, 2023, https://www.nytimes.com/2023/04/25/arts/unesco-world-heritage-site.html.

32. McGuire, "What's at Stake in Designating Japan's Sacred Mountains," pp. 339–40.

33. McGuire, "What's at Stake in Designating Japan's Sacred Mountains," p. 345.

34. World Heritage Committee, Decision 37 COM 8B.29.

35. Noguchi, *Sekai isan ni sarete Fujisan wa naite iru*, pp. 7, 43–44, 63–64, 93, 100–102, 109, 229.

36. Jones, Beeton, and Cooper, "World Heritage Listing as a Catalyst for Collaboration," pp. 224–27.

37. Kankyōshō (Ministry of the Environment), "Fuji Hakone Izu Kokuritsu Kōen, gaiyō, keikakusho," accessed September 20, 2024, https://www.env.go.jp/park/fujihakone/intro/index.html.

38. World Heritage Committee, Decision 40 COM 7B.39, "Fujisan, sacred place and source of artistic inspiration (Japan) (C 1418)," UNESCO World Heritage Convention, https://whc.unesco.org/en/decisions/6704.

39. For details on the composition and work of the council, see Fujisan World Cultural Heritage Council web site, https://www.fujisan-3776.jp/en/.

40. Government of Japan, *State of Conservation Report*, "Vision and Strategies: Fujisan, Sacred Place and Source of Artistic Inspiration," October 2015, pp. 47–48, 65, 85–86, 100, 104, 114–15; "Fujisan: Comprehensive Preservation and Management Plan," January 2016, pp. 203–4, 238, 244–49.

41. Government of Japan, *State of Conservation Report*, "Vision and Strategies," p. 52; "Fujisan: Comprehensive Preservation and Management Plan," p. 234.

42. Government of Japan, *State of Conservation Report*, "Vision and Strategies," pp. 40, 45, 50; "Fujisan: Comprehensive Preservation and Management Plan," pp. 221–24, 228.

43. Council for the Promotion of the Proper Use of Mt. Fuji, "Official Web Site for Mt. Fuji Climbing," accessed September 20, 2024, https://www.fujisan-climb.jp/en/plan-your-trip-index.html.

44. "Arupinisuto Noguchi Ken, Sekai isan ni natte Fujisan wa 'yoi hōkō e'," *Sankei shinbun*, September 20, 2022, https://www.sankei.com/article/20220920-6QFLRWO3TVM7BD3XZLHJPFD6VQ/.

45. Kankyōshō (Ministry of the Environment), *2021 nen kaki no Fujisan tozanshasū ni tsuite*, September 30, 2021, https://www.env.go.jp/park/fujihakone/sfujihakone_R3.pdf.

46. Kankyōshō (Ministry of the Environment), *2023 nen kaki no Fujisan tozanshasū ni tsuite*, September 19, 2023, https://www.env.go.jp/park/fujihakone/data/files/fujihakone_shosai_R5.pdf.

47. "Mt. Fuji Climbers Will Be Asked to Wait When Trails Become Crowded," *The Asahi Shimbun*, August 10, 2023, https://www.asahi.com/ajw/articles/14978598.

48. Council for the Promotion of the Proper Use of Mt. Fuji, "[Yoshida Trail] 2024 Restrictions Overview and FAQ," accessed January 14, 2025, https://www.fujisan-climb.jp/en/news/240417_yoshida_trail_faq.html.

49. Kankyōshō (Ministry of the Environment), *2024 nen kaki no Fujisan tozanshasū ni tsuite*, September 30, 2024, https://www.env.go.jp/park/fujihakone/data/files/fujihakone_shosai_R6.pdf.

50. Yamanaka-mura no Rekishi Hensan Iinkai, *Yamanaka-mura no rekishi*, pp. 260–64; Shizuoka Chiri Kyōiku Kenkyūkai, *Fujisan sekai isan e no michi*, pp. 199–201.

51. Shizuoka Chiri Kyōiku Kenkyūkai, *Fujisan sekai isan e no michi*, p. 197.

52. Kankyōshō (Ministry of the Environment), "Kokuritsu kōen riyōshasū," https://www.env.go.jp/park/doc/data/natural/naturalpark_04.pdf.

53. Shizuoka Chiri Kyōiku Kenkyūkai, *Fujisan sekai isan e no michi*, p. 203.

54. UNESCO, UNESCO World Heritage Convention, "Fujisan, Sacred Place and Source of Artistic Inspiration," Video, accessed September 20, 2024, https://whc.unesco.org/en/list/1418/video/.

55. UNESCO, UNESCO World Heritage Convention, "Fujisan, Sacred Place and Source of Artistic Inspiration," Gallery, accessed September 20, 2024, https://whc.unesco.org/en/list/1418/gallery/.

56. UNESCO, UNESCO World Heritage Convention, "Fujisan, Sacred Place and Source of Artistic Inspiration," Description, accessed September 20, 2024, https://whc.unesco.org/en/list/1418/.

57. UNESCO, UNESCO World Heritage Convention, "Fujisan, Sacred Place and Source of Artistic Inspiration," accessed September 20, 2024, https://whc.unesco.org/en/list/1418/.

58. Fujisan World Cultural Heritage Council, "Outstanding Universal Value of Mt. Fuji," accessed September 20, 2024, https://www.fujisan-3776.jp/en/value/index.html.

59. Marcon, *The Knowledge of Nature*, pp. 5–7.

60. The literary and cultural critic Raymond Williams called nature "perhaps the most complex word in the language." Williams, *Keywords*, p. 219.

61. Marcon, *The Knowledge of Nature*, pp. 17–20; Thomas, *Reconfiguring Modernity*, p. 7.

62. Government of Japan, *State of Conservation Report: Fujisan, Sacred Place and Source of Artistic Inspiration*, November 2018, "Appendix: Implementation Report," p. 77, https://whc.unesco.org/en/list/1418/documents/.

63. A prominent critic of the use of the term "nature" to distinguish nonhumans from humans is Bruno Latour, who has proposed employing the word "collective" to indicate a "procedure for *collecting* associations of humans and nonhumans" that dispenses with the nature/humanity divide. Latour, *Politics of Nature*, p. 238.

64. Cronon, "The Trouble with Wilderness," pp. 80–81.

65. White, *The Organic Machine*, p. 112.

66. See, for example, Koyama, *Fujisan daifunka ga sematte iru*.

67. As part of the planning effort, a delegation from the Tokyo fire department visited the US Pacific Northwest in the spring of 2023 to investigate the response to the eruption of Mount Saint Helens in 1980 as well as preparations that communities in Washington and Oregon have made for future eruptions of Mount Saint Helens and other local volcanoes. Amagasa Masaaki, Ōba Ryōsuke, Okado Kōji, and Satō Mizuki, interview by author, Portland, Oregon, April 23, 2023.

BIBLIOGRAPHY

Abe Hajime. "Kindai Nihon no kyōkasho to Fujisan." In *Fujisan to Nihonjin*, edited by Seikyūsha Henshūbu. Tokyo: Seikyūsha, 2002.

Adachi, Noboru, Ken-ichi Shinoda, and Masami Izuho. "Further Analyses of Hokkaido Jomon Mitochondrial DNA." In Kaifu et al., *Emergence and Diversity.*

Aizawa, Koki, Ryokei Yoshimura, and Naoto Oshiman. "Splitting of the Philippine Sea Plate and a Magma Chamber Beneath Mt. Fuji." *Geophysical Research Letters* 31, no. 9 (May 2004): 1–4. https://doi.org/10.1029/2004GL019477.

Alcock, Rutherford. *The Capital of the Tycoon: A Narrative of a Three Years' Residence in Japan.* 2 vols. 1863. Reprint, Boston: Adamant Media, 2005.

Amano Ken'ichi. "Yokonosannōbara iseki no hakkutsu chōsa—Fujisan Hōei daifunka no higai to fukkō." *Nihon kōkogaku,* no. 44 (October 2017): 91–98.

Amano Kiyoko. "Kodaijin no Fujisan kan—hi no yama, hi no moto no shizume." In Amano and Sawato, *Fujisan to Nihonjin no shinsei.*

Amano Kiyoko and Sawato Hirosato, eds. *Fujisan to Nihonjin no shinsei.* Tokyo: Iwata Shoin, 2007.

Ambros, Barbara. *Emplacing a Pilgrimage: The Ōyama Cult and Regional Religion in Early Modern Japan.* Cambridge, MA: Harvard University Press, 2008.

Andrei, Talia. "*Sankei Mandara*: Layered Maps to Sacred Places." *Cross-Current: East Asian History and Culture Review* 6, no. 2 (November 2017): 365–403. https://doi.org/10.1353/ach .2017.0014.

Anma Sō. "Fujisan de hassei suru rahāru to surasshu rahāru." In Nihon Kazan Gakkai, *Fuji kazan.*

Aoki, Yosuke, Kae Tsunematsu, and Mitsuhiro Yoshimoto. "Recent Progress of Geophysical and Geological Studies of Mt. Fuji Volcano, Japan." *Earth-Science Reviews* 194 (July 2019): 264–282.

Aoyagi Shūichi. *Fugaku tabi hyakkei: Kankō chiikishi no kokoromi.* Tokyo: Kadokawa Shoten, 2002.

Arai Hakuseki. *Told Round a Brushwood Fire: The Autobiography of Arai Hakuseki.* Translated by Joyce Ackroyd. Tokyo: University of Tokyo Press, 1979.

Arakawa Hidetoshi. *Nihon kishōgakushi.* Tokyo: Kawade Shobō, 1941.

Aramaki Shigeo. "Fuji kazan no tokushoku." In Fujisan Sekai Bunka Isan Tōroku Suishin Shizuoka Yamanashi Ryōken Gōdō Kaigi and Fujisan o Sekai Isan ni Suru Kokumin Kaigi, *Fujisan: Shinkō to geijutsu no minamoto/Mt. Fuji: The Wellspring of Our Faith and Arts.*

Asahi Shigeaki. *Tekiroku Ōmu rōchūki: Genroku bushi no nikki*. Vol. 2, edited by Tsukamoto Manabu. Tokyo: Iwanami Shoten, 1995.

Baba Hiroomi. "Genroku dai jishin to Hōei Fujisan funka, sono 2: Sagami kuni Odawara hanryō muramura no nengu waritsukejō bunseki kara." *Bunmei*, no. 21 (January 2016): 1–21. https://www.u-tokai.ac.jp/uploads/2021/03/03-5.pdf.

Barnes, Gina L. *Archaeology of East Asia: The Rise of Civilization in China, Korea and Japan*. Oxford: Oxbow Books, 2015.

Barnes, Gina L. "A Hypothesis for Early Kofun Rulership." *Japan Review: Journal of the International Research Center for Japanese Studies* 27 (November 2014): 3–29. https://nichibun.repo.nii.ac.jp/records/7171.

Barnes, Gina L. *State Formation in Japan: Emergence of a 4th-Century Ruling Elite*. London: Routledge, 2007.

Barnes, Gina L. *Tectonic Archaeology: Subduction Zone Geology in Japan and its Archaeological Implications*. Oxford: Archaeopress, 2022.

Batdorff, Kara. "Mapping a Mountain: Mt. Fuji Land Cover Transitions over the 20th Century." Senior thesis, Lewis and Clark College, 2014.

Bell, Jon. *On Mount Hood: A Biography of Oregon's Perilous Peak*. Seattle: Sasquatch Books, 2013.

Berger, Klaus. *Japonisme in Western Painting from Whistler to Matisse*. Cambridge: Cambridge University Press, 1993.

Bernstein, Andrew. *Modern Passings: Death Rites, Politics, and Social Change in Imperial Japan*. Honolulu: University of Hawai'i Press, 2006.

Bernstein, Andrew. "Sacred Space on Postwar Fuji." In Wigen, Sugimoto, and Karacas, *Cartographic Japan*.

Bernstein, Andrew. "Weathering Fuji: Marriage, Meteorology, and the Meiji Bodyscape." In Miller, Thomas, and Walker, *Japan at Nature's Edge*.

Bernstein, Andrew. "Whose Fuji? Religion, Region, and State in the Fight for a National Symbol." *Monumenta Nipponica* 63, no. 1 (Spring 2008): 51–99. https://doi.org/10.1353/mni.0.0001.

Berry, Mary Elizabeth. *Japan in Print: Information and Nation in the Early Modern Period*. Berkeley: University of California Press, 2006.

Black, John R. *Young Japan: Yokohama and Yedo; A Narrative of The Settlement and the City from the Signing of the Treaties in 1858, to the Close of the Year 1879, with a Glance at the Progress of Japan During a Period of Twenty-One Years*. 2 vols. London: Trubner, 1881.

Bleed, Peter, and Akira Matsui. "Why Didn't Agriculture Develop in Japan? A Consideration of Jomon Ecological Style, Niche Construction, and the Origins of Domestication." *Journal of Archaeological Method and Theory* 17, no. 4 (July 2010): 356–70. https://doi.org./10.1007/s10816-010-9094-8.

Bodart-Bailey, Beatrice M. *The Dog Shogun: The Personality and Policies of Tokugawa Tsunayoshi*. Honolulu: University of Hawai'i Press, 2006.

"Bokusen." *Encyclopedia of Shinto*. In *Kokugakuin Digital Museum*, edited by Kokugakuin University. https://d-museum.kokugakuin.ac.jp/eos/detail/?id=8969.

Breen, John, and Mark Teeuwen, eds. *Shinto in History: Ways of the Kami*. Honolulu: University of Hawai'i Press, 2000.

Brown, Philip C. "Constructing Nature." In Miller, Thomas, and Walker, *Japan at Nature's Edge*.

Butchōji Gorō. *Fuji ni tatsu hito—iriai tōshi, Takamura Fujiyoshi no han seki.* Sakura-shi: Chōkōsha Shuppan, 2000.

Caldecott, Leonie. "At the Foot of the Mountain: The Shibokusa Women of Kita Fuji." In *Keeping the Peace,* edited by Lynne Jones. London: The Women's Press, 1983.

Carter, Caleb Swift. *A Path into the Mountains: Shugendō and Mount Togakushi.* Honolulu: University of Hawai'i Press, 2022.

Castiglioni, Andrea. "Devotion in Flesh and Bone: The Mummified Corpses of Mount Yudono Ascetics in Edo-Period Japan." *Asian Ethnology* 78, no. 1 (2019): 26–51. https://asianethnology.org/articles/2167.

Castiglioni, Andrea, Fabio Rambelli, and Carina Roth, eds. *Defining Shugendō: Critical Studies on Japanese Mountain Religion.* London: Bloomsbury, 2022.

Chamberlain, Basil Hall. *Japanese Things: Being Notes on Various Subjects Connected with Japan.* Rutland, VT: Charles E. Tuttle Company, 1971.

Chūjō Junko. "Miyako no Yoshika saku 'Fujisanki' ni tsuite—Chūgoku rikuchō bungaku to no kanren kara." *Kodai bunka* 33, no. 8 (August 1981): 31–42.

Cioc, Mark. *The Rhine: An Eco-Biography, 1815–2000.* Seattle: University of Washington Press, 2002.

Clancey, Gregory. *Earthquake Nation: The Cultural Politics of Japanese Seismicity, 1868–1930.* Berkeley: University of California Press, 2006.

Clark, Peter U., et al. "The Last Glacial Maximum." *Science* 325, no. 5941 (August 7, 2009): 710–14. https://doi.org/10.1126/science.1172873.

Clark, Timothy. *100 Views of Mount Fuji.* Trumbull, CT: Weatherhill, 2001.

Cooper, Tim Ervin. "Tokugawa Yoshimune versus Tokugawa Muneharu: Rival Visions of Benevolent Rule." PhD diss., University of California, Berkeley, 2010. https://escholarship.org/uc/item/4nn6p372.

Cronon, William. "The Trouble with Wilderness; or, Getting Back to the Wrong Nature." In *Uncommon Ground: Rethinking the Human Place in Nature,* edited by William Cronon. New York: W. W. Norton, 1996.

Date Mitsuyoshi. *Nihon shūkyō seido shiryō ruiju kō.* Rev. ed. Kyoto: Rinsen Shoten, 1974.

Dauvergne, Peter. *Shadows in the Forest: Japan and the Politics of Timber in Southeast Asia.* Cambridge, MA: MIT Press, 1997.

Davis, Natasha. "Paperpower: The Japanese Paper Industry and the Environment in Australia and Japan." *Japanese Studies Bulletin* 15, no. 1 (1995): 1–25. https://doi.org/10.1080/10371399508571518.

Debarbieux, Bernard, and Gilles Rudaz. *The Mountain: A Political History from the Enlightenment to the Present.* Translated by Jane Marie Todd. Chicago: University of Chicago Press, 2015.

Earhart, H. Byron. *Mount Fuji: Icon of Japan.* Columbia: University of South Carolina Press, 2011.

Ehlers, Maren A. *Give and Take: Poverty and the Status Order in Early Modern Japan.* Cambridge, MA: Harvard University Asia Center, 2018.

Endō Hideo. "Fujisan shinkō no hassei to Sengen shinkō no seiritsu." In *Fuji Sengen shinkō,* edited by Hirano Eiji. Tokyo: Yūzankaku Shuppan, 1987.

Endō Hideo. "Fuji shinkō no seiritsu to Murayama shugen." In *Fuji, Ontake to chūbu reizan,* edited by Suzuki Shōei. Tokyo: Meicho Shuppan, 1978.

Endō Kunihiko. "Fuji goko no nazo—Yamanakako o rei to shite." In Nihon Daigaku Bun-rigakubu Chikyū Shisutemu Kagaku Kyōshitsu, *Fujisan no nazo o saguru.*

Endō, Shūsaku. *Volcano.* Translated by Richard A. Schuchert. New York: Taplinger, 1980.

Esper, Jan, et al. "European Summer Temperature Response to Annually Dated Volcanic Erup-tions Over the Past Nine Centuries." *Bulletin of Volcanology* 75, no. 7 (June 14, 2013): 1–14. https://doi.org/10.1007/s00445-013-0736-z.

Farris, William Wayne. *A Bowl for a Coin: A Commodity History of Japanese Tea.* Honolulu: University of Hawai'i Press, 2019.

Farris, William Wayne. *Japan's Medieval Population: Famine, Fertility, and Warfare in a Transfor-mative Age.* Honolulu: University of Hawai'i Press, 2006.

Fedman, David. *Seeds of Control: Japan's Empire of Forestry in Colonial Korea.* Seattle: University of Washington Press, 2020.

Figal, Gerald. *Civilization and Monsters: Spirits of Modernity in Meiji Japan.* Durham, NC: Duke University Press, 1999.

Fleming, James Rodger. *Meteorology in America, 1800–1870.* Baltimore: Johns Hopkins Univer-sity Press, 1990.

Forestry Agency, Ministry of Agriculture, Forestry and Fisheries, Japan. *Annual Report on Forest and Forestry in Japan, Fiscal Year 2022 (Summary).* https://www.maff.go.jp/e/data/publish/attach/pdf/index-193.pdf.

Fujii Toshitsugu. "Fuji kazan no magumagaku." In Nihon Kazan Gakkai, *Fuji kazan.*

Fujinomiya Shishi Hensan Iinkai, ed. *Fujinomiya shishi.* Vol. 1. Fujinomiya-shi: Fujinomiya-shi, 1971.

Fujioka, Nobuko. "Vision or Creation? Kojima Usui and the Literary Landscape of the Japanese Alps." *Comparative Literature Studies* 39, no. 4 (2002): 282–92. https://doi.org/10.1353/cls.2002.0028.

Fuji Sabo Work Office. *Kuzure (Collapse): Mt. Fuji Ohsawa Failure.* Fujinomiya-shi: Fuji Sabo Work Office, Ministry of Land, Infrastructure and Transport, n.d.

Fuji Sanchō Shiyūka Hantai Dōmei. *Fuji sanchō haraisage mondai no shinsō.* Pamphlet. February 1953.

Fujisan Hongū Sengen Taisha, ed. *Kokuyū keidaichi jōyo shinsei fukyoka shobun torikeshi seikyū jiken.* Fujinomiya-shi: Fujisan Hongū Sengen Taisha, 1993.

Fujisan Sekai Bunka Isan Tōroku Suishin Shizuoka Yamanashi Ryōken Gōdō Kaigi and Fujisan o Sekai Isan ni Suru Kokumin Kaigi, eds. *Fujisan: Shinkō to geijutsu no minamoto/Mt. Fuji: The Wellspring of Our Faith and Arts.* Tokyo: Shōgakukan, 2009.

Fuji Shiritsu Hakubutsukan, ed. *Fuji no cha.* Hito to tomo ni ikiru shokubutsu 1. Fuji-shi: Fuji Shiritsu Hakubutsukan, 2006.

Fuji Shiritsu Hakubutsukan, ed. *Fuji-shi no seishigyō.* Fuji-shi: Fuji Shiritsu Hakubutsukan, 1997.

Fuji Shiritsu Hakubutsukan, ed. *Kashima kome to mizu: Fujikawa karyū no kome zukuri.* Fuji-shi: Fuji Shiritsu Hakubutsukan, 1998.

Fuji Shiritsu Hakubutsukan, ed. *Ukishimanuma to kome zukuri.* Fuji-shi: Fuji Shiritsu Hakubut-sukan, 1984.

Fuji Shishi Hensan Iinkai, ed. *Fuji shishi.* 3rd ed. 2 vols. Fuji-shi: Fuji-shi, 1982.

Fuji Shishi Hensan Iinkai, ed. *Yoshiwara shishi.* 3 vols. Fuji-shi: Fuji-shi, 1972–78.

Fujiyoshida-shi Hoka Nikason Onshi Ken'yū Zaisan Hogo Kumiai, ed. *Onshirin kumiaishi.* 4 vols. Fujiyoshida-shi: Fujiyoshida-shi Hoka Nikason Onshi Keny'ū Zaisan Hogo Kumiai, 1997–2001.

Fujiyoshida-shi Kyōiku Iinkai Rekishi Bunkaka, ed. *Yoshida no himatsuri.* Fujiyoshida-shi: Fujiyoshida Kyōiku Iinkai, 2005.

Fujiyoshida Shishi Hensan Iinkai, ed. *Fujiyoshida shishi shiryōhen.* Vol. 2, *Kodai, chūsei.* Fujiyoshida-shi: Fujiyoshida-shi, 1992.

Fujiyoshida Shishi Hensan Iinkai, ed. *Fujiyoshida shishi shiryōhen.* Vol. 3, *Kinsei 1.* Fujiyoshida-shi: Fujiyoshida-shi, 1994.

Fujiyoshida Shishi Hensan Iinkai, ed. *Fujiyoshida shishi shiryōhen.* Vol. 4, *Kinsei 2.* Fujiyoshida-shi: Fujiyoshida-shi, 1994.

Fujiyoshida Shishi Hensan Iinkai, ed. *Fujiyoshida shishi shiryōhen.* Vol. 5, *Kinsei 3.* Fujiyoshida-shi: Fujiyoshida-shi, 1997.

Fujiyoshida Shishi Hensan Iinkai, ed. *Fujiyoshida shishi shiryōhen.* Vol. 7, *Kin, gendai 2.* Fujiyoshida-shi: Fujiyoshida-shi, 1995.

Fujiyoshida Shishi Hensan Iinkai, ed. *Fujiyoshida shishi tsūshihen.* Vol. 1, *Genshi, kodai, chūsei.* Fujiyoshida-shi: Fujiyoshida-shi, 2000.

Fujiyoshida Shishi Hensan Iinkai, ed. *Fujiyoshida shishi tsūshihen.* Vol. 2, *Kinsei.* Fujiyoshida-shi: Fujiyoshida-shi, 2001.

Fujiyoshida Shishi Hensan Iinkai, ed. *Fujiyoshida shishi tsūshihen.* Vol. 3, *Kin, gendai.* Fujiyoshida-shi: Fujiyoshida-shi, 1999.

Galton, Francis. *The Art of Travel: or, Shifts and Contrivances Available in Wild Countries.* Introduction by Dorothy Middleton. 1855. Reprint, London: Phoenix Press, 1971.

George, Timothy S. *Minamata: Pollution and the Struggle for Democracy in Postwar Japan.* Cambridge, MA: Harvard University Asia Center, 2001.

Gluck, Carol. *Japan's Modern Myths: Ideology in the Late Meiji Period.* Princeton, NJ: Princeton University Press, 1985.

Gojira tai hedora. Directed by Banno Yoshimitsu. Tōhō Kabushiki Kaisha, 1971.

Gotenba Shishi Hensan Iinkai, ed. *Gotenba shishi.* Vol. 9, *Tsūshihen 2.* Gotenba-shi: Gotenba-shi, 1983.

Gottardo, Marco. "On the Identity of Mt. Fuji's Deity: A Study on the Role of Benzai-ten in the Development of the Fuji Cult." *Tamagawa Daigaku Bungakubu kiyō,* no. 57 (2016): 47–60.

Government of Japan. *Nomination of Fujisan for Inscription on the World Heritage List,* January 2012. https://whc.unesco.org/uploads/nominations/1418.pdf.

Government of Japan. *State of Conservation Report: Fujisan, Sacred Place and Source of Artistic Inspiration,* January 2016. https://whc.unesco.org/document/139890.

Grapard, Alan. "The Economics of Ritual Power." In Breen and Teeuwen, *Shinto in History: Ways of the Kami.*

Guth, Christine M. E. *Hokusai's Great Wave: Biography of a Global Icon.* Honolulu: University of Hawai'i Press, 2015.

Guth, Christine M. E. "Hokusai's Great Waves in Nineteenth-Century Japanese Visual Culture." *Art Bulletin* 93, no. 4 (December 2011): 468–85. https://doi.org/10.1080/00043079.2011.10786019.

Habu, Junko. *Ancient Jomon of Japan.* Cambridge: Cambridge University Press, 2004.

Hall, John W. "Rule by Status in Tokugawa Japan." *Journal of Japanese Studies* 1, no. 1 (Autumn 1974): 39–49. https://doi.org/10.2307/133436.

Hardacre, Helen. *Shinto: A History.* Oxford: Oxford University Press, 2017.

Harris, Townsend. *The Complete Journal of Townsend Harris: First American Consul General and Minister to Japan.* Introduction and notes by Mario Emilio Cosenza. Garden City, NY: Doubleday, Doran, 1930.

Hashimoto Eikichi. *Fuji sanchō.* Tokyo: Kamakura Bunko, 1948.

Havlick, David G. *Bombs Away: Militarization, Conservation, and Ecological Restoration.* Chicago: University of Chicago Press, 2018.

Hayase Shinzo. "Manila Hemp in World, Regional, National, and Local History." *Journal of Asia-Pacific Studies* (Waseda University), no. 31 (March 2018): 171–88. https://doi.org/10 .57278/wiapstokyu.31.0_171.

Hayashi Takeshi. "Fujisan hokubu ni okeru chikasui ryūdō kikō no kaimei no kadai to tenbō." *Chigaku zasshi* 129, no. 5 (October 2020): 677–95. https://doi.org/10.5026 /jgeography.129.677.

Hayashi Yutaka and Koyama Masato. "Hōei yonen Fujisan funka ni sakidatte hassei shita jishin no kibo no suitei." *Rekishi jishin,* no. 18 (2002): 127–32. https://www.histeq.jp/kaishi_18 /20-Hayashi.pdf.

Hearn, Lafcadio. *Exotics and Retrospectives.* Boston: Little, Brown, 1898.

Hellyer, Robert I. *Green with Milk and Sugar: When Japan Filled America's Tea Cups.* New York: Columbia University Press, 2021.

Hesselink, Reinier H. "The Assassination of Henry Heusken." *Monumenta Nipponica* 49, no. 3 (Autumn 1994): 331–51. https://doi.org/10.2307/2385453.

Heusken, Henry. *Japan Journal: 1855–1861.* Translated and edited by Jeannette C. van der Corput and Robert A. Wilson. New Brunswick, NJ: Rutgers University Press, 1964.

Hiraga Gennai. *Fūrai sanjin shū.* Vol. 55 of *Nihon koten bungaku taikei,* edited by Nakamura Yukihiko. Tokyo: Iwanami Shoten, 1961.

Hirano Eiji. *Fuji shinkō to Fujikō.* Tokyo: Iwata Shoin, 2004.

Hirano, Kyoko. *Mr. Smith Goes to Tokyo: Japanese Cinema under the American Occupation, 1945– 1952.* Washington, DC: Smithsonian Institution Press, 1992.

Hirata Daiji et al. "Puroto Izu-Mariana tōko no shōtotsu fuka tekutonikusu—rebyū." *Chigaku zasshi* 119, no. 6 (December 2010): 1125–60. https://doi.org/10.5026/jgeography.119.1125.

"Hitachi Province Gazetteer (*Hitachi Fudoki*)." Translated by Edwin Cranston. In Shirane, *Traditional Japanese Literature: An Anthology.*

Hōjō Hiroshi. *Yamanashi-ken iriai tōsōshi.* Tokyo: Ochanomizu Shobō, 1998.

Horiuchi Tatsuzō. "'Fujimi jūsanshū yochi no zenzu' ihan saikō." *Kochizu kenkyū,* no. 310 (September 2002): 9–16.

Ikeya Nobuyuki. "Marine Transport of Obsidian in Japan During the Upper Paleolithic." In Kaifu et al., *Emergence and Diversity.*

Ikku Jippensha. *Shank's Mare: A Translation of the Tokaido Volumes of Hizakurige, Japan's Great Comic Novel of Travel & Ribaldry.* Translated by Thomas Satchell. 1960. Reprint, Boston: Tuttle, 2001.

Inobe Shigeo. *Fuji no rekishi.* Vol. 1 of *Fuji no kenkyū,* edited by Kanpei Taisha Sengen Jinja Shamusho. Tokyo: Kokon Shoin, 1928.

Inoue Kimio. "Fujisan Hōei funka (1707) go no chōkikan ni oyonda dosha saigai." In Nihon Kazan Gakkai, *Fuji kazan.*

Inoue Kimio. "Genroku jishin (1703) to Fujisan Hōei funka (1707) ni yoru dosha saigai to fukkō katei—Kanagawa-ken Yamakita-chō ni okeru saikin no shiryōgaku, kōkogaku teki seika ni yoru saikentō." *Rekishi jishin,* no. 20 (2005): 247–55. https://www.histeq.jp/kaishi_20/34-Inoue.pdf.

International Council of Monuments and Sites. *Fujisan (Japan), no. 1417,* March 6, 2013. https://whc.unesco.org/document/152623.

Itō Kenkichi. *Fujisan oshi.* Tokyo: Zufu Shuppan, 1968.

Iwase, Akira, Keiichi Takahashi, and Masami Izuho. "Further Study on the Late Pleistocene Megafaunal Extinction in the Japanese Archipelago." In Kaifu et al., *Emergence and Diversity.*

Iwashina Koichirō. *Fujikō no rekishi: Edo shomin no sangaku shinkō.* Tokyo: Meicho Shuppan, 1983.

Izuho, Masami, and Yousuke Kaifu. "The Appearance and Characteristics of the Early Upper Paleolithic in the Japanese Archipelago." In Kaifu et al., *Emergence and Diversity.*

Izuho, Masami, et al. "The Upper Paleolithic of Hokkaido: Current Evidence and its Geochronological Framework." In Ono and Izuho, *Environmental Changes and Human Occupation.*

Jaffe, Richard M. *Neither Monk nor Layman: Clerical Marriage in Modern Japanese Buddhism.* Princeton, NJ: Princeton University Press, 2001.

Jamieson, Alan J., et al. "Bioaccumulation of Persistent Organic Pollutants in the Deepest Ocean Fauna." *Nature Ecology & Evolution* 1, no. 3 (February 13, 2017): 1–4. https://doi.org/10.1038/s41559-016-0051.

Janković, Vladimir. *Reading the Skies: A Cultural History of English Weather, 1650–1820.* Chicago: University of Chicago Press, 2001.

Jasny, Aaron. "In Praise of the Peaks: Science, Art, and Nature in Kojima Usui's Mountain Literature." PhD diss., Washington University in St. Louis, 2019. Arts & Sciences Electronic Theses and Dissertations. https://doi.org/10.7936/7gyc-9x32.

Jones, Thomas, Sue Beeton, and Malcolm Cooper. "World Heritage Listing as a Catalyst for Collaboration: Can Mount Fuji's Trail Signs Point the Way for Japan's Multi-purpose National Parks?" *Journal of Ecotourism* 17, no. 3 (August 2018): 220–38. https://doi.org/10.1080/14724049.2018.1503769.

Jones, Thomas E., Yang Yang, and Kiyotatsu Yamamoto. "Assessing the Recreational Value of World Heritage Site Inscription: A Longitudinal Travel Cost Analysis of Mount Fuji Climbers." *Tourism Management* 60 (June 2017): 67–78. http://dx.doi.org/10.1016/j.tourman.2016.11.009.

Kaifu, Yousuke, et al., eds. *Emergence and Diversity of Modern Human Behavior in Paleolithic Asia.* College Station: Texas A&M Press, 2015.

Kaigo Tokiomi, ed. *Nihon kyōkasho taikei kindaihen.* Vol. 5, *Kokugo* 2. Tokyo: Kōdansha, 1964.

Kaigo Tokiomi, ed. *Nihon kyōkasho taikei kindaihen.* Vol. 6, *Kokugo* 3. Tokyo: Kōdansha, 1964.

Kaigo Tokiomi, ed. *Nihon kyōkasho taikei kindaihen.* Vol. 7, *Kokugo* 4. Tokyo: Kōdansha, 1963.

Kaigo Tokiomi, ed. *Nihon kyōkasho taikei kindaihen.* Vol. 8, *Kokugo* 5. Tokyo: Kōdansha, 1964.

Kaigo Tokiomi, ed. *Nihon kyōkasho taikei kindaihen.* Vol. 25, *Shōka.* Tokyo: Kōdansha, 1965.

Kaigo Tokiomi, ed. *Nihon kyōkasho taikei kindaihen.* Vol. 26, *Zuga.* Tokyo: Kōdansha, 1966.

Kamigaito Ken'ichi. *Fujisan: sei to bi no yama.* Tokyo: Chūō Kōron Shinsha, 2009.

Karatani Kōjin. *Origins of Modern Japanese Literature*. Durham, NC: Duke University Press, 1993.

Kashiwazaki, Chikako. "The Politics of Legal Status: The Equation of Nationality with Ethnonational Identity." In *Koreans in Japan: Critical Voices from the Margin*, edited by Sonia Ryang. London: Routledge, 2000.

Kawabata Yasunari. *The Tale of the Bamboo Cutter*. Translated by Donald Keene. Tokyo: Kodansha International, 1998.

Kawai Yasuyo. "Fujikō kara mita seichi Fujisan no fūkei—Tōkyō-to 23 ku no Fujizuka no rekishiteki henyō o tsūjite." *Chirigaku hyōron* 74A, no. 6 (June 2001): 349–66. https://doi.org/10.4157/grj1984a.74.6_349.

Kawamura, Yoshinari, and Ryohei Nakagawa. "Terrestrial Mammal Faunas in the Japanese Islands During OIS3 and OIS2." In Ono and Izuho, *Environmental Changes and Human Occupation*.

Ketelaar, James Edward. *Of Heretics and Martyrs in Meiji Japan: Buddhism and Its Persecution*. Princeton, NJ: Princeton University Press, 1990.

Kijima, Yoko, Takeshi Sakurai, and Keijiro Otsuka. "Iriaichi: Collective versus Individualized Management of Community Forests in Postwar Japan." *Economic Development and Cultural Change* 48, no. 4 (July 2000): 867–86. https://doi.org/10.1086/452481.

Kim Tal-su. "In the Shadow of Mount Fuji" (*Fuji no mieru mura de*). Translated by Sharalyn Orbaugh. In *Into the Light: An Anthology of Literature by Koreans in Japan*, edited by Melissa L. Wender. Honolulu: University of Hawai'i Press, 2011.

Kinsei Shiryō Kenkyūkai, ed. *Edo machibure shūsei*. 22 vols. Tokyo: Hanawa Shobō, 1994–2012.

Kitahara Itoko. *Nihon shinsaishi*. Tokyo: Chikuma Shobō, 2016.

Kitahara Itoko et al. *Nihon rekishi saigai jiten*. Tokyo: Yoshikawa Kōbunkan, 2012.

Kobayashi Issa. *Issa haikushū*. Edited by Ogiwara Seisensui. Tokyo: Iwanami Shoten, 1976.

Kojiki. Translated by Donald L. Philippi. Tokyo: University of Tokyo Press, 1968.

Kojima Usui. *Fujisan*. In Shimada et al., *Kojima Usui zenshū*, vol. 4.

Kojima Usui. *Kikōbun ron*. In Shimada et al., *Kojima Usui zenshū*, vol. 5.

Kokkai (National Diet of Japan), Shūgiin (House of Representatives). *Dai jūkyū kai Kokkai Shūgiin Gyōsei Kansatsu Tokubetsu Iin kaigiroku dai ni gō, dai san gō*, December 17–18, 1953. https://kokkai.ndl.go.jp/pdf/101904280X00219531217; https://kokkai.ndl.go.jp/pdf/101904280X00319531218.

Kokinshū: A Collection of Poems Ancient and Modern. Translated and annotated by Laurel Rasplica Rodd with Mary Catherine Henkenius. Boston: Cheng & Tsui, 1996.

Kokin Wakashū: The First Imperial Anthology of Japanese Poetry, With Tosa Nikki and Shinsen Waka. Translated by Helen Craig McCullough. Stanford, CA: Stanford University Press, 1985.

Kondō Seiichi. *Fujisan: Sekai isan e no michi*. Tokyo: Mainichi Shinbunsha, 2014.

Kotō, Bunjirō. "The Scope of the Vulcanological Survey of Japan." In *Publications of the Earthquake Investigation Committee in Foreign Languages*, no. 3. Tokyo: The Earthquake Investigation Committee, 1900.

Koyama Masato. *Fujisan daifunka ga sematte iru*. Tokyo: Gijutsu Hyōronsha, 2011.

Koyama Masato. *Fujisan funka to hazādo mappu: Hōei funka no 16 nichikan*. Tokyo: Kokon Shoin, 2009.

Koyama Masato. "Fujisan no rekishi funka sōran." In Nihon Kazan Gakkai, *Fuji Kazan*.

Koyama Yasunori. "East and West in the Late Classical Age." In Piggott, *Capital and Countryside in Japan, 300–1180*.

Krauss, Ellis S., and Bradford L. Simcock. "Citizens' Movements: The Growth and Impact of Environmental Protest in Japan." In Steiner, Krauss, and Flanagan, *Political Opposition and Local Politics in Japan*.

Kubota Jun. *Fujisan no bungaku*. Tokyo: Bungei Shunjū, 2004.

Kudo, Yuichiro. "Absolute Chronology of Archaeological and Paleoenvironmental Records from the Japanese Islands, 40–15 ka BP." In Ono and Izuho, *Environmental Changes and Human Occupation*.

Kudo, Yuichiro, and Fujio Kumon. "Paleolithic Cultures of MIS 3 to MIS 1 in Relation to Climate Changes in the Central Japanese Islands." *Quaternary International* 248 (March 7, 2011): 22–31. https://doi.org/10.1016/j.quaint.2011.02.016.

Kumar, Amit, and Dharm Dutt. "A Comparative Study of Conventional Chemical Deinking and Environment-Friendly Bio-Deinking of Mixed Office Wastepaper." *Scientific African* 12 (July 2021): 1–10. https://doi.org/10.1016/j.sciaf.2021.e00793.

Kuroita Katsumi and Kokushi Taikei Henshūkai, eds. *Tokugawa jikki*. Vol. 6. Tokyo: Yoshikawa Kōbunkan, 1965.

Kurosawa, Takafumi, and Tomoko Hashino. "From the Non-European Tradition to a Variation on the Japanese Competitiveness Model: The Modern Japanese Paper Industry Since the 1870s." In *The Evolution of the Global Paper Industry, 1800–2050, a Comparative Analysis*, edited by Juha-Antti Lamberg et al. Dordrecht: Springer Science+Business Media Dordrecht, 2012.

Kuzmin, Yaroslav V., and Michael D. Glascock, eds. *Crossing the Straits: Prehistoric Obsidian Source Exploitation in the North Pacific Rim*. Oxford: Archaeopress, 2010.

Kuzmin, Yaroslav V., and Michael D. Glascock. "Introduction: Obsidian Sourcing in the North Pacific Rim Region and Beyond It." In Kuzmin and Glascock, *Crossing the Straits*.

Kuznetsov, Anatoly. "Some Issues on the Origin of Microblade Industries in Northeast Asia During OIS2." In Ono and Izuho, *Environmental Changes and Human Occupation*.

Latour, Bruno. *Politics of Nature: How to Bring the Sciences Into Democracy*. Translated by Catherine Porter. Cambridge, MA: Harvard University Press, 2004.

Levy, Ian Hideo. *The Ten Thousand Leaves: A Translation of the Man'yōshū, Japan's Premier Anthology of Classical Poetry*. Princeton, NJ: Princeton University Press, 1987.

Lewis, Jack G. "Civic Protest in Mishima: Citizens' Movements and the Politics of the Environment in Contemporary Japan." In Steiner, Krauss, and Flanagan, *Political Opposition and Local Politics in Japan*.

Lindström, Kati. "Universal Heritage Value, Community Identities and World Heritage: Forms, Functions, Processes and Context at a Changing Mt. Fuji." *Landscape Research* 44, no. 3 (March 2019): 278–91. https://doi.org/10.1080/01426397.2019.1579899.

Lippit, Yukio. *Painters of the Realm: The Kano House of Painters in 17th-Century Japan*. Seattle: University of Washington Press, 2012.

Lu, David J. *Japan: A Documentary History*. Vol. 1, *The Dawn of History to the Late Tokugawa Period*. Armonk, NY: M. E. Sharpe, 1997.

Makimura, Yasuhiro. *Yokohama and the Silk Trade: How Eastern Japan Became the Primary Economic Region of Japan, 1843–1893*. Lanham, MD: Lexington Books, 2017.

Marcon, Federico. *The Knowledge of Nature and the Nature of Knowledge in Early Modern Japan.* Chicago: University of Chicago Press, 2015.

Marra, Michele. *The Aesthetics of Discontent: Politics and Reclusion in Medieval Japanese Literature.* Honolulu: University of Hawai'i Press, 1991.

Marshak, Stephen. *Earth: Portrait of a Planet.* 4th ed. New York: W. W. Norton, 2012.

Martini, Malgorzata. "Katsushika Hokusai and Mount Fuji." In Martini, *Mount Fuji: Hokusai and Hiroshige.*

Martini, Malgorzata. *Mount Fuji: Hokusai and Hiroshige, Japanese Landscape Prints from the Collection of Feliks Manggha Jasieński.* Krakow: Manggha Museum of Japanese Art and Technology, 2012.

Martini, Malgorzata. "Utagawa Hiroshige and Mount Fuji." In Martini, *Mount Fuji: Hokusai and Hiroshige.*

Maruyama Hiroshi. *Kindai Nihon kōenshi no kenkyū.* Kyoto: Shibunkaku Shuppan, 1994.

"Maruyamakyō." *Encyclopedia of Shinto.* In *Kokugakuin Daigaku Digital Museum,* edited by Kokugakuin University. https://museum.kokugakuin.ac.jp/eos/detail/?id=9809.

Matsuo Basho. *Basho: The Complete Haiku.* Edited and translated by Jane Reichhold. New York: Kodansha USA, 2013.

"Matsuo Bashō, Two Haiku." Translated by Peter MacMillan. *Mānoa* 29, no. 2 (2017): 12. https://doi.org/10.1353/man.2017.0017.

Matsushima Jin. "Man'en gannen no deipuromateikku gifuto—Kanō Tōsen Nakanobu hitsu [Fuji hikaku zu] no seisaku to igi." *Kokka* 128, no. 8 (March 2023): 5–26.

Matsushima, Jin. "Mount Fuji and the Tokugawa Shogunate." In *Multidisciplinary Studies of the Environment and Civilization: Japanese Perspectives,* edited by Yoshinori Yasuda and Mark J. Hudson. London: Routledge, 2018.

McClain, James L., and John M. Merriman. "Edo and Paris: Cities and Power." In *Edo and Paris: Urban Life and the State in the Early Modern Era,* edited by James L. McClain, John M. Merriman, and Ugawa Kaoru. Ithaca, NY: Cornell University Press, 1994.

McGuire, Mark Patrick. "What's at Stake in Designating Japan's Sacred Mountains as UNESCO World Heritage Sites? Shugendo Practices in the Kii Peninsula." *Japanese Journal of Religious Studies* 40, no. 2 (2013): 323–54. https://doi.org/10.18874/jjrs.40.2.2013.323-354.

McKean, Margaret A. "Defining and Dividing Property Rights in the Japanese Commons." Paper presented at the annual meeting of the International Association for the Study of Common Property, Winnipeg, Canada, September 26–29, 1991.

McKean, Margaret A. *Environmental Protest and Citizen Politics in Japan.* Berkeley: University of California Press, 1981.

McKean, Margaret A. "Management of Traditional Common Lands (*Iriaichi*) in Japan." In *Making the Commons Work: Theory, Practice, and Policy,* edited by Daniel W. Bromley and David Feeny. San Francisco: ICS Press, 1992.

Miller, Ian Jared, Julia Adeney Thomas, and Brett L. Walker, eds. *Japan at Nature's Edge: The Environmental Context of a Global Power.* Honolulu: University of Hawai'i Press, 2013.

Mills, D. E. "*Soga Monogatari, Shintōshū* and the Taketori Legend." *Monumenta Nipponica* 30, no. 1 (Spring 1975): 37–68. https://doi.org/10.2307/2383695.

Miraculous Stories from the Japanese Buddhist Tradition: The "Nihon ryōiki" of the Monk Kyōkai. Translated by Kyoko Motomochi Nakamura. Richmond, UK: Curzon Press, 1997.

Miura, Takashi. *Agents of World Renewal: The Rise of Yonaoshi Gods in Japan.* Honolulu: University of Hawai'i Press, 2020.

Miyachi Naokazu and Hirono Saburō. *Sengen jinja no rekishi.* Vol. 2 of *Fuji no kenkyū,* edited by Kanpei Taisha Sengen Jinja Shamusho. Tokyo: Kokon Shoin, 1929.

Miyaji Naomichi. "Kako ichi man sen nenkan no Fuji kazan no funkashi to funshutsuritsu, funka kibo no sui'i." In Nihon Kazan Gakkai, *Fuji kazan.*

Miyaji, Naomichi, et al. "High-Resolution Reconstruction of the Hoei Eruption (AD 1707) of Fuji Volcano, Japan." *Journal of Volcanology and Geothermal Research* 207, no. 3 (October 2011): 113–29. https://doi.org/10.1016/j.jvolgeores.2011.06.013.

Miyaji Naomichi and Koyama Masato. "Fujisan 1707 nen funka (Hōei funka) ni tsuite no saikin no kenkyū seika." In Nihon Kazan Gakkai, *Fuji kazan.*

Miyaji Naomichi, Togashi Shigeko, and Chiba Tatsurō. "Fuji kazan tōshamen de 2,900 nen mae ni hassei shita santai hōkai." *Kazan* 49, no. 5 (October 2004): 237–42.

Miyake Okiko. "Iwaya Sazanami." In *The Oxford Encyclopedia of Children's Literature,* edited by Jack Zipes. Oxford: Oxford University Press, 2006. Oxford Reference Online.

Miyako no Yoshika. *Fujisanki.* In *Nihon koten bungaku taikei.* Vol. 69, *Kaifūsō bunka shūreishū honchō monzui,* edited by Kojima Noriyuki. Tokyo: Iwanami Shoten, 1964.

Miyanaga, Kentaro, and Daisaku Shimada. "'The Tragedy of the Commons' by Underuse: Toward a Conceptual Framework Based on Ecosystem Services and *Satoyama* Perspective." *International Journal of the Commons* 12, no. 1 (April 2018): 332–51. https://doi.org/10.18352/ijc.817.

Miyata Noboru. "Fuji shinkō to Miroku." In *Sangaku shūkyō to minkan shinkō no kenkyū,* edited by Sakurai Tokutarō. Tokyo: Meicho Shuppan, 1976.

Miyata Noboru. "Taketori monogatari to Fuji shinkō." *Kokubungaku* 38, no. 4 (April 1993): 28–31.

Miyazaki Fumiko. "An Artist's Rendering of the Divine Mount Fuji." In Wigen, Sugimoto, and Karacas, *Cartographic Japan.*

Miyazaki Fumiko. "Daikyō senpu no jidai no shinshūkyō—Jikkōsha no bai." In *Meiji Nihon no seijika gunzō,* edited by Fukuchi Atsushi and Sasaki Takashi. Tokyo: Yoshikawa Kōbunkan, 1993.

Miyazaki Fumiko. "Female Pilgrims and Mt. Fuji: Changing Perspectives on the Exclusion of Women." *Monumenta Nipponica* 60, no. 3 (Autumn 2005): 339–91. https://doi.org/10.1353/mni.2005.0034.

Miyazaki Fumiko. "The Formation of Emperor Worship in the New Religions: The Case of Fujidō." *Japanese Journal of Religious Studies* 17, nos. 2–3 (June–September 1990): 281–314. https://doi.org/10.18874/jjrs.17.2-3.1990.281-314.

Miyazaki Fumiko. "Gaikokujin no Fuji tozan ni tai suru bakumatsu Nihonjin no taiō." *Fujisan bunka kenkyū,* nos. 9–10 (2008): 13–32.

Miyazaki Fumiko. "Minshū no shūkyō undo." In *Kaikoku,* Kōza Nihon kinseishi 7, edited by Aoki Michio and Kawachi Hachirō. Tokyo: Yūhikaku, 1985.

Miyazaki Fumiko. "Networks of Believers in a New Religion: Female Devotees of Fujidō." In *Women and Networks in Nineteenth-Century Japan,* edited by Bettina Gramlich-Oka, Anne Walthall, Miyazaki Fumiko, and Sugano Noriko. Ann Arbor: University of Michigan Press, 2020.

Mizoguchi, Kōji. *The Archaeology of Japan: From the Earliest Rice Farming Villages to the Rise of the State.* Cambridge: Cambridge University Press, 2013.

Moerman, D. Max. "The Archeology of Anxiety, An Underground History of Heian Religion." In *Heian Japan, Centers and Peripheries*, edited by Mikael Adolphson, Edward Kamens, and Stacie Matsumoto. Honolulu: University of Hawai'i Press, 2007.

Nagahara Keiji. *Fujisan Hōei daibakuhatsu*. Tokyo: Shūeisha, 2004.

Nagashima Taigyō. *Fujisan shinkei no zu*, 1847. Reprinted in *Fujisan shinkei no zu: Edo jidai sankei emaki*, edited by Okada Hiroshi. Tokyo: Meicho Shuppan, 1985.

Naitō Chisō. *Tokugawa jūgodaishi*. Vol. 3. Reprint, Tokyo: Shin Jinbutsu Ōraisha, 1985.

Nakada Setsuya, Yoshimoto Mitsuhiro, and Fujii Toshitsugu. "Sen Fuji kazangun." In Nihon Kazan Gakkai, *Fuji kazan*.

Nakajima, Ryota, et al. "Occurrence and Levels of Polybrominated Diphenyl Ethers (PBDEs) in Deep-Sea Sharks from Suruga Bay, Japan." *Marine Pollution Bulletin* 176 (March 2022): 1–6. https://doi.org/10.1016/j.marpolbul.2022.113427.

Nakaoka Hirofumi et al. "HLA idenshi takei kara mita Nihonjin shūdan no kongōteki kigen." *Nihon soshiki tekigōsei gakkai shi* 21, no. 1 (April 2014): 37–44. https://doi.org/10.12667/mhc.21.37.

Naruse Fujio. "Fujisan no kaiga." In *Nihon no bi: Fuji*, edited by Suzuki Susumu. Tokyo: Bijutsu Nenkansha, 2000.

Natsume Sōseki. *Sanshirō*. Translated by Jay Rubin. Seattle: University of Washington Press, 1977.

Nelson, Lindsay. *Circulating Fear: Japanese Horror, Fractured Realities, and New Media*. Lanham, MD: Lexington Books, 2021.

Nenzi, Laura. *Excursions in Identity: Travel and the Intersection of Place, Gender, and Status in Edo Japan*. Honolulu: University of Hawai'i Press, 2008.

Nihon Daigaku Bunrigakubu Chikyū Shisutemu Kagaku Kyōshitsu, ed. *Fujisan no nazo o saguru: Fuji kazan no chikyū kagaku to bōsaigaku*. Tokyo: Tsukiji Shokan, 2006.

Nihongi: Chronicles of Japan from the Earliest Times to A.D. 697. 2 vols. Translated by W. G. Aston. Rutland, VT: Charles E. Tuttle, 1972.

Nihon Kazan Gakkai, ed. *Fuji kazan*. Fujiyoshida: Yamanashi-ken Kankyō Kagaku Kenkyūjo, 2007.

Nippon Gakujutsu Shinkōkai. *The Manyōshū: One Thousand Poems Selected and Translated from the Japanese*. Tokyo: Iwanami Shoten, 1940.

Nishida Kaoru. "Kawaguchi-mura ni okeru Fujisan oshi no seiritsu to sono katsudō." In *Fujisan oshi no rekishiteki kenkyū*, edited by Kōshū Shiryō Chōsakai. Tokyo: Yamakawa Shuppansha, 2009.

Nitta Hitoshi. "Shinto as a 'Non-religion': The Origins and Development of an Idea." In Breen and Teeuwen, *Shinto in History: Ways of the Kami*.

Nitta Jirō. *Fuyō no hito*. 1971. Reprint, Tokyo: Bungei Shunjū, 1975.

Noguchi Ken. *Sekai isan ni sarete Fujisan wa naite iru*. Tokyo: PHP Kenkyūjo, 2014.

Nonaka Chiyoko. *Fuyō nikki*. In Ōmori, *Fuji annai, Fuyō nikki*.

Nonaka Itaru. *Fuji annai*. In Ōmori, *Fuji annai, Fuyō nikki*.

Ochiai Naobumi. *Takane no yuki*. Tokyo: Meiji Shoin, 1896.

Odawara-shi, ed. *Odawara shishi shiryōhen, kinsei*. Vol. 2. Odawara-shi: Odawara-shi, 1989.

Odawara-shi, ed. *Odawara shishi tsūshihen*. Vol. 2, *Kinsei*. Odawara-shi: Odawara-shi, 1999.

Oda Zenkō. *Fujisan ga sekai isan ni naru hi*. Tokyo: PHP Kenkyūjo, 2006.

Ogawa Satoshi et al. "Fuji santai o riyō shita kaki jiyū tairyūken ni okeru gasujō suigin no kansoku: 2014 nen kaki shūchū kansoku kekka." *Taiki kankyō gakkai shi* 50, no. 2 (March 2015): 100–106. https://doi.org/10.11298/taiki.50.100.

Ó Gráda, Cormac. *Black '47 and Beyond: The Great Irish Famine in History, Economy, and Memory*. Princeton, NJ: Princeton University Press, 2000.

Okada, Richard. "'Landscape' and the Nation-State: A Reading of *Nihon fūkei ron*." In *New Directions in the Study of Meiji Japan*, edited by Helen Hardacre with Adam L. Kern. Leiden: Brill, 1997.

Okamoto Bun'ichi. *Fugaku roppyaku kei*. Shizuoka-shi: Wagendō Shūkokan, 1978.

Ōkurashō Kanzaikyoku, ed. *Shaji keidaichi shobunshi*. Tokyo: Ōkurashō Zaimu Kyōkai, 1954.

Okuwaki Kazuo. "Fujisan no shinkō to 'ohachi meguri': Shōfukuji han 'happa kyūson zu' o chūshin ni." *Fujiyoshida-shi Rekishi Minzoku Hakubutsukan kenkyū kiyō* 1 (March 2015): 76–99.

Ōmori Hisao, ed. *Fuji annai, Fuyō nikki*. Tokyo: Heibonsha, 2006.

Ono, Akira, and Masami Izuho. *Environmental Changes and Human Occupation in East Asia During OIS3 and OIS2*. Oxford: Archaeopress, 2012.

Ono, Akira, and Masayoshi Yamada. "The Upper Palaeolithic of the Japanese Islands: An Overview." *Archeometriai Mühely* 9, no. 4 (2012): 219–28.

Ono Shin'ichi. "Tokugawa Tsunayoshi to dojō hiryō gaku." *Nōgyō to kankyō*, no. 101 (September 2008). https://www.naro.affrc.go.jp/archive/niaes/magazine/101/mgzn10108.html.

Ooms, Herman. *Imperial Politics and Symbolics in Ancient Japan: The Tenmu Dynasty, 650–800*. Honolulu: University of Hawaii Press, 2008.

Orwell, George. *Animal Farm and 1984*. Orlando, FL: Harcourt, 2003.

Ōtani Masayuki. "Kinmeisui to Fujikō." *Fūzoku shigaku*, no. 16 (2001): 38–55.

Oyama Chōshi Hensan Senmon Iinkai, ed. *Oyama chōshi*. Vol. 2, *Kinsei shiryōhen*. Oyama-chō: Oyama-chō, 1991.

Oyama Chōshi Hensan Senmon Iinkai, ed. *Oyama chōshi*. Vol. 6, *Genshi, kodai, chūsei tsūshihen*. Oyama-chō: Oyama-chō, 1996.

Oyama Chōshi Hensan Senmon Iinkai, ed. *Oyama chōshi*. Vol. 7, *Kinsei tsūshihen*. Oyama-chō: Oyama-chō, 1998.

Pan-Pacific Science Congress. *Guide-book Excursion C-4: The Lake District Around Mt. Fuji*. Tokyo: Department of Education, 1926.

Parker, Geoffrey. *Global Crisis: War, Climate Change and Catastrophe in the Seventeenth Century*. New Haven, CT: Yale University Press, 2013.

Pearson, Richard. "Debating Jomon Social Complexity." *Asian Perspectives* 46, no. 2 (Fall 2007): 361–88. https://doi.org/10.1353/asi.2007.0015.

Pearson, Richard. "New Perspectives on Jomon Society." In *Bulletin of the International Jomon Culture Conference*, edited by Richard Pearson. Vol. 1. Tokyo: International Jomon Culture Conference, 2004.

Perry, Matthew Calbraith, and Francis L. Hawks. *Narrative of the Expedition of an American Squadron to the China Seas and Japan*, edited by Sidney Wallach. New York: Coward McCann, 1952.

Piggott, Joan. *The Emergence of Japanese Kingship*. Stanford, CA: Stanford University Press, 1997.

Piggott, Joan R., ed. *Capital and Countryside in Japan, 300–1180: Japanese Historians in English.* Ithaca, NY: East Asia Program, Cornell University, 2006.

Plutschow, Herbert E. *Chaos and Cosmos: Ritual in Early and Medieval Japanese Literature.* Leiden: E. J. Brill, 1990.

Ravina, Mark. *Land and Lordship in Early Modern Japan.* Stanford, CA: Stanford University Press, 1999.

Ravina, Mark. *To Stand with the Nations of the World.* Oxford: Oxford University Press, 2017.

Roberts, Luke S. *Performing the Great Peace: Political Space and Open Secrets in Tokugawa Japan.* Honolulu: University of Hawai'i Press, 2012.

Rosenfield, John M. "Nihonga and Its Resistance to 'the Scorching Drought of Modern Vulgarity.'" In *Births and Rebirths in Japanese Art,* edited by Nicole Coolidge Rousmaniere. Leiden: Hotei Publishing, 2001.

Ruoff, Kenneth J. *Imperial Japan at its Zenith: The Wartime Celebration of the Empire's 2,600th Anniversary.* Ithaca, NY: Cornell University Press, 2010.

Saigyō: Poems of a Mountain Home. Translated by Burton Watson. New York: Columbia University Press, 1991.

Saitō Haruo. "Seigyō no ba to shite no Fujisan." In *Kochizu de tanoshimu Fujisan,* edited by Ōtaka Yasumasa. Nagoya: Fūbaisha, 2020.

Sakio, Hitoshi, and Takehiro Masuzawa. "Advancing Timberline on Mt. Fuji between 1978 and 2018." *Plants* 9, no. 11 (November 2020): 1537–51. https://doi.org/10.3390/plants9111537.

Sato, Hiroyuki. "Trap-Pit Hunting in Late Pleistocene Japan." In Kaifu et al., *Emergence and Diversity.*

Sawada, Janine Anderson. *Faith in Mount Fuji: The Rise of Independent Religion in Early Modern Japan.* Honolulu: University of Hawai'i Press, 2022.

Sawada, Janine Anderson. "Mind and Morality in Nineteenth-Century Japanese Religions: Misogi-Kyō and Maruyama-Kyō." *Philosophy East and West* 48, no. 1 (January 1998): 108–41. https://doi.org/10.2307/1399927.

Sawada, Janine Anderson. "Sexual Relations as Religious Practice in the Late Tokugawa Period: Fujidō." *Journal of Japanese Studies* 32, no. 2 (Summer 2006): 341–66. https://doi.org/10.1353/jjs.2006.0063.

Sawato Hirosato. "Fuji shinkō girei to Edo bakufu no Fujikō torishimarirei." *Hōsei Daigaku Bungakubu kiyō,* no. 47 (2001): 147–78.

Scott, James. *Seeing Like a State: How Certain Schemes to Improve the Human Condition Have Failed.* New Haven, CT: Yale University Press, 1999.

Screech, Timon. *The Lens Within the Heart: The Western Scientific Gaze and Popular Imagery in Later Edo Japan.* Honolulu: University of Hawai'i Press, 2002.

Screech, Timon. *Tokyo Before Tokyo: Power and Magic in the Shogun's City of Edo.* London: Reaktion Books, 2020.

Seo, Kami, and Jonathan Taylor. "Forest Resource Trade Between Japan and Southeast Asia: The Structure of Dual Decay." *Ecological Economics* 45 (April 2003): 91–104. https://doi.org/10.1016/S0921-8009(03)00005-3.

Shiga Shigetaka. *Nihon fūkeiron.* 1894. Reprint, Tokyo: Iwanami Shoten, 2001.

Shikazono Naotatsu, Arakawa Takayuki, and Nakano Takanori. "Fujisan nanroku no chikasui suiritsu, ryūdō to chisso osen." *Chigaku zasshi* 123, no. 3 (June 2014): 323–42. https://doi.org/10.5026/jgeography.123.323.

Shimada, Daisaku. "Multi-level Natural Resources Governance Based on Local Community: A Case Study of Semi-natural Grassland in Tarōji, Nara, Japan." *International Journal of the Commons* 9, no. 2 (September 2015): 486–509. https://doi.org/10.18352/ijc.510.

Shimada, Kazutaka. "Pioneer Phase of Obsidian Use in the Upper Palaeolithic and the Emergence of Modern Human Behavior in the Japanese Islands." In Ono and Izuho, *Environmental Changes and Human Occupation.*

Shimada Tatsumi et al., eds. *Kojima Usui zenshū.* 14 vols. Tokyo: Taishūkan Shoten, 1979–87.

Shinno Toshikazu. "Journeys, Pilgrimages, Excursions: Religious Travels in the Early Modern Period." Translated by Laura Nenzi. *Monumenta Nipponica* 57, no. 4 (Winter 2002): 187–206. https://www.jstor.org/stable/3096735.

Shirane, Haruo, ed. *Early Modern Japanese Literature: An Anthology, 1600–1900.* New York: Columbia University Press, 2002.

Shirane, Haruo, ed. *Traditional Japanese Literature: An Anthology, Beginnings to 1600.* New York: Columbia University Press, 2007.

Shirane, Haruo, and Tomi Suzuki, eds. *Inventing the Classics: Modernity, National Identity, and Japanese Literature.* Stanford, CA: Stanford University Press, 2000.

"Shishino Nakaba." *Encyclopedia of Shinto.* In *Kokugakuin Digital Museum,* edited by Kokugakuin University. https://d-museum.kokugakuin.ac.jp/eos/detail/?id=8714.

Shizaki Daisaku. *Fujisan sokkōjo monogatari.* Tokyo: Seizandō Shoten, 2002.

Shizuoka Chiri Kyōiku Kenkyūkai. *Fujisan sekai isan e no michi: Sanroku ni ikiru hitobito no sugata o otte.* Tokyo: Kokon Shoin, 2000.

Shizuoka-ken, ed. *Shizuoka kenshi tsūshihen.* Vol. 1, *Genshi, kodai.* Shizuoka-ken: Shizuoka-ken, 1994.

Sigurdsson, Haraldur. "Introduction." In *Encyclopedia of Volcanoes,* edited by Haraldur Sigurdsson. New York: Academic Press, 2000.

Skabelund, Aaron Herald. *Inglorious, Illegal Bastards: Japan's Self-Defense Force during the Cold War.* Ithaca, NY: Cornell University Press, 2022.

Smith, Henry D., ed. *Hiroshige: One Hundred Famous Views of Edo.* New York: George Braziller, 1986.

Smith, Henry D. "Hokusai and the Blue Revolution in Edo Prints." In *Hokusai and His Age: Ukiyo-e Painting, Printmaking, and Book Illustration in Late Edo Japan,* edited by John T. Carpenter. Amsterdam: Hotei Publishing, 2005.

Smith, Henry D. *Hokusai: One Hundred Views of Mt. Fuji.* New York: George Braziller, 1988.

Smits, Gregory. *Seismic Japan: The Long History and Continuing Legacy of the Ansei Edo Earthquake.* Honolulu: University of Hawai'i Press, 2013.

Starr, Frederick. *Fujiyama: The Sacred Mountain of Japan.* Chicago: Covici-McGee, 1924.

Steiner, Kurt, Ellis S. Krauss, and Scott C. Flanagan, eds. *Political Opposition and Local Politics in Japan.* Princeton, NJ: Princeton University Press, 1980.

Stockdale, Jonathan. *Imagining Exile in Heian Japan: Banishment in Law, Literature, and Cult.* Honolulu: University of Hawai'i Press, 2015.

Sugano Masao. "Mushimaro no Fuji no yama no uta." In *Seminā man'yō no kajin to sakuhin*. Vol. 7, *Yamabe Akihito, Takahashi Mushimaro*, edited by Kōnoshi Takamitsu and Sakamoto Nobuyuki. Osaka: Izumi Shoin, 2001.

Sugimoto Yūki. "Sankeidō to shite no Kamakura kaidō." *Yamanashi Kenritsu Fujisan Sekai Isan Sentā kenkyū kiyō* 2 (March 2018): 12–22. https://www.fujisan-whc.jp/archive/documents/worldheritagefujisan02_2018.pdf.

Sugiyama Kōhei and Kaneko Takayuki. "Jōmon jidai kōhanki no Izu, Hakone, Fujisan no funka katsudō to shūraku dōtai." *Kōkogaku kenkyū* 60, no. 2 (September 2013): 34–54.

Susono Shishi Hensan Senmon Iinkai, ed. *Susono shishi*. Vol. 8, *Tsūshihen* 1. Susono-shi: Susono-shi, 2000.

Susono Shishi Hensan Senmon Iinkai, ed. *Susono shishi*. Vol. 9, *Tsūshihen* 2. Susono-shi: Susono-shi, 2001.

Suzuki Junko. "Seeking Accuracy: The First Modern Survey of Japan's Coast." In Wigen, Sugimoto, and Karacas, *Cartographic Japan*.

Suzuki Kōshirō. "Yunesuko no tsuika kankoku ni miru Fujisan no sekai bunka isan to shite no kadai." *Chigaku zasshi* 124, no. 6 (December 2015): 995–1014. https://doi.org/10.5026/jgeography.124.995.

Suzuki Masataka. "A Critical History of the Study of Shugendō and Mountain Beliefs in Japan." In Castiglioni, Roth, and Rambelli, *Defining Shugendō*.

Suzuki Masataka. "Meiji ishin to shugendō." *Shūkyō kenkyū* 92, no. 2 (September 2018): 131–57. https://doi.org/10.20716/rsjars.92.2_131.

Suzuki Masataka. *Nyōnin kinsei*. Tokyo: Yoshikawa Kōbunkan, 2002.

Takada Akira et al. *Geological Map of Fuji Volcano*. 2nd ed. Tsukuba: Geological Survey of Japan, 2016. https://gbank.gsj.jp/volcano/Act_Vol/fujisan/map/volcmap-s.html.

Takagi Ichinosuke et al., eds. *Nihon koten bungaku taikei*. Vols. 4–7, *Man'yōshū* 1–4, edited by Takagi Ichinosuke, Gomi Tomohide, and Ōno Susumu. Tokyo: Iwanami Shoten, 1957–62.

Takahashi, Keiichi, and Masami Izuho. "Formative History of Terrestrial Fauna of the Japanese Islands During the Plio-Pleistocene." In Ono and Izuho, *Environmental Changes and Human Occupation*.

Takahashi Masaki et al. "Fujisan jōgan funka to Aokigahara yōgan." In Nihon Kazan Gakkai, *Fuji kazan*.

Takahashi, Shin, et al. "Contamination and Specific Accumulation of Organochlorine and Butyltin Compounds in Deep-sea Organisms Collected from Suruga Bay, Japan." *Science of the Total Environment* 214, nos. 1–3 (June 1998): 49–64. https://doi.org/10.1016/S0048-9697(98)00088-6.

Takahashi Tsuyoshi and Kanayama Yoshiaki. "Fujisan ni tai suru Jōmonjin no ishikika ni tsuite." In Amano and Sawato, *Fujisan to Nihonjin no shinsei*.

Takebayashi Seizō. "Nihon zenkoku no mitate Fuji." In Watanabe and Sano, *Fujisan o shiru jiten*.

Takeda Sachiko. "Roads in the *Tennō*-Centered Polity." In Piggott, *Capital and Countryside in Japan, 300–1180*.

Takeda Yūkichi and Satō Kenzō, eds. *Kundoku Nihon sandai jitsuroku*. Kyoto: Rinsen Shoten, 1986.

Takemi Momoko. "'Menstruation Sutra' Belief in Japan." Translated by W. Michael Kelsey. *Japanese Journal of Religious Studies* 10, no. 2–3 (June–September 1983): 229–46. https://doi.org/10.18874/jjrs.10.2-3.1983.229-246.

Takeuchi, Melinda. "Making Mountains: Mini-Fujis, Edo Popular Religion and Hiroshige's *One Hundred Famous Views of Edo*." *Impressions: The Journal of the Ukiyo-e Society of America* 24 (2002): 24–45. https://www.jstor.org/stable/42597928.

Takeya Yukie. *Fujisan no seishinshi: Naze Fujisan o sanpō ni egaku no ka*. Sagamihara-shi: Seizansha, 1998.

"The Tale of the Fuji Cave." Translated by Keller R. Kimbrough. In *Monsters, Animals, and Other Worlds: A Collection of Short Medieval Japanese Tales*, edited by R. Keller Kimbrough and Haruo Shirane. New York: Columbia University Press, 2018.

"The Tales of Ise (Ise Monogatari, ca. 947)." Translated by Jamie Newhard and Lewis Cook. In Shirane, *Traditional Japanese Literature: An Anthology*.

Tamamuro Fumio. "Local Society and the Temple-Parishioner Relationship within the Bakufu's Governance Structure." *Japanese Journal of Religious Studies* 28, nos. 3–4 (Fall 2001): 261–92. https://doi.org/10.18874/jjrs.28.3-4.2001.261-292.

Tanaka Eriko, Morita Kei, and Sano Mitsuru. "Yama no kashi'iki kara mita Fujisan no nichijōsei to shōchōsei ni kan suru bunseki." *Fujigaku kenkyū* 6, no. 2 (March 2008): 72–77. https://doi.org/10.57325/fujisan.6.2_72.

Tanaka Kei. "Fujisan no kashi'iki—4000 man nin ga miru yama." In Watanabe and Sano, *Fujisan o shiru jiten*.

Tange Kenzō. "Dai Tōa kensetsu kinen eizō keikaku: Kyōgi sekkei tōsen zuan." *Kenchiku zasshi* 56, no. 693 (December 1942): 963–74.

Tani Yūji. *Fujisan wa naze Fujisan ka: watakushi no Nihon sanmei tanken*. Tokyo: Yama to Keikokusha, 1983.

Tani Yūji. *Kurofune Fujisan ni noboru! Bakumatsu gaikō ibun*. Tokyo: Dōhōsha, 2001.

Teeuwen, Mark. "Kokugaku vs. Nativism." *Monumenta Nipponica* 61, no. 2 (Summer 2006): 227–42. https://doi.org/10.1353/mni.2006.0023.

Thal, Sarah. *Rearranging the Landscape of the Gods: The Politics of a Pilgrimage Site in Japan, 1573–1912*. Chicago: University of Chicago Press, 2005.

Thomas, Jolyon Baraka. *Faking Liberties: Religious Freedom in American-Occupied Japan*. Chicago: University of Chicago Press, 2019.

Thomas, Julia Adeney. *Reconfiguring Modernity: Concepts of Nature in Japanese Political Ideology*. Berkeley: University of California Press, 2001.

Thomas, Julia Adeney, Mark Williams, and Jan Zalasiewicz, eds. *The Anthropocene: A Multidisciplinary Approach*. Cambridge: Polity Press, 2020.

Toby, Ronald P. *Engaging the Other: "Japan" and Its Alter Egos, 1500–1850*. Leiden: Brill, 2019.

Togashi Shigeko and Takahashi Masaki. "Fujisan no maguma no kagaku sosei to gansekigakuteki tokuchō: Maguma no jitai e no seiyaku jōken." In Nihon Kazan Gakkai, *Fuji kazan*.

Tōkyō-shi, ed. *Tōkyō shishikō sangyōhen*. Vol. 16. Tokyo: Tōkyō-to, 1972.

Tōkyō-shi, ed. *Tōkyō shishikō sangyōhen*. Vol. 41. Tokyo: Tōkyō-to, 1997.

Tōkyō-shi, ed. *Tōkyō shishikō shigaihen*. Vol. 9. Tokyo: Tōkyō-to, 1995.

Tōkyō Shiyakusho, ed. *Tōkyō shishikō shūkyōhen*. Vol. 2. Tokyo: Tōkyō-shi, 1936.

Torii Kazuyuki et al. *Nihon no kokoro: Fuji no bi ten*. Nagoya: NHK Nagoya Hōsōkyoku, 1998.

Tosaki Yūki and Asai Kazuyoshi. "Fujisan no chikasui nendai." *Chigaku zasshi* 126, no. 1 (February 2017): 89–104. https://doi.org/10.5026/jgeography.126.89.

Totman, Conrad D. *Early Modern Japan*. Berkeley: University of California Press, 1993.

Totman, Conrad D. *Japan: An Environmental History*. London: I.B. Tauris, 2016.

Totman, Conrad D. *Japan's Imperial Forest, Goryōrin, 1889–1946*. Folkestone, UK: Global Oriental, 2007.

Traganou, Jilly. *The Tōkaidō Road: Traveling and Representation in Edo and Meiji Japan*. London: RoutledgeCurzon, 2004.

Tsuchi Ryūichi. "Fujisan no chikasui, yūsui." In Nihon Kazan Gakkai, *Fuji kazan*.

Tsutsui, William M. "Landscapes in the Dark Valley: Toward an Environmental History of Wartime Japan." In *Natural Enemy, Natural Ally: Toward an Environmental History of War*, edited by Richard P. Tucker and Edmund Russell. Corvallis: Oregon State University Press, 2004.

Tsutsumi Takashi. "MIS3 Edge-Ground Axes and the Arrival of the First *Homo Sapiens* in the Japanese Archipelago." *Quaternary International* 248 (January 2012): 70–78. https://doi.org/10.1016/j.quaint.2011.01.030.

Tsutsumi, Takashi. "Prehistoric Procurement of Obsidian from Sources on Honshu Island (Japan)." In Kuzmin and Glascock, *Crossing the Straits*.

Tyler, Royall. "'The Book of the Great Practice': The Life of the Mt. Fuji Ascetic Kakugyō Tōbutsu Kū." *Asian Folklore Studies* 52, no. 2 (1993): 251–331. https://doi.org/10.2307/1178159.

Tyler, Royall. "The Tokugawa Peace and Popular Religion: Suzuki Shōsan, Kakugyō Tōbutsu, and Jikigyō Miroku." In *Confucianism and Tokugawa Culture*, edited by Peter Nosco. Princeton, NJ: Princeton University Press, 1984.

Uematsu Shōhachi. "Fuji shinkō to kami Kaguya-hime: sono seiritsu to shūen." *Fujigaku kenkyū* 7, no. 2 (October 2010): 5–8. https://doi.org/10.57325/fujisan.7.2_5.

UNESCO. *UNESCO 2013*. Paris: UNESCO, 2014. UNESDOC Digital Library, https://unesdoc.unesco.org/ark:/48223/pf0000227146.

Vaporis, Constantine Nomikos. *Breaking Barriers: Travel and the State in Early Modern Japan*. Cambridge, MA: Harvard University Press, 1994.

Vlastos, Stephen. *Peasant Protests and Uprisings in Tokugawa Japan*. Berkeley: University of California Press, 1990.

Wada Yūji. "Nonaka Itaru-shi no Fujisan kansokujo." *Taiyō*, January 1896.

Waley, Arthur. *The Nō Plays of Japan*. New York: Grove Press, 1957.

Walker, Brett L. *Toxic Archipelago: A History of Industrial Disease in Japan*. Seattle: University of Washington Press, 2010.

Walthall, Anne. "Popular Movements in Early Modern Japan: Petitions, Riots, Martyrs." In *The New Cambridge History of Japan*. Vol. 2, *Early Modern Japan in Asia and the World, c. 1580–1877*, edited by David L. Howell. Cambridge: Cambridge University Press, 2023.

Walthall, Anne. *The Weak Body of a Useless Woman: Matsuo Taseko and the Meiji Restoration*. Chicago: University of Chicago Press, 1998.

Watai Hideyo. "Fuji sanroku ni okeru kodai no iseki bunpu." *Fujigaku kenkyū* 6, no. 2 (March 2008): 66–71. https://doi.org/10.57325/fujisan.6.2_66.

Watai Hideyo. "Kōkogakuteki na kenchi kara mita Sengen jinja no hajimari." *Fujigaku kenkyū* 8, no. 2 (March 2011): 23–30. https://doi.org/10.57325/fujisan.8.2_23.

Watai Hideyo. "Reimeiki no Fujisan shinkō." *Fujigaku kenkyū* 12, no. 1 (December 2014): 45–49. https://doi.org/10.57325/fujisan.12.1_45.

Watai Hideyo. "Suruga ni okeru shuchōbo no hensan to Fujisan funka." *Fujigaku kenkyū* 10, no. 1 (November 2012): 1–8. https://doi.org/10.57325/fujisan.10.1_1.

Watanabe Michihito. *The Natural Features of Mount Fuji.* Vol. 2, *Satoyama Environments in Crisis.* Nagano-shi: Hōzuki Shoseki, 2018.

Watanabe Sadamoto. "The Nature and Scenery of Mt. Fuji Area." In Fujisan Sekai Bunka Isan Tōroku Suishin Shizuoka Yamanashi Ryōken Gōdō Kaigi and Fujisan o Sekai Isan ni Suru Kokumin Kaigi, *Fujisan: Shinkō to geijutsu no minamoto/Mt. Fuji: The Wellspring of Our Faith and Arts.*

Watanabe Sadamoto and Sano Mitsuru, eds. *Fujisan o shiru jiten.* Tokyo: Nichigai Asoshiētsu, 2013.

Weston, Walter. *Mountaineering and Exploration in the Japanese Alps.* London: John Murray, 1896.

Weston, Walter. *The Playground of the Far East.* London: John Murray, 1918.

White, Richard. *The Organic Machine: The Remaking of the Columbia River.* New York: Hill and Wang, 1995.

Wigen, Kären. "Discovering the Japanese Alps: Meiji Mountaineering and the Quest for Geographical Enlightenment." *Journal of Japanese Studies* 31, no. 1 (2005): 1–26. https://doi.org/10.1353/jjs.2005.0031.

Wigen, Kären, Sugimoto Fumiko, and Cary Karacas, eds. *Cartographic Japan: A History in Maps.* Chicago: University of Chicago Press, 2016.

Williams, Raymond. *Keywords: A Vocabulary of Culture and Society.* Rev. ed. New York: Oxford University Press, 1985.

Wilson, Noell. *Defensive Positions: The Politics of Maritime Security in Tokugawa Japan.* Cambridge, MA: Harvard University Press, 2015.

Wilson, Roderick I. *Turbulent Streams: An Environmental History of Japan's Rivers, 1600–1930.* Leiden: Brill, 2021.

Winichakul, Thongchai. *Siam Mapped: A History of the Geo-Body of a Nation.* Honolulu: University of Hawai'i Press, 1994.

Witze, Alexandra. "Geologists Reject the Anthropocene as Earth's New Epoch—after 15 Years of Debate." *Nature* 627, no. 8003 (March 6, 2024): 249–50. https://doi.org/10.1038/d41586-024-00675-8.

Woodard, William P. *The Allied Occupation of Japan 1945–1952 and Japanese Religions.* Leiden: Brill, 1972.

Yamakawa Shūji. "Fujisan no kasagumo—Fujisan kishōgaku nyūmon." In Nihon Daigaku Bunrigakubu Chikyū Shisutemu Kagaku Kyōshitsu, *Fujisan no nazo o saguru.*

Yamamoto Shin'ya et al. "Eruptive History of Mt. Fuji over the Past 8000 Years Based on Integrated Records of Lacustrine and Terrestrial Tephra Sequences and Radiocarbon Dating." *Quaternary Science Advances* 12 (October 2023): 1–9. https://doi.org/10.1016/j.qsa.2023.100091.

Yamamoto Shin'ya et al. "Fujisan hokuroku, Kawaguchiko no kotei yūsui to mizu no kigen." *Chigaku zasshi* 129, no. 5 (October 2020): 665–76. https://doi.org/10.5026/jgeography.129.665.

Yamamoto Takahiro et al. "Shinki Fuji kazan kōka kasaibutsu no saikisai to funshutsuryō no mitsumori." *Chishitsu chōsa kenkyū hōkoku* 71, no. 6 (2020): 517–80. https://doi.org/10.9795/bullgsj.71.517.

Yamanaka-mura no Rekishi Hensan Iinkai. *Yamanaka-mura no rekishi.* Yamanakako-mura: Sengen Jinja Yūchi Iriaiken Yōgo Iinkai, 2002.

Yamanashi Kenritsu Fujisan Sekai Isan Sentā. *Yōgan dōketsu o meguru shinkō.* Fujikawaguchiko-machi: Yamanashi Kenritsu Fujisan Sekai Isan Sentā, 2019. https://www.fujisan-whc.jp/archive/documents/r1kikakuten2-leaflet.pdf.

Yamanashi Nichinichi Shinbunsha, ed. *Narusawa sonshi.* Vol. 1. Narusawa-mura: Narusawa-mura Yakuba, 1988.

Yamawaki Takumi et al. "Fuji santai o mochiita kaki jiyū tairyūken ni okeru unsuichū kihatsusei yūkikagōbutsu no kansoku." *Taiki kankyō gakkai shi* 55, no. 5 (August 2020): 191–203. https://doi.org/10.11298/taiki.55.191.

Yasumaru Yoshio and Itō Kenkichi, eds. "Fujikō." In *Nihon shisō taikei.* Vol. 67, *Minshū shūkyō no shisō,* edited by Murakami Shigeyoshi and Yasumaru Yoshio. Tokyo: Iwanami Shoten, 1971.

Yi Yang-ji. "Fujisan." In *Yi Yang-ji zenshū.* Tokyo: Kōdansha, 1993.

Yokoi Haruno. "Fuji tozan annai." *Rekishi chiri* 36, no. 1 (1920): 100–11.

Yokoi Haruno. "Sekaiteki kōen to shite no Fuji." *Teikokumin* 7 (1918): 43–46.

Yonemoto, Marcia. *Mapping Early Modern Japan: Space, Place, and Culture in the Tokugawa Period (1603–1868).* Berkeley: University of California Press, 2003.

Yoneyama, Lisa. *Hiroshima Traces: Time, Space, and the Dialectics of Memory.* Berkeley: University of California Press, 1999.

Yoshida Keiji. "Fujisan no sabō jigyō." *Chishitsu to chōsa,* no. 140 (August 2014): 17–20. https://www.zenchiren.or.jp/jgca/pdf/jgca140.pdf.

Page numbers in italics indicate maps, figures, or plates.